Phenotypic Evolution:
A Reaction Norm Perspective

Phenotypic Evolution: A Reaction Norm Perspective

CARL D. SCHLICHTING
University of Connecticut

MASSIMO PIGLIUCCI
University of Tennessee, Knoxville

Sinauer Associates, Inc., *Publishers*
Sunderland, Massachusetts

THE COVER: Developmental reaction norms for the seasonal forms of the satyrine butterfly, *Bicyclus anynana*. The curves show the change in size (*y*-axis) of the large forewing eyespot of the female butterfly through ontogeny (*x*-axis) in the wet and dry season environments (*z*-axis). The first pair of photos (upper and lower) shows no difference in expression of the gene *Distal-less* at the fifth instar. The second pair reveals that *Distal-less* expression is greater in the wet season form shortly after pupation. The third pair shows the major difference in phenotypes of adult females.

The photos of *Distal-less* expression are reprinted with permission from *Nature* 384:236–242 (Brakefield et al., 1996), © 1996 Macmillan Magazines Limited. We thank Paul Brakefield and Sean Carroll for providing originals. The photos of the adult females are courtesy of Paul Brakefield, University of Leiden.

For information or to order, address:
Sinauer Associates, Inc., PO Box 407
23 Plumtree Road, Sunderland, MA, 01375 U.S.A.
FAX: 413-549-1118
Internet: publish@sinauer.com; http://www.sinauer.com

Library of Congress Cataloging-in-Publication Data
Schlichting, Carl.
    Phenotypic evolution: a reaction norm perspective / Carl D. Schlichting,
  Massimo Pigliucci.
        p.   cm.
    Includes bibliographical references and index.
    ISBN 0-87893-799-4
      1. Adaptation (Biology)   2. Evolution (Biology)   3. Phenotype.
  I. Pigliucci, Massimo, 1964–   .   II. Title.
  QH546.S36   1998
  576.5'3—dc21   98–7457
                                                                  CIP

Printed in U.S.A.

5 4 3 2 1

Contents

Preface

Our purpose in writing this book is to accumulate evidence from a disparate set of fields and weave these pieces into an idea that had been evolving in our brains for many years. Phenotypes are the complex result of the interactions of three phenomena: phenotypic plasticity (the genetically mediated response to external environmental changes), epigenetics (the genetically mediated response to internal environmental changes), and allometry (the way in which multitudes of internal and external responses are coordinated to yield a coherent whole). We would feel satisfied if the reader will consider these relationships and think about their evolutionary implications.

We have vacillated between the urge to be comprehensive versus merely representative in our citations. We started out by trying to read everything relevant, but the explosion of literature in many of the interconnected fields of inquiry covered by this book soon made it clear that such an ambition was naive. We apologize to authors for any omissions—we know just how annoying it can be to read something that has not adequately investigated antecedent work. Our notions of what to include (and exclude) from this book have been influenced by several recent books: Brian Hall's *Evolutionary Developmental Biology*, David Rollo's *Phenotypes*, and Rudoph Raff's *The Shape of Life*.

We single out for special thanks our colleagues Cindi Jones and Kurt Schwenk at the University of Connecticut: during numerous long and interesting discussions, they were instrumental in helping us to focus many of the ideas in this book. Immense gratitude is also due to those who provided comments on one or more chapters: Jennifer Butler, Jan Conn, Jeff Conner, Cindi Jones, Rick Miller, Courtney Murren, Sonia Sultan, and Günter Wagner. Thanks also go to the members of YUCEE (Yale-UConn Evolution Ensemble) and to Hilary Callahan, Mark Camara, Carolyn Wells, and Mitch Cruzan, who discussed these topics on several occasions. We also acknowledge the countless colleagues who provided references and preprints/reprints of their work.

Our research on phenotypic plasticity and reaction norm evolution has been supported in part by the National Science Foundation, the University of Connecticut Research Foundation, and the University of Tennessee.

Carl Schlichting & Massimo Pigliucci
April, 1998

CHAPTER 1

Phenotypic Evolution and Its Many Facets

Our ignorance of how genotypes produce phenotypes is, I believe, the greatest gap in our understanding of the evolutionary process and it is a huge gap indeed. (Futuyma, 1979, p. 438)

Although written nearly 20 years ago, we find this quote fully apropos today.[1] How evolutionary processes produce phenotypic change is one of the most perplexing issues to face evolutionary biologists. A vague sense of dissatisfaction permeates the field despite recent advances in both phenotypic and molecular evolution. Evolutionary biologists are finding disparate areas of inquiry difficult to merge and are still struggling with the integration of developmental processes into the neo-Darwinian synthesis. Some argue that the synthesis itself is incomplete because macroevolutionary phenomena are not adequately explained (or are, in fact, not

---

[1] Although we could find no comparable statement in Futuyma (1997).

explained at all). The origin of novel phenotypes is still shrouded in mystery, and the process of the formation of new species remains elusive. Furthermore, there is even disagreement between the theory and the observation of genetic variation in natural populations—Is there more variation than there should be?

Although we cannot offer a panacea for all these difficulties, we will suggest that the adoption of a different perspective is necessary in order to make headway on these issues. We agree with others that the integration of development into the evolutionary synthesis has been slow and largely one-sided (the molecular genetics is rapidly being elucidated, but epigenetic phenomena remain largely enigmatic). However, we also perceive an additional lacuna: the synthesis has inadequately incorporated the ecological context, particularly the multifaceted ramifications of genotype by environment interactions. In their analysis of the contributions of various biological disciplines to the evolutionary synthesis, Mayr and Provine (1980) included no coverage of ecology.

We propose the concept of a **developmental reaction norm** as a framework to facilitate connections among the various theories on the evolution of phenotypes. The developmental reaction norm is the set of ontogenies that can be produced by a single genotype when it is exposed to internal or external environmental variation. We propose that natural selection operates on developmental reaction norms (as opposed to operating on particulate traits). Our interest is the elucidation of the developmental programs involved in the *epigenetic* transition from genotype to phenotype (Figure 1.1), and our goal is to ultimately tie the patterns and processes of evolution together. In this book we will weave together approaches gleaned from theoretical and experimental studies on the genetics, ontogeny, plasticity, and allometry of the construction of phenotypes, with the intention of enhancing the understanding of the processes underlying phenotypic evolution.

Charlesworth (1990a) enumerated three classes of explanation for adaptive phenotypic evolution. Class 1 represents a Darwin/Fisher view in which complex adaptations arise through the incremental addition of modifications to existing structures. Class 2 is a Goldschmidtian perspective in which a single mutation can produce a range of concurrent changes, resulting in a coordinated, functional phenotype. And class 3 is the perspective of Wright and Haldane, which incorporates a major role for the random process of drift, with the potential to move a population across a valley separating two adaptive peaks.[2] Charlesworth concluded that Goldschmidt was refuted both empirically and theoretically, and that "there is no evidence that compels us to accept a class 3 over a class 1 explanation for any evolutionary process involving morphological

---

[2] We note that, in his description of class 3, Charlesworth left out a vital component of Wright's conception, namely the importance of universal epistasis—the pervasive interrelationships among genes and gene products.

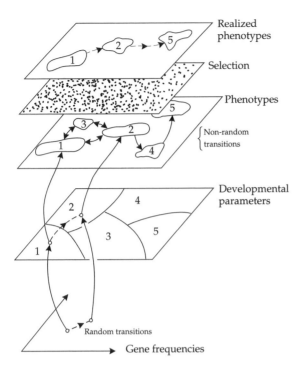

Realized
phenotypes

Selection

Phenotypes

Non-random
transitions

Developmental
parameters

Random transitions

Gene frequencies

Figure 1.1 Diagrammatic representation of the genotype to phenotype map. Selection acts directly on phenotypes, and this action is then translated into changes in gene frequencies through the filter of developmental parameters. The realized phenotype is the resultant of the directional forces of selection and the constraints of the genetic/developmental system. (From Oster and Alberch, 1982.)

change." The position we take throughout this book is that the mounting evidence from modern molecular and developmental genetics supports the Wright/Haldane paradigm of extensive gene-gene interaction in building phenotypic architecture (and perhaps even a modified Goldschmidtian perspective). These discoveries critically undermine aspects of the Fisherian paradigm of gradual evolution by alterations in many genes with small effects ("mass selection"). This leaves evolutionary biologists with the complex task of incorporating what once was the black box of development into evolutionary theory, and coming to a mechanistic understanding of genotype by environment interaction through a creative merger of ecological and molecular perspectives on how organisms live and evolve.

FISHER VERSUS WRIGHT?

The two classical approaches to modeling phenotypic evolution were already in place by the early 1930s (Provine, 1971). Fisher's view of the additivity of gene action and the foundations for population and quantitative genetics were first presented in 1918, and fully elucidated in his book, *The Genetical Theory of Natural Selection* (Fisher, 1930). Wright's view that phenotypic expression was strongly dependent on both pleiotropic and epistatic gene effects was incorporated in his shifting balance theory (Wright, 1931a). Our expectation would be that, over 60 years later, one point of view (or perhaps a synthesis of both) would prevail among evolutionary biologists. However, as pointed out by Wade (1992), there is a curious mixing of gene-level and phenotype-level views by modern evolutionary biologists—Fisher's additive genetic model is embraced for the operational description of phenotypic evolution, while Wright's (1969) notion of universal pleiotropy/epistasis is regarded as a more likely model for the genetic architecture of organisms.

Fisher's conception of the genetic architecture of traits was based on the idea of genes as independent factors—their effects could be added together to produce the phenotype. This particulate approach was embraced as he built his "genetical theory" of evolution by natural selection. Fisher was strongly influenced in this endeavor by the success of physicists in building mechanistic models[3] (Depew and Weber, 1995). His view of the additivity of genes enabled him to devise the conventional statistical approach to partitioning their effects into "additive" and "non-additive" components. Because Fisher believed that the majority of genes had additive effects, the projection of the mechanistic basis of gene action (additivity) onto a statistical representation was direct, and posed no particular analytical problems. His development of analysis of variance flowed directly from this view of genes (Provine, 1971). The variance in phenotypic expression could be partitioned into main effects—the linear or "additive" components—and interaction effects—the non-linear or "non-additive" components. In the application of these methods to quantitative genetics, only the statistically linear effects, that is, those presumed by Fisher to be due to the additivity of gene effects, would be important for determining responses to selection.

In Fisher's view, natural selection was the preeminent evolutionary force driving phenotypic change; random processes would only rarely play an important role. Adaptive evolution of phenotypes proceeded on a global landscape with a single fitness peak as the ultimate objective. Evolution in very large populations would utilize the available additive genetic variation to move up the slope. Fisher's "fundamental theorem of natural selection" therefore predicts that evolution will continue until the (additive) genetic variation for fitness is exhausted.[4]

Wright's work on physiological genetics led him to a divergent view of gene action. He was convinced that not only were there manifold effects of single genes (pleiotropy), but that systems of interacting genes (epistasis) were pervasive. Furthermore, he believed these interactions would be context dependent,

---

[3] In fact, Fisher may represent one of the earliest cases of physics envy in a biologist, a malady that afflicted other early biologists, such as Lotka, and appears occasionally to this day.

[4] See Chapter 4 in Hartl and Clark (1989) for a discussion of restrictions on the applicability of Fisher's fundamental theorem, and Frank (1997) for an examination of just how "fundamental" it is.

with the effect of any particular allele determined by the background of alleles present at other loci (Wright, 1931a,b). The importance of genetic background was convincingly demonstrated (e.g., in *Drosophila* by Haldane, 1932a, Table V, p. 56; and in *Gossypium* by Harland, 1936), and has evolved into the theory of coadapted gene complexes (Dobzhansky, 1951; Allard et al., 1972). A representative example of the complexity of such hierarchical genetic systems is depicted for the *Drosophila* gene *Deformed* in Figure 1.2.

In contrast to Fisher, Wright's view of the fitness landscape permits multiple peaks of various heights, such that adaptive evolution is local rather than global, with smaller populations climbing the closest hill. However, this scenario results in many populations achieving lower relative fitness than can be attained globally. How can phenotypes move from these local optima? The shifting balance theory (Wright, 1931a, 1932) emerged as an answer, with epistasis playing a central role in the depiction of how changes in phenotypic states (i.e., peak shifts) could occur. Movement from lower to higher peaks would be possible with the exchange of alleles between the somewhat differentiated subpopulations, creating new genetic combinations that could propel a population in a new direction. Epistatic effects would result from new associations of alleles at different loci *across* rather than within subpopulations. Wright did not dismiss the prominence of natural selection, but saw it as only one among several potent evolutionary forces.

Wright's shifting balance theory, although intuitively appealing, has garnered limited empirical support (e.g., Wade and Goodnight, 1991; King, 1993; Moore and Tonsor, 1994; Coyne et al., 1997), while results of other studies have not corroborated the model's predictions (e.g., Cohan et al., 1989). Other investigations have questioned its generality on theoretical grounds. For example, Nei (1987) pointed out that the level of migration necessary for peak shifts is large enough to effectively limit differentiation of local populations, and Gavrilets (1996) argued that the final phase of the process can occur only with a much more restricted set of parameter values than previously suggested. Recent theoretical investigations have examined how peak shifts might arise with and without the conditions of the shifting balance (Barton, 1992; Phillips, 1993; Price et al., 1993; Wagner et al., 1994; Whitlock, 1997).

Fisher argued for a smooth adaptive landscape as a consequence of the large variety of phenotypic characteristics that are targets of selection—this leads to a very high dimensionality of the phenotype space and concomitantly fewer discontinuities in the space. Wright's rugged landscape with multiple peaks was envisioned as a consequence of context-dependent interactions of genes (e.g., coadaptation), but could also result from an effectively low dimensionality of those aspects of the phenotype that are under strong selection at any particular time.

A continuing source of confusion arising from the difference between Fisher's and Wright's conceptions is that they imply different meanings for the term

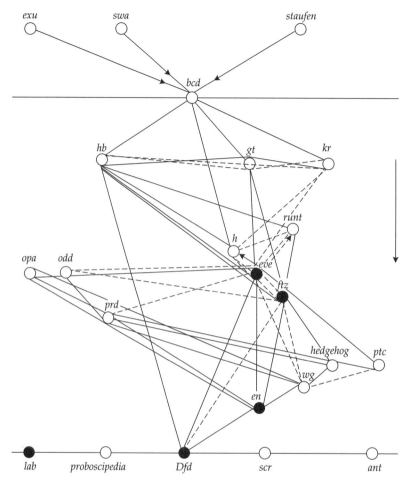

Figure 1.2 Network of interacting genes that either directly or indirectly regulate expression of the *Deformed* (*Dfd*) gene (bottom center) in the development of the *Drosophila* body plan. Regulation proceeds from top to bottom; dashed lines represent negative regulation, solid lines positive regulation. Black circles are possible autoregulatory genes. There are a variety of other genes not shown that are also activated or repressed by this network. (From Burstein, 1995.)

epistasis. Wade (1992) provides a clear account of the distinction: In Fisher's model epistasis is a statistical phenomenon referring to non-linear sources of variation in phenotypes, whereas Wright uses the term in the mechanistic sense of an actual interaction between the products of different genes. Fisher's view is embodied in quantitative genetic applications in which epistasis is a property of populations, with no necessary reflection at the gene level. Wright's view is based on the individual and not on populations, and is represented in standard biology and genetics texts, and embodied in the evolutionary literature on metabolic pathways and control theory (e.g., Kacser and Burns, 1981).

The success of quantitative genetics in plant and animal breeding (Falconer, 1989; Comstock, 1996; Evans and Marshall, 1996) has led to its enthusiastic employment for describing phenotypic evolution in natural populations (e.g., Lande, 1982; Lofsvold, 1986; Shaw et al., 1995b). However, problems associated with applying a strictly statistical approach to understanding mechanisms of biological evolution have been pointed out repeatedly (historical perspective provided by Wade, 1992; see also Ward, 1994; Schlichting and Pigliucci, 1995a; Pigliucci and Schlichting, 1997; and below). The most common criticisms are leveled precisely at the conflation of additive and epistatic sources of variation into a statistical "additive" genetic variance. The crux of this debate is whether the evolution of phenotypic traits, whose "additive" variation is governed partially or wholly by epistatic (in the mechanistic sense) gene systems, will necessarily follow a somewhat different long-term evolutionary trajectory than traits controlled by genes with strictly additive effects.

Finally, we submit that there is another key distinction between Wright and Fisher, one that can be appreciated by inspecting their blind spots. Wright championed the importance of the genetic context as a determinant of gene expression, yet virtually ignored the ecological context of organismal performance. Fisher, as noted above, thought genetic context unimportant, yet appears to have explicitly included the role of changing environments in the formulation of his fundamental theorem (see Frank and Slatkin, 1992). We suggest that, to a certain extent, the limitations encountered by both theoretical traditions can be traced to these curious omissions.

OPEN QUESTIONS ON THE EVOLUTION OF PHENOTYPES

Remarkable progress has been made in understanding microevolutionary phenomena due to the revelations of protein electrophoresis and its successors (e.g., Hamrick and Godt, 1996), as well as to the increasing sophistication of population and quantitative genetic models of the processes of evolution (Pollak et al., 1977; Weir et al., 1988; Boake, 1994). There remains, however, a substantial corpus of unresolved or partially resolved issues related to phenotypic evolution in general, and to adaptive phenotypic change in particular. These issues are micro- and macroevolutionary in nature, and are especially vexing in relation to the generation and integration of new features at higher taxonomic levels. They will surface repeatedly throughout the book as links underlying the main discourse.

How Do We Connect Observed Patterns to Underlying Processes? Mechanistic Biology and the Comparative Method

One of the ongoing tensions among evolutionary biologists derives from disagreements concerning the relationship between evolutionary patterns and processes. The linking of pattern and process is in fact wickedly difficult—historical patterns typically yield little insight into the processes that gave rise to them, and, conversely, we lack methods for reliably projecting processes onto future patterns (Rieppel and Grande, 1994).

Although it is difficult to confidently form a picture of the importance of recent trends in science, it seems clear that at least two major avenues of evolutionary research have opened in the past decade—the molecular renaissance in developmental biology (e.g., Akam et al., 1994b), and the inclusion of explicit phylogenetic comparisons in ecological and evolutionary studies (i.e., the "comparative method," Harvey and Pagel, 1991). These represent, respectively, a process- and a pattern-oriented approach to evolutionary issues, with but a few attempts to combine the two (e.g., Sommer and Sternberg, 1994; Warren et al., 1994; Wray and Bely, 1994; Wake and Hanken, 1996; McEdward and Janies, 1997).[5] A third route has developed around the investigation of reaction norm evolution, with recent studies moving toward an integration of patterns and processes (Markwell and Osterman, 1992; Schmitt et al., 1995).

What Are the Limits to Variability?
The Role of Constraints in Phenotypic Evolution

This issue has a venerable history in evolutionary biology. There has been substantially more effort spent on conceptual and semantic wrangling about constraints than on their empirical investigation, and a peculiar tendency toward impenetrability or mysticism in some of the arguments. At one end of the spectrum lies the implausible stance that there are no limits to variation and evolution, while at the other end are process structuralists who hold that innate (environment-independent) potentialities principally dictate the alternatives available to developmental programs (Ho and Saunders, 1979; Goodwin, 1989; Smith, 1992). The research conducted between these extremes must focus on what determines the limits and potentials for natural selection (e.g., Are there limits to mutational expression?).

Wagner (1995) and Wagner and Altenberg (1996) have argued persuasively for distinguishing the observable differences among individuals in a population (e.g., the actual number of different alleles at a particular locus), which they call *variation*, from the potential or propensity of a trait to change (e.g., the rate and type of mutations that can be produced, or the new combinations of genes that can be created by recombination), referred to as *variability*.[6] Short-term responses to selection are limited by available variation, while changes over the long

---

[5]Somewhat less successful attempts exist as well. Cheetham et al. (1994) applied quantitative genetic techniques to paleontological data to illuminate pattern/process issues, and Reyment and Kennedy (1991) used multivariate statistics to characterize phenotypic plasticity in ammonites. The conclusions of these studies can neither be supported nor refuted by their data: neither genetic nor environmental influences on phenotypes can be extracted from rocks.

[6]See Thoday (1953) for an early call for this distinction.

term depend on variability. Variability, in turn, is a reflection of fundamental genetic/epigenetic constraints. However, the central question remains—How does the construction of developmental systems limit phenotypic expression?[7]

How Are Complex Phenotypes Coordinated?
The Evolution of Epigenetic Systems

This essentially translates into a question concerning the evolution of pleiotropy, epistasis, and, especially, gene regulation. Several workers have probed the conceptual issues of how character coupling can arise, with one focus being the integration of functionally related characters (Frazzetta, 1975; Riedl, 1978; Cheverud, 1984, 1988; Wagner, 1986, 1988a, 1995; Wagner and Altenberg, 1996). There has also been substantial attention devoted to the statistical theory of the evolution of correlated characters (Riska, 1986; Clark, 1987a; Arnold, 1989; Wagner, 1989c; Houle, 1991; Price and Langen, 1992; Gromko, 1995). Unfortunately, there have been relatively few studies dealing directly with the evolution of the mechanistic bases of such coupling (e.g., Bush and Paigen, 1992; Carroll, 1994; Cohan et al., 1994; Doyle, 1994; Roff, 1995; Zera et al., 1997a,b). In the majority of cases of pleiotropy and epistasis the genetic or biochemical structure of the underlying pathways remains unknown.

How Do Organisms Deal with Changing Environments?
Genetic Variation and Phenotypic Plasticity

For those evolutionists inclined to a "selectionist" point of view, the role of genetic variability had been envisioned as an adaptive one. What, though, is one to make of the "vast stores of genetic variation within populations [that are] difficult to reconcile with balancing selection" (Clegg, 1997)? Phenotypic plasticity is increasingly viewed as an important facet of adaptation to varying conditions. The old (and "not dead yet") paradigm in studies in this area is that genetic variation (or even heterozygosity) and plastic response represent alternative and mutually exclusive means of dealing with environmental variability (e.g., Lewontin, 1957; Marshall and Jain, 1968; Pederson, 1968; Gillespie and Turelli, 1989). A related idea is that phenotypic plasticity acts as a buffer *against* natural selection (Wright, 1931a; Grant, 1977; Stearns, 1982; Sultan, 1987; Levin, 1988) allowing genetic variation to be maintained. There is also a third possibility, that of "adaptive coin-flipping" (Cooper and Kaplan, 1981), in which variable offspring are produced in an environment that is changing unpredictably. Although there has been an explosion of work on these topics in recent years (Hoffman et al., 1995), the key ideas on genetic polymorphism were formulated by Levene (1953) and Haldane and Jayakar (1963), and those on plasticity by

---

[7]Debate over the form and function of "constraints" on phenotypic evolution has increased, in part, due to overuse of the term (Antonovics and van Tienderen, 1991; Perrin and Travis, 1992; Schwenk 1995).

Levins (1963, 1968) and Bradshaw (1965). In this area in particular, empirical studies are vastly outnumbered by theoretical approaches.

How Do Novel Phenotypes Arise?
Micro- versus(?) Macroevolution

Many of the issues surrounding phenotypic novelty have a close connection with the question posed above, namely, How can any major change be integrated into an existing coordinated system? (Mayr, 1960; Muller, 1990). The "ancient" battles (Goldschmidt, 1940; Wright, 1941) still rage over the ability of microevolutionary theory to account for macroevolutionary events and patterns (Gould, 1980a, 1982; Charlesworth et al., 1982; Coyne and Charlesworth, 1997; Eldredge and Gould, 1997). The limiting factor to understanding novelty stems from the relatively large role that chance appears to play in the generation of new forms. Attempts at a more pluralistic approach have been made (e.g., contributions in Wake and Roth, 1989, and Nitecki, 1990), and new synthetic views have been proposed (Bonner, 1988; West-Eberhard, 1989; Orr and Coyne, 1992).

THEORETICAL APPROACHES

To better understand the current state of evolutionary theory as it applies to phenotypic evolution, we must briefly consider the major efforts at modeling evolution in recent years. Far from being a comprehensive survey, the following pages are intended simply as a reminder to the reader about the state of the art, and as a pointer to the primary literature.

Quantitative Genetics

Quantitative genetics was initially developed by R. A. Fisher, Sewall Wright, and J. B. S. Haldane, as a tool for understanding the inheritance of quantitative traits and their response to selection (Fisher, 1918; Wright, 1921; Haldane, 1932a). The success and influence of this approach have, deservedly, been enormous (Falconer and Mackay, 1996). Applications of these techniques have been, until fairly recently, largely limited to plant and animal breeding, or to artificial selection on laboratory populations. Lande promoted the explicit usage of quantitative genetics for the study of evolution in natural populations (Lande, 1979, 1982; Lande and Arnold, 1983).

The foundations of quantitative genetics were laid when Fisher recognized that the gulf between the biometricians (studying continuous characters) and the Mendelians (focusing on discrete characters) could be bridged by postulating a system with many genes acting in a Mendelian fashion.[8] A gene system with

---

[8] As noted by Provine (1971), Yule and Weinberg had actually pointed this out some time before (in the years 1902 and 1909–1910, respectively). See Provine, and Depew and Weber (1995) for comprehensive accounts of this debate.

small additive effects would produce the continuously varying distributions of phenotypic values on which the Biometricians had focused. This coup erased the major point of contention between the two groups (if not the contention itself), and set the stage for a rapid succession of contributions by Wright (e.g., heritability) and Haldane (e.g., examination of selection intensity) over the next decade. Fisher also recognized the inherent statistical tractability of such a system, thereby making possible the study of quantitative genetics using nested analysis of variance (Becker, 1984; Falconer, 1989).

The simple equation

$$R = h^2 S \qquad \text{(Equation 1.1)}$$

(where R is the observed or predicted **response to selection**, h^2 is the **heritability** of the trait, and S is the **selection differential**), actually describes both the *selection* and *transmission* phases of the evolutionary process. Measures of the strength of selection on the phenotypic characteristics are represented by S, and the capacity for genotypic response is summarized by h^2. Heritability refers to the proportion of the total variation that is due to additive genetic effects, and the selection differential records the effectiveness of a particular **intensity of selection** (i) given the available phenotypic variation (σ_P)

$$S = i\sigma_P \qquad \text{(Equation 1.2)}$$

Although quantitative genetics was enormously successful for evaluating evolutionary change under artificial selection regimes (i.e., with known S), the practical problems associated with imposing (or inferring) breeding designs in nature prohibited its use for wild populations (many studies have estimated heritabilities and genetic correlations from wild populations under laboratory conditions). Lande (1979) saw a way around this restriction: By limiting the focus to the first part of the process, it is not particularly difficult to gain estimates of the intensity of natural selection or to identify the form of its effect (i.e., directional, stabilizing, disruptive) on a phenotypic character. Even better, Lande and Arnold (1983) proposed to generalize the approach for response to selection from one correlated character (e.g., Young and Weiler, 1960) to a matrix of correlated characters

$$\Delta \bar{z} = GP^{-1}S \qquad \text{(Equation 1.3)}$$

where $\Delta \bar{z}$ is the change in the vector of mean phenotypes across one generation, G and P are, respectively, the additive genetic and total phenotypic variance-covariance matrices, and S is the vector of selection differentials on the traits. GP^{-1} is then the multivariate analogue of heritability—the relative proportion of total phenotypic variance-covariance that is additive for the set of traits measured. For the response to selection as a function of the contributions of the various characters to fitness (w), this relationship can be rewritten as

$$\nabla \ln \overline{w} = P^{-1}S = G^{-1}\Delta \overline{z} \qquad \text{(Equation 1.4)}$$

where $\nabla \ln \overline{w}$ is the gradient operator representing the partial derivative of each trait with respect to fitness. ∇ produces a vector that will move fitness most quickly up an adaptive peak (i.e., the steepest ascent; Hartl and Clark, 1989). Multiple regression techniques applied to a sample of individuals from a population can estimate selection's intensity and form, given certain assumptions (e.g., Rausher, 1992; Culver et al., 1994; Miller et al., 1994; Conner et al., 1996).

Endler (1986) and Mitchell-Olds and Shaw (1987) extensively reviewed the statistical assumptions and potential pitfalls of the multiple regression approach, in particular the importance of unmeasured traits. Wade and Kalisz (1990) reminded users that this technique can directly illuminate only the magnitude and direction of selection intensity, and cannot be used to infer the causes of selection. All of these authors recommended experimental approaches such as direct manipulation to complement the results from multivariate regression. Several authors suggested alternatives to the regression approach. Schluter and Nychka (1994) pointed out that the parametric analysis of Lande and Arnold can produce misinterpretations of the form of selection. Crespi and Bookstein (1989) proposed path analysis as a means of taking into account the inherent hierarchical nature of intertrait relationships.

Modelers of phenotypic evolution have applied the multiple regression approach to subjects ranging from phenotypic plasticity to allometry, maternal effects, and development (Via and Lande, 1985; Zeng, 1988; Kirkpatrick and Lande, 1989; Kirkpatrick and Lofsvold, 1992). Although the quantitative genetic model itself cannot be inherently right or wrong (i.e., it is mathematically consistent), there has been considerable discussion on the relevance of some of its assumptions and on its restricted applicability to evolution in natural populations (Mitchell-Olds and Rutledge, 1986; Barker and Thomas, 1987; Barton and Turelli, 1989; Riska, 1989; Pigliucci and Schlichting, 1997). Some of the points that have been discussed are:

1. Infinite effective population size. Barker and Thomas (1987) pointed out that natural populations are probably more commonly quite small rather than extraordinarily large.

2. Constancy of quantitative genetic parameters. Although most models assume that these will remain constant over evolutionary time, numerous authors have pointed out the likelihood that phenotypic and genetic variances and covariances will change with environment and following selection (Schlichting, 1986; Turelli, 1988; Wilkinson et al., 1990; Jernigan

et al., 1994; Windig, 1994; Blows and Sokolowski, 1995; Schlichting and Pigliucci, 1995a; Shaw et al., 1995a; Bennington and McGraw, 1996).

3. Statistically additive and epistatic variation can be interconverted (Bryant et al., 1986; Goodnight, 1987, 1988; Bryant and Meffert, 1992; Cheverud and Routman, 1996). A focus on additive variance is, therefore, likely to underestimate the potential for evolutionary response.

4. Absence of genotype by environment interaction. Empirical work suggests that G by E interactions will be present more often than they are not (Schlichting, 1986; Scheiner, 1993a; Asins et al., 1994).

5. Absence of mechanistic detail. Due to the lack of direct correspondence between statistical and mechanistic properties of biological systems, statistical descriptions of populations are not an adequate substitute for knowledge of the genetic control of trait expression (Mitchell-Olds and Rutledge, 1986; Houle, 1991; Ward, 1994; Gromko, 1995; Schlichting and Pigliucci, 1995a; Cheverud and Routman, 1995; Comes, 1998). This will lead to unreliability in long-term projections (Barton and Turelli, 1989; Leroi et al., 1994a; Turelli and Barton, 1994; Pigliucci and Schlichting, 1997).

Optimality Models

Originally constructed for use in economics, these models address adaptive evolution in terms of phenotypes and selective forces based on twin propositions—that fitness (or some trait correlated with it) should be maximized, and that there are inherent tradeoffs between components of fitness (Seger and Stubblefield, 1996). Optimality approaches incorporate two different kinds of "constraints." They stipulate the phenotypes on which selection can operate (the **strategy set**), and they specify the underlying factors that are assumed to be invariant (i.e., What is the currency of tradeoffs?). For example, in the evolution of offspring size, one can assume that size is free to vary, but that the total amount of resources available for reproduction is fixed. Because these are equilibrium models, they are only able to postulate factors responsible for the maintenance of a trait, not to trace its evolutionary history (Schmid-Hempel, 1990). [However, dynamic modeling has recently been used to address problems in behavioral ecology (Clark, 1993).]

There are two classes of optimality models—frequency-independent (simple optimization) and frequency-dependent (competitive optimization or game theory). In the case of frequency independence, an optimization model maximizes fitness for all individuals in the population; there are no influences due to the phenotypes of other individuals. For example, in studying leaf shape evolution, a model could take into account temperature, availability of water, and transpiration rates, but ignore competition for light from neighboring plants. Frequency-dependent models use the mathematical tools of game theory to lead to the definition of the so-called evolutionarily stable strategies (ESS). In this case, the strategy (behavioral or phenotypic) of an organism, depends on those of other organisms. An ESS is defined as "a strategy such that, if all members of a population adopt it, then no mutant strategy could invade the population under the influence of natural selection" (Maynard Smith, 1982).

Optimization and game theoretical models have been widely used to provide insights into the characteristics and behaviors of organisms. Schmid-Hempel (1990) provides a detailed review specifically on the use of optimization models. Parker and Maynard Smith (1990) and, more recently, Seger and Stubblefield (1996) comment on the applicability of optimality models in evolutionary biology. These models have continued to be applied successfully to problems of phenotypic evolution (Ashman and Schoen, 1994; Parker and Simmons, 1994; Lucas and Howard, 1995).

There have been numerous criticisms leveled at optimality modeling (Hoekstra, 1988; Parker and Maynard Smith, 1990; Schmid-Hempel, 1990; Roff, 1994):

1. The particulate approach ignores the potential contributions of other unmeasured traits, and thus any additional tradeoffs that may be in force. This criticism has been answered in part by extending models to multiple factors (Lloyd and Venable, 1992), or by applying methods of dynamic modeling or programming based on simulations that make use of iterative techniques applied to non-linear differential equations (Mangel and Clark, 1988; Mangel and Ludwig, 1992; Clark, 1993; Rand et al., 1994).

2. The genetic details of the phenotypic expression are ignored, that is, the mechanisms underlying tradeoffs are unknown (Weissing, 1996). These details can be examined, however, such as in the studies by Riechert and Maynard Smith (1989).

3. The populations may not be in equilibrium.

4. Genetic variation for the traits in question is simply assumed to be present.

5. de Laguerie et al. (1993) allowed tradeoffs of physiological response to richness of the environment to be non-linear and found ESS calculations to be dependent on the resource level. In these models, selection can act in different directions depending on the quality of the environment.

Several authors have explicitly compared quantitative genetic and optimality models, and have found the two to be equivalent under some conditions (Charlesworth, 1990b; Abrams et al., 1993; Roff, 1994). Because these two classes of models have somewhat different aims, it is not unreasonable to choose one approach over the other where appropriate, or even to investigate the results of both approaches (Moore and Boake, 1994; Weeks and Meffe, 1996).

Complexity Theory

Recent years have seen the creation and rise to fame of a theoretical framework proposed to deal with systems displaying emergent properties—chaos theory and its latest generalization, complexity theory (Holland, 1988; Bak and Chen, 1991; Kauffman, 1993; Perry, 1995; Ferriere and Fox, 1995). The basic tenet of this approach is that complex systems cannot be understood by dissecting them into elemental constituents. To do so leaves out one very basic property, namely, the interactions among the parts, which, in turn, explain the complex behavior of these systems, sometimes referred to as "emergent properties." A fundamental corollary of complexity/chaos theory is that the emergent properties of a complex system may be understood as general patterns even though the specific long-term behavior of such systems might be unpredictable *in principle* (i.e., not simply because of a lack of computing power or time: Wolfram, 1984; Langton, 1986; Arthur, 1988; Kauffman, 1988).

More specifically, complexity theorists classify systems into three broad classes: subcritical, critical, and supercritical (Bak and Chen, 1991). Subcritical systems are characterized by a very limited number of interactions among parts, and display static or very simple behaviors. These situations can be dealt with using classical analytical approaches. Supercritical systems include a very high number of interconnections among parts, leading to highly complex and mostly unpredictable behaviors. The study of these phenomena (such as weather patterns) formed the foundation for the development of chaos theory in the early 1960s. Critical systems are potentially the most relevant from a biological standpoint. They display complex interactions, but the degree of connectivity among their parts is intermediate, and the behavior, although usually predictable, may be punctuated by spurts of quasi-chaotic changes that account for the emergent properties (Kauffman and Levin, 1987; Kauffman, 1991).

The models proposed have been applied to a range of difficult or previously intractable problems, from evolution under natural selection (Kauffman and Smith, 1986; Ferriere and Fox, 1995), to the mechanics of language and learning (Holland, 1986), to the stability (or lack thereof) of economic webs (Arthur, 1988; Holland, 1988; Kauffman, 1988). This flexibility has attracted both interest and criticism to the whole endeavor. The interest is aroused by the promise of a unified theory to explain why a deep understanding of so many complex natural phenomena has proved elusive to modern scientific practice. Criticisms of complexity theory attack this promise to explain everything, and anticipate its falling seriously short of its goals, as happened to the closely related catastrophe theory of René Thom in the 1970s (Bazin and Saunders, 1978; Waddington, 1979). In fact, like catastrophe theory, most of the models proposed by complexity theory "explain" natural phenomena only by analogy, and are largely untestable by the usual scientific methods (e.g., they generally do not make quantitative predictions). On the other hand, it may truly be that such systems are beyond the reach of deterministic science, and that we will have to be content to understand general patterns without being able to predict specific outcomes. It will be fascinating to follow the development of this field either into a fundamental rethinking of modern science, or into another blind alley, of interest only to historians and philosophers of science (Depew and Weber, 1995).

EMPIRICAL APPROACHES

The experimental study of adaptive phenotypic evolution (Rose and Lauder, 1996) also has a rich and controversial literature. In the following sections we will briefly discuss some of the major avenues of investigation that experimental biologists are using to tackle the problem. This is meant as a summary and introduction to the literature and not as a review. Later we will discuss some specific applications of empirical studies in more detail.

Comparative Studies

While comparative studies have been part of the backbone of evolutionary biology since its beginning (after all, *The Origin* is but a long and very detailed comparative study . . .), the modern approach makes explicit use of phylogenetic information (Harvey and Pagel, 1991; Silvertown and Dodd, 1996; Titus and Larson, 1996; Ward and Seely, 1996). A large set of methods has been developed not only to reconstruct phylogenies but also to use them in studies of the evolution of morphological, life history, and behavioral characters (Felsenstein, 1985, 1988; Gittleman and Kot, 1990; Garland et al., 1992; Gittleman and Decker, 1994; Martins, 1994, 1996). Furthermore, and almost as important from the point of view of the practicing biologist, there is ample software available to carry out such analyses.[9]

The advantage of comparative studies, of course, is that they provide the researcher with a "long view." There is virtually no limit to how far back in time one can go to follow the evolution of a trait of interest. Several cautions must be heeded. First, conclusions from comparative studies are only as good as the phylogeny, and it is often difficult to find a reliable phylogeny for the desired group and taxonomic level that one is interested in. Second, comparative studies are all about patterns, not processes. Therefore, inferences about processes are dubious and require multiple concordant approaches to be validated. Third, there is still discussion about the characteristics and relative efficacy of differ-

---

[9]Software, for a variety of platforms, is available at no cost (e.g., http://evolution.uoregon.edu/~COMPARE/).

ent comparative methods (see references above). Fourth, special attention must be paid to the relationships between the extent of the taxonomic sample, the depth of the phylogeny, and the evolutionary lability of the trait. As an example consider behavioral characters that are known to evolve rapidly. If one focuses on a deep phylogeny and samples just a few taxa from each branch, the conclusion may be reached that there is random evolution of the trait, but such a conclusion is more likely the consequence of superficial sampling that fails to uncover regular patterns embedded within subclades.

Finally, phylogenetic information is only part of the story, and it is the easier part to read. Westoby et al. (1995) pointed out that most phylogenetic "corrections" actually eliminate valuable evolutionary information, by assuming that there is no overlap between the phylogenetic and the ecological signals (i.e., that there is no "phylogeny by ecology" interaction). On the contrary, there are good reasons to expect that closely related species will colonize similar habitats, thereby confounding historical and selective signals. There is no shortcut to solving this problem—detailed information about the ecology of the taxa being included in the phylogeny is essential. We needn't underscore the fact that there is a dearth of such information, a void that is more time-consuming to fill than are attempts to reconstruct the phylogeny itself.

Evolution in the Field

Experiments in the field make it possible to identify the existence of ecotypes (Turesson, 1922), study the extent of local adaptation, and even estimate quantitative genetic parameters such as heritabilities (Riska et al., 1989; Mitchell-Olds and Bergelson, 1990; Bennington and McGraw, 1996; Weigensberg and Roff, 1996). The appeal of experiments with natural populations is that these are as close as a biologist can get to "real" situations. The difficulty they often present is that reality may be much too complex; it may become impossible to distinguish among many competing causal hypotheses. However, in such cases, field experiments can provide preliminary clues for narrowing the numbers of interacting factors, and can thus guide laboratory manipulations that reduce these factors to a logistically feasible subset.

There are several different types of experiments that fall into this category: measurements of natural selection (Endler, 1986), mark-recapture studies, classical transplant trials (Clausen et al., 1940, 1948), and environmental modification studies (pros: Partridge and Harvey, 1988; cons: Reznick, 1992, and Rose et al., 1996). These are most often carried out on plants or on animals that are relatively easy to keep track of (Davies and Snaydon, 1972, 1976; Brodie, 1992; Mitchell, 1993; Godfray, 1995; Kadmon, 1995; LeRoy Poff and Ward, 1995; Campbell, 1996, 1997; Reznick and Travis, 1996). The studies of Reznick and coworkers have shown guppies (*Poecilia reticulata*) to be a particularly good species on which to use an environmental modification approach (e.g., Carvalho et al., 1996; Reznick et al., 1997).

Phenotypic Manipulation

Many distinct techniques can be grouped under this heading (Sinervo and Basolo, 1996). While on the one hand some of the criticisms of this approach are on target—for example, it may not be possible to predict the response of a population to selection based on its response to environmental manipulation—on the other hand, phenotypic manipulation by alteration of the environment *does* indeed provide a wealth of information on the response of a genotype to multiple environments. We believe a genotype's developmental reaction norm *is* relevant to the genotype's past selective history and future potential.

Manipulation of the phenotype can be achieved either directly or by genetic engineering. Sinervo et al. (1992) chose the direct approach, adding or removing yolk from lizard eggs to examine the relationship between egg size and success. They found that fecundity selection favored production of large clutches of smaller eggs, but this was balanced by survival selection favoring larger offspring and thus fewer, larger eggs. However, the largest offspring were not always favored—an intermediate optimum was seen in five of eight comparisons. A genetic engineering approach was taken by Schmitt and coworkers (Schmitt et al., 1995; see also Chapter 3) who substituted altered genes into a constant background and then compared the reaction norms of the controls and mutants.

Another approach is to perform environmental manipulations aimed at triggering a response that should be maladaptive only under certain conditions. The performance and fitness of the typical and "wrong" phenotypes are then compared (Dudley and Schmitt, 1996). Genetic and environmental "phenotypic" manipulations represent an elegant approach to merging mechanistic and ecological information. Of course, there are limits to what can be achieved by altering genes directly—the commonality of extensive pleiotropic and epistatic effects makes it impossible to track causal pathways from the single gene all the way to the complex phenotype. Furthermore, only some organisms and environmental cues are suitable to the manipulations needed for these studies.

Classical Quantitative Genetics

The empirical quantitative genetic approach to studying phenotypic variation of natural populations is well enough established that it is unnecessary to summarize it here (see Falconer and Mackay, 1996; Lynch and Walsh, 1997; Roff, 1997). It relies on field or laboratory estimates of heritabilities, genetic variances, and covariances, as well as regression of trait means against fitness parameters to gauge the amount and distribution of genetic variation for quantitative characters, to detect forms of selection, and to predict possible evolutionary responses.

The advantages and disadvantages of empirical quantitative genetics mirror those already discussed above for its theoretical counterpart. The simplicity and statistical tractability of the approach can be counteracted by the narrow scope of the applicability of the conclusions due to the difficulty of generalizing across populations, environments, or evolutionary time.

Laboratory Evolution

Laboratory evolution and cage experiments are aimed at finding out what kind of phenotypes evolve under pre-set circumstances. This approach is very different from artificial selection in that the experimenter is not selecting trait values defined *a priori*. Instead, the goal is to create a relatively well-defined mini-universe, and to step aside and let "natural" selection take over (Rose et al., 1996; Bell, 1997). Recent, intensive studies with *Drosophila* (Chippindale et al., 1993, 1994, 1996, 1997; Leroi et al., 1994a,b,c) and *E. coli* (Bennett et al., 1992; Bennett and Lenski, 1993, 1996, 1997; Leroi et al., 1994d; Travisano et al., 1995; Mongold et al., 1996; Travisano, 1997) are representative and have led to interesting insights into tradeoffs, life history evolution, and even long-term evolutionary trajectories (at least for bacteria).

An advantage of laboratory evolution is that it represents an intermediate between field experiments and artificial selection. It retains the *a priori* unpredictability of the former, and the potential for controlling many environmental variables typical of the latter. Of course, there are several disadvantages as well (Rose et al., 1996). First of all, it is not applicable to many organisms, leading to questions about the generality of the conclusions. Second, "cryptic" selection (via other, unwanted selection agents) may alter, or render unintelligible, the experimental results. Third, both maternal effects and past evolutionary history are hard to control for, since the laboratory cage will almost always be a rather novel environment for the experimental population (Service and Rose, 1985). Finally, unaccounted genotype by environment interactions (i.e., genetic variation for plasticity) can lead to spurious or uninterpretable results. Of course, if the environment is properly manipulated, such interactions can also lead to interesting insights into the genetic architecture of the population under study.

Artificial Life

A parallel approach to laboratory evolution has been developed in artificial life studies (Langton, 1994). Entire microcosms have been created and monitored during their "natural" progressions (Kauffman, 1993). For example, Ray (1994) has developed a computer world called Tierra, in which organismal and ecological complexity can be seen to evolve. Taylor and Jefferson (1994) provide a general introduction to the types of problems for which artificial life approaches might be useful (see also Mitchell, 1994; Toquenaga and Wade, 1996). How much these experiments can tell us about biological evolution (which is assumed to be a particular example of a general category of systems capable of evolving) remains to be seen.

We believe that all of the above approaches can yield valuable insights into phenotypic evolution. Although some theoretical methods may have limited applicability (or at least should have limited generalizations), each of the empirical ways of cutting the phenotypic pie seem worthy of investigation. They provide us with a series of different instruments to look at overlapping parts of the same problem. For example, one could envision beginning the study of a suspected adaptation with a comparative approach, which would tell the investigator how common that adaptation is, how frequently it evolved, and if it is subject to phylogenetic inertia. A quantitative genetic study of natural populations might then reveal current genetic variation or selection on that trait. Field investigations could point to ecological factors that determine the value of the trait, and phenotypic manipulations may help to uncover its mechanistic basis and to understand its functional ecology. Finally, laboratory evolution studies could test the predictions of functional hypotheses by subjecting a population to a determined set of conditions that should favor the evolution of the structure under investigation.

A DIFFERENT VIEW: THE DEVELOPMENTAL NORM OF REACTION

The battle between Fisher and Wright over primacy of their world views continued to be waged among their supporters, and is still with us today. Fisher's particulate view of genes combined with the supremacy of natural selection stands in opposition to Wright's view of a network of gene interactions and a fundamental role for random processes in evolutionary change. Depew and Weber (1995) present a superb account of these skirmishes. E. B. Ford and his school of ecological genetics sought (successfully) to demonstrate that many phenotypic differences among populations are adaptive. The culmination of Fisher's particulate view of genes is to be found in Richard Dawkins's concept of the selfish gene, in which the gene itself, and not the individual organism, is the object of selection. Dobzhansky and his protégés (most notably Richard Lewontin) held to the view that gene effects were context-dependent, and, although not denying the role of selection, subscribed to a pluralism of evolutionary forces. Depew and Weber point to Kimura's neutral theory as one hyperextension of the Wrightian tradition, in which random events predominate in weaving the evolutionary fabric of genes (and phenotypes?), with other forces (e.g., selection) playing relatively minor roles.

We favor the Wright side of the arguments for the importance of epistasis and genetic drift (and stochastic events in general), but we do not believe that it is sufficient. Specifically, we believe that the role of the environment, and particularly the interaction of genotypes with environmental factors, has been glaringly underestimated (see Sultan, 1992 and Frank and Slatkin, 1992 for a view of Fisher's contribution). This oversight, combined with the enduring necessity for the incorporation of development into the evolutionary synthesis, has provided the catalyst for our pursuit of the integration of genetic, environmental, and developmental perspectives on the evolution of the phenotype (Pigliucci et al., 1996).

Our approach to investigating the thicket of phenotypic complexity is to adopt a comprehensive view of the components of phenotypic expression. Various aspects of phenotypes have been investigated throughout history, with most observations focusing on variation in single characters of adult organisms in a particular environment. But there have also been three major departures from that approach: (1) the study of character change during development (ontogeny), (2) the evaluation of associations between pairs or groups of characters (allometry), and (3) the recognition that organismal form is altered by changes in the environment (plasticity). We see these dynamic concepts as necessary extensions of the predominantly univariate, developmentally static view of phenotypes, and propose that they represent the cardinal units of phenotypic expression.

Ontogenetic Trajectories and Development

The ontogeny of an organism includes all developmental events, both quantitative and qualitative, occurring from the single cell through to the adult stages. As a field of a study, developmental biology is obviously much older than evolutionary biology. Its practitioners have been interested not only in documenting patterns, but in identifying general rules for the production of organismal form, as well as establishing evolutionary relationships. As statistical sophistication has increased, the depiction of the course of development as a trajectory has been reduced to a set of numerical coefficients to represent the beginnings and endings of particular processes, and the rates at which such processes proceed (Alberch et al., 1979). But there is clearly a lot more to the integration of development into evolutionary theory than a series of summary coefficients. There is considerable excitement over the recent contributions of molecular developmental genetic studies in model systems such as *Drosophila*, *Caenorhabditis*, *Arabidopsis*, mice, and sea urchins. An enormous amount of information is being uncovered on cell fate, pattern formation, developmental timing, and the roles of major regulators of morphogenesis such as homeotic genes.

Phenotypic Complexity and Multivariate Allometries

Although D'Arcy Thompson clearly had the big picture in focus when he penned *On Growth and Form* (Thompson, 1917), the statistical concept of allometry is ultimately due to Huxley's interest (1932) in the changes of one character relative to a second one, or relative to some measure of overall organ/body size. In Huxley's view, allometry referred specifically to instances in which the rate of growth of one feature does not equal that of the second feature (i.e., a regression of one trait on the other yields an allometric coefficient not equal to one). Isometry represents the special case where the rates of increase of the two features is the same (i.e., the allometric coefficient equals one).

Huxley's description of developmental allometry has since been expanded (Cock, 1966; Gould, 1966; Cheverud, 1982a) to include regression analyses based on measures of multiple individuals taken at a specific developmental stage (static allometry), and regressions of species means along a phylogeny (evolutionary allometry). Allometric associations have been used to infer shared developmental mechanisms and to examine evolutionary forces underlying the observed patterns. The empirical study of allometry allows us to determine how characters are correlated and therefore integrated into the whole phenotype. In subsequent sections, we broaden the scope of the term "allometry" to include the interrelationships among all categories of traits—morphological, physiological, and behavioral—and we will use this viewpoint in the context of both developmental and environmental phenomena.

Environmental Modifications: Reaction Norms and Phenotypic Plasticity

Environmentally induced changes in phenotypes may take many forms, ranging from changes in physiology, to alterations of morphological structure, to shifts in behavioral repertoires. All of these represent instances of phenotypic plasticity—cases in which a particular genotype has the capacity to produce a different phenotype in response to a change in the environment. The term **norm of reaction** is more inclusive because it encompasses all possibilities of plastic responses, including the particular case of no change of a trait in a new environment. As we have come to expect, Darwin probably recognized the potential of plasticity as an adaptive mechanism, but later biologists deserve the credit for formulating the theory of the evolution of reaction norms, namely Schmalhausen (1949, although he was mostly ignored), Levins (1963, 1968), and Bradshaw (1965, 1973). More recent authors appear to have supplied some of the stimulus for turning the study of reaction norms into the cottage industry it is today (e.g., Via and Lande, 1985; Schlichting, 1986; Sultan, 1987; papers in *BioScience* vol. 39 no. 7).

Developmental Reaction Norms

Once these three components of the phenotype are recognized, the task becomes one of integrating them. Our approach is centered on the concept of the **developmental reaction norm** (**DRN**) (Pigliucci and Schlichting, 1995a; Pigliucci et al., 1996). The developmental reaction norm encompasses: (1) the processes that alter the phenotype throughout the ontogenetic trajectory; (2) the recognition that different aspects of the phenotype are (and must be) correlated; and (3) the ability of a genotype to produce phenotypes in different environments. The rarefied definition of the DRN is: *the complete set of multivariate ontogenetic trajectories that can be produced by a single genotype exposed to all biologically relevant environments* (Figure 1.3). A practical definition is, by necessity, somewhat less stringent: *the set of (multivariate) ontogenetic trajectories produced by a genotype (or sibship) in response to naturally occurring (or experimentally imposed) environmental variation* (Figure 1.4).

We view the DRN as a framework for interconnections among various approaches to understanding the evolution of phenotypes through the elucidation of the developmental programs involved in the epigenetic transition from genotype to phenotype. We see it tying pattern and process together (e.g., Figure 1.1).

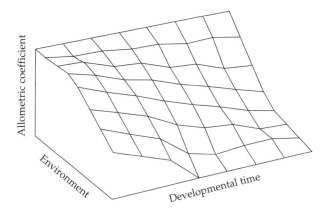

Figure 1.3 A three-dimensional representation of the developmental reaction norm, depicting the dynamics of the relationship between two traits through ontogeny and in different environments. The allometric coefficient in this case is high early in development and tends to decline through ontogeny; it is also dependent on the environment—in some environments the relationship between the traits stays uniformly strong.

We also wish to make the following point very clear from the outset—that we certainly do *not* expect all future studies to incorporate all developmental stages and to grow organisms under multiple environmental conditions. When possible, or appropriate, these variables should be investigated and all results placed in a phylogenetic context. What we are proposing is that the *perspective* of the developmental reaction norm should be consistently applied to any study of variation and evolution in phenotypes. By remembering that organismal form is the result of developmental processes determined by the interplay between a particular genotype and its environment, we become less likely to "wax poet-

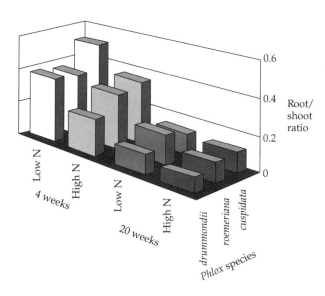

Figure 1.4 A developmental reaction norm representing the effects of nutrients and ontogenetic stage on root/shoot ratios in three species of *Phlox*. Schlichting (unpublished) took sequential harvests of replicate individuals of the three species reported in Schlichting and Levin (1984). Root/shoot ratios declined substantially (~3–4 fold) from 4 weeks up to the final harvests in both the high nutrient (control) and low nutrient treatments. However, the differences between species and treatments are most pronounced at 4 weeks.

ic" about adaptive scenarios, less likely to ascribe phenotypes directly to genes, and more likely to appreciate the scope of an organism's capacity for coping with environmental heterogeneity (Levins and Lewontin, 1985). We will show that this perspective can simplify some seemingly complex phenomena, and that it has the potential to complexify some that are apparently simple. The bottom line may be that the entire developmental reaction norm is impossible to include in a single empirical study. But its greater importance may be to alter the frame of mind that one brings to experimental studies and theoretical modeling—its effect on the fundamental perceptions that guide our attempts and inform our conclusions.

We propose the following as the core concepts arising from this perspective.

THE DEVELOPMENTAL NORM OF REACTION IS ITSELF THE OBJECT OF NATURAL SELECTION We maintain that it is not particular genes (with minor exceptions), nor the discrete phenotypes produced in different environments, that are the objects of selection: *it is the capacity of organisms to deal with environmental contingency that is the focus of selective forces in nature* (McNamara and Houston, 1996). This position is in opposition to the Fisherian tradition (and neo-Fisherians such as Dawkins and Lande), and in agreement with the Dobzhansky-Lewontin-Levins-Gould view of the irreducibility of whole organisms and the inextricability of organism from environment. Organisms grow, develop, and reproduce in heterogeneous environments; the abilities to process environmental information and to respond appropriately to changes are paramount in the adaptive evolutionary process. Organisms lacking this capacity are long since extinct! Even the simplest extant bacteria have marvelously complicated networks of environmentally determined responses; in fact, the very first model of gene regulation (Jacob and Monod, 1961) was derived from the plastic response of the *lac* operon to changes in food supply! In the case of the *lac* operon it is an *external* environmental factor that effects a regulatory shift. We expressly extend the concept of developmental reaction norm to encompass adjustments made in response to changes in the *internal* environment as well. Thus, "typical" developmental sequences (e.g., cell fate determination) and plastic macroenvironmental responses are simply manifestations of different scales of "environmental" effects along a continuum (Rachootin and Thomson, 1981). Nijhout (1990) explicitly adopted such a perspective with his suggestion that all gene expression is ultimately environment-dependent.[10]

---

[10] We offer, as an example of the artificial nature of this dichotomy, the report by Bry et al. (1996) that members of the indigenous bacterial flora are essential for completion of mammalian intestinal epithelial cell differentiation. Should we classify this as internal or external environmental influence? We maintain that the distinction is irrelevant.

BOTH THE TYPICAL SEQUENCE OF DEVELOPMENT OF COMPLEX PHENOTYPES AND THE CAPACITY TO RESPOND TO ENVIRONMENTAL VARIABILITY (INTERNAL OR EXTERNAL) REQUIRE FLEXIBLE SYSTEMS OF BALANCED REGULATORY CONTROL The unfolding of ontogeny from the egg to the adult is an innately hierarchical process that continues to astonish us with its intricacy. However, the accuracy with which it sails towards its ultimate destination is not necessarily a reflection of an invariant sequence of steps—all sorts of environmental insults and genetic exigencies can be absorbed as numerous course corrections are made. The obvious answer to the question "How is such buffering achieved?" is "By means of regulatory control." It has been far less obvious how such regulatory systems can be constructed to enable the production of a bacterium, much less a 300-foot *Eucalyptus*, a 50 ton dinosaur, or a cellular slime mold.

Evidence is accumulating that duplication of genetic material has allowed divergence of both gene function and specificity of control. Even the tens of thousands of genes in vertebrate genomes are composed of derivatives of only about 1000 gene families (Chothia, 1994). The diversification of duplicated gene functions has been especially well documented (e.g., Hunkapillar and Hood, 1986; Doyle, 1994; Holland et al., 1994; Henikoff et al., 1997; Tatusov et al., 1997; Mermall et al., 1998). For example, genes of the MADS-box family have been implicated in affecting the identity of plant meristems and floral organs (Coen and Meyerowitz, 1991a; Meyerowitz, 1994a; Motte et al., 1998; see Chapter 8). There have been three or more major duplications in this gene family (Figure 1.5) that date to the mid- to late-Paleozoic, before flowering plants (Purugganan et al., 1995; Munster et al., 1997; Purugganan, 1997).

One interesting finding has been that of redundant duplicated systems (Gralla, 1991; Thomas, 1993; Pickett and Meeks-Wagner, 1995). Some of these have been discovered when mutants of genes of known function have little or no effect (Davis and Capecchi, 1996). These redundant genes perform similar functions to their antecedents, and multiple copies of regulatory elements may allow for a separate or more precise control of expression. Smith (1990) proposed that differential regulation of genes within multigene families represents the molecular basis of phenotypic plasticity.

THE INNATE COMPLEXITY OF GENETIC SYSTEMS NECESSARILY LEADS TO EMERGENT PROPERTIES ARISING FROM EPIGENETIC PROCESSES THAT INTEGRATE THE OUTPUT COMPONENTS OF MYRIAD LOCAL GENETIC PROGRAMS INTO A FUNCTIONAL GLOBAL PHENOTYPE Wright, with his typical insight, perceived how the control of complex developmental programs would be achieved:

> Within the organism as a more or less integrated reaction system, there is a hierarchy of subordinate reaction systems, each with considerable independence, as shown by capacities for self differentiation. (1941, pp. 168–169)

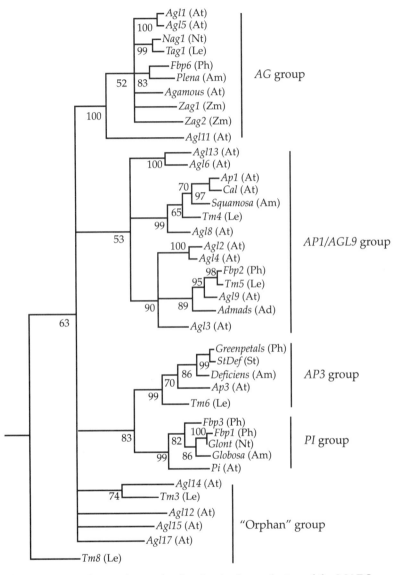

Figure 1.5 Evidence of gene duplication in the evolution of the MADS-box plant gene family, involved in development of the flower. Groups: *AG, agamous; AP, apetala; PI, pistillata.* Abbreviations in parentheses identify various species of flowering plants. (From Purugganan et al., 1995.)

This perspective has been reified in complexity theory as the "critical" N–K system referred to above—systems of numerous (N) elements are loosely bound together by an intermediate number (K) of connections (Bak and Chen, 1991).

Examination of any biochemistry text, with its diagrams of enzymatic mediation of metabolic pathways, will reinforce the imagery of networks and interaction.

Current research focusing on cell signaling and transduction pathways is pinpointing the importance of regulatory cascades in development (Simon et al., 1992; Egan and Weinberg, 1993; Kimble, 1994; Klauck et al., 1996). How, though, do these connections and pathways arise and diversify? (Rollo, 1994). Stated in another way, how has the epigenetic system evolved? We have already alluded to the work of theorists such as Cheverud and Wagner on the evolution of pleiotropy and trait coupling, and we will return to these themes in Chapters 7 and 9. Of immediate interest is the concept of emergence. Colloquially defined as "the whole is greater than the sum of its parts," the biological demonstration of emergence is not exactly straightforward.[11] Perhaps our difficulty with it stems from the fact that the origin of life itself represents transcendence of the emergence boundary.

It seems clear that, no matter how strong our belief in the power of reductionism as an explanatory scheme, the nature of the phenotype of any organism cannot be mechanistically deduced, even if we possess a complete DNA sequence of its genome (Levins and Lewontin, 1985; Nijhout, 1990). The elucidation of the utterly fascinating (and mind-numbing) gymnastics that comprise transcription and translation (Lewin, 1997) has definitively crushed that hope. Thus "emergence" arises somewhere between DNA and the phenotype. This black box, often referred to as epigenetics, is only now being perforated with numerous small holes, shedding some dim light on portions of its contents.

The concept of emergence in epigenetic systems also has a parallel incarnation in complexity theory. Kauffman (1993) has suggested that only "critical" systems have the appropriate combination of numbers of genes and loose connectivity to display emergent properties, and that "critical" biological systems are likely to be the most successful (evolutionarily). Ultimately, the emergent properties of epigenetic systems may frustrate attempts to construct a detailed flow chart of the sequence of events linking genotype and phenotype. However, given the ubiquity of emergence in living systems, the effort to recognize the processes by which it is exhibited may be greatly repaid.

WHAT WILL FOLLOW

Our goal in this book is to demonstrate the utility of the perspective of the developmental reaction norm for gaining insight into processes of phenotypic evolution. Before that, however, we will provide and discuss some background on the antecedents to our approach. In Chapter 2 we recognize the contributions

---

[11] We can grasp the full meaning of an "emergent property" by considering a very familiar simple example—the binding of hydrogen and oxygen to form water. The physical-chemical properties of water cannot be understood by simply adding together the characteristics of the two types of atoms. The key element is in the *interaction* between the atoms.

of a cadre of accomplished researchers who focused on the evolutionary importance of reaction norms or development (or both), with special attention on
Goldschmidt, Waddington, Schmalhausen, and the euphonious Clausenkeck-
andhiesey. In Chapters 3, 4, and 5 we provide historical background and an
overview of the current state of affairs for each of the three major phenotypic
components we unite in the DRN—phenotypic plasticity in Chapter 3, allometry
in Chapter 4, and ontogeny in Chapter 5. Obviously, these chapters could be the
subject of three individual books; our intention is to provide a brief tour of the
fields as they have been historically, and are currently, involved in the investigation of phenotypic evolution.

Starting with Chapter 6 we begin the process of applying reaction norm
thinking to problems in phenotypic evolution. Chapter 6 examines the topic of
constraints. More than enough has been written on this topic, but obviously not
enough to dissuade us from applying a DRN perspective to it. We cut through
the terminological logjam in an attempt to achieve a clearer focus on what exactly are the factors limiting phenotypic evolution. In Chapter 7 we investigate the
topic of how the phenotype is integrated, moving from a historical review to a
detailed scrutiny of developmental and environmental influences on the interrelationships of traits. Chapter 8 tackles the slippery subject of epigenetics: What
is it, and what are the components of developmental systems that can evolve?
We very briefly review the explosion in molecular developmental genetics and
assess the impacts of these findings as well as point to further questions of direct
interest to evolutionists. Chapter 9 seeks to bring it all back together. Starting
with the three-way consolidation of allometry, plasticity, and ontogeny in the
DRN, we proceed to a discussion of the theory of evolution in heterogeneous
environments, and the objects of natural selection in such environments. We end
chapter 9 with an exploration of the question of how developmental reaction
norms evolve. In Chapter 10 we discuss the impact of the DRN perspective on
the entrenched view of how phenotypic evolution proceeds, address the question of the origin of novel phenotypes, and then summarize the conceptual
implications of the DRN and suggest areas where it might be effectively applied.

CHAPTER 2

A Brief History of an Alternative Evolutionary Synthesis

All of the components of the developmental reaction norm—changes in the ontogenetic trajectory, correlations among characters, and responses to different environmental conditions—were being investigated by the early 1900s. For a variety of reasons these subjects have rarely been discussed within a common framework and have not been successfully incorporated within the "modern synthesis." We believe this failure of integration presents many problems for the current paradigm. We suggest that there were elements of an "alternative synthesis" developing in the 1930s and 1940s, but that a series of complex historical, sociological, and scientific factors stalled its development. We envision a synthetic view that includes major roles for developmental biology and the study of environmental influences—not mere bit parts—on the evolutionary stage. This chapter will briefly outline the contributions of earlier workers to such a unification, and discuss some of the opposition to their viewpoints.

We identify three distinct historical phases of this prior alternative and unfinished synthesis. The first phase coincided with the rediscovery of Mendel's work and the birth of modern genetics. This phase saw the shaping of the concepts of genotype and phenotype by Johannsen (1911) and the definition of the reaction norm by Woltereck (1909). The second phase was the concurrent formation of a theoretical framework for macroevolution by Goldschmidt, Schmalhausen, and Waddington that was antithetical to, or at least outside of, the developing neo-Darwinian synthesis. The third phase, which partially overlapped the second, consisted of empirical investigations of the ideas put forward by Schmalhausen (and Baldwin; see below). The most notable of these are the benchmark works of Waddington (1942, 1957), and Clausen, Keck, and Hiesey (Clausen et al., 1940, 1945, 1948; Clausen, 1951; Clausen and Hiesey, 1958)—contributions that to this day are unmatched in scope or detail, but are generally unmentioned in most considerations of the development of the synthesis. Finally, we recognize the contemporary renaissance in studies on development and plasticity as the fourth and, hopefully, ultimate stage.

THE PRECURSORS: BALDWIN, WOLTERECK, AND JOHANNSEN

Around the turn of the century, and prior to the rediscovery of Mendel's work, Darwinism was in a period of eclipse (Bowler, 1983). The challenge of Lamarckism was stronger than ever, and Darwin's inability to propose a mechanism for the inheritance of phenotypic changes represented a major challenge for contemporary evolutionary biologists. James Mark Baldwin, a comparative developmental psychologist, published in 1896 a fundamental paper in *The American Naturalist* that single-handedly initiated a new line of thinking. This paper claimed that a straightforward application of Darwin's principle of natural selection was sufficient to explain the origin of phenotypic novelties without any recourse to Lamarck, initiating a series of theoretical and experimental investigations throughout the early part of the twentieth century.

Baldwin's "new factor in evolution" (Baldwin, 1896) arose from the realization that individuals differ not only in their phenotypic attributes (as Darwin knew well), but in the way those attributes are altered by changing environmental circumstances (in their reaction norms). He argued, from Darwinian principles, that those organisms that are more "adaptable" (i.e., plastic) to new circumstances are bound to leave more progeny, to breed among themselves, and to therefore shape the direction of evolutionary change in the population.

The "Baldwin effect" is a clever Darwinian reinterpretation of situations that might otherwise appear Lamarckian.[1] Baldwin went on to elaborate the idea, writing *Development and Evolution* (Baldwin, 1902), which was widely read by contributors to both the orthodox (e.g., Simpson, 1953) and alternative modern syntheses. It is important to emphasize, however, that Baldwin's arguments were entirely intuitive, based on abstract reasoning without empirical evidence. Such experimental support only came much later, as we shall see, thanks mostly to Waddington, and to others catalyzed by his work.

Independent from Baldwin, there was an early debate within the new discipline of genetics that would have great consequences not just for our understanding of the evolution of reaction norms, but for the concept of genotype itself. R. Woltereck (1909) proposed the term reaction norms (initially called "phenotypic curves") to describe a striking phenomenon he observed in clones of the microscopic crustacean *Daphnia*. He found that during a season, samples of *Daphnia* from a given pond would be characterized by markedly different morphologies, with the head of the organism changing in both size and shape. To determine if differences between strains were constant, and therefore within the realm of genetics, Woltereck plotted the phenotypic trait head size against a measure of the environment for each clone of *Daphnia* studied. The results were the first published reaction norms, and showed differences in response of head size to food availability. This seemed a perfect example of the distinction between phenotype and genotype that was then being developed by W. Johannsen (1911).

However, in the course of his investigations Woltereck also found an unexpected phenomenon that shook his faith in the usefulness of the concept of genotype proposed by Johannsen. He discovered some genetically distinct clones of *Daphnia* that displayed identical, or very similar, phenotypes under certain environmental conditions (Figure 2.1). Woltereck failed to grasp the evolutionary relevance of this observation; he concluded that there was no distinction between genotype and phenotype, because differences between purportedly distinct genotypes could disappear "simply" due to environmental influences.[2]

Johannsen (1911) was quick to praise Woltereck's experimental approach, pointing to the close similarity between the ideas of genotype and of reaction norms:

> We do not know a "genotype," but we are able to demonstrate "genotypical" differences or accordances. . . . The very appropriate German term

---

[1] At least two other scientists reached conclusions similar to those of Baldwin more or less concurrently; namely, Conway Lloyd Morgan, another comparative psychologist, and Henry Fairfield Osborn, a biologist (see Gottlieb, 1992, for a thorough discussion).

[2] This marked the beginning of a long history of either misunderstanding or neglecting the role of environmental effects in genetics and evolution. The view of plasticity as unimportant in evolution or as a source of experimental "noise" continued through the modern synthesis and well into the 1980s, and the notion that it is non-genetic can still be found today.

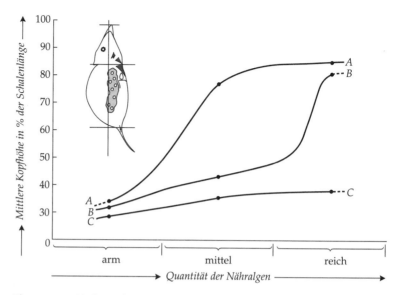

Figure 2.1 Woltereck's original representation of reaction norms in clones from three populations (A, B, and C) of *Hyalodaphnia*, recording changes in head size relative to the rest of the body (*y*-axis) as a function of food availability (low, intermediate, and high algae, *x*-axis). (Redrawn from Woltereck, 1909.)

"Reaktionsnorm" used by Woltereck is, as may be seen, nearly synonymous with "genotype," in so far as the "Reaktionsnorm" is the sum total of the potentialities of the zygotes in question. That these potentialities are partly separable ("segregating" after hybridization) is adequately expressed by the "genotype" as composed of "genes." The "Reaktionsnorm" emphasizes the diversity and still the unity in the behavior of the individual organism. . . . Thence, the notion of "Reaktionsnorm" is fully compatible with the genotype-conception.

However, Johannsen went on to criticize Woltereck's misunderstanding of the genotype concept:

Woltereck might apply [to the discrepancy between his observations and his deductions] the famous words from Harvey's times: *"video sed non credo"* [I see, yet I do not believe]. The phenotype-curves of the Daphnias in question sometimes show rather constant differences between the pure strains compared; but mostly this is not the case. Especially under extreme conditions . . . some of the curves are confluent. So the differences between the phenotype-curves may vary considerably or may even vanish entirely. . . . But when Woltereck thinks that these facts are inconsistent with the existence of constant differences between the genotypes, he shows himself to have totally misunderstood the question! Of course the *phenotypes* of the special characters, *i.e., the reactions of the genotypical constituents*, may under different conditions exhibit all possible forms of transition or transgression—this has nothing at all to do with constancy or inconstancy of genotypical differences. . . . Every student of genetics ought to know this.

Since *we* are amazed at how many students of genetics still do not fully realize the consequences of Johannsen's point, we are more forgiving toward Woltereck, who, after all, was just then being exposed to the novel ideas of a genotype mapping to a phenotype, and to a genotype's capability of responding to environmental influences.

THE THEORETICAL FRAMEWORK: GOLDSCHMIDT AND SCHMALHAUSEN

Several decades passed between Baldwin's intuition, Johannsen's and Woltereck's formalization of genotypes and reaction norms, and the first coordinated and complete theoretical attacks on the foundations of the maturing neo-Darwinian paradigm. As the major monographs of the modern synthesis were published (Dobzhansky, 1937; Huxley, 1942; Mayr, 1942; Simpson, 1944), two very influential "alternative" books also appeared. The German geneticist Richard Goldschmidt published *The Material Basis of Evolution* in 1940, while the Russian Ivan Ivanovich Schmalhausen published his *Factors of Evolution* in 1946 (in Russian, translated to English in 1949, largely through the efforts of Theodosius Dobzhansky). Both authors expressed dissatisfaction toward the developing Dobzhansky-Mayr-Huxley-Simpson synthesis of Darwin's ideas, population genetics, systematics, and paleontology. For Goldschmidt, the synthesis was simply unable to explain macroevolutionary phenomena, including the origin of species and of phenotypic novelties, and he advocated radical revisions. Schmalhausen's position was more moderate (which possibly contributed to a greater acceptance of his ideas)—that development was the missing link in the synthetic view, and that it should be brought to the forefront, thereby augmenting and completing the synthesis. Waddington's ideas on development and evolution appeared in definitive form (*The Strategy of the Genes*, 1957) much later than Goldschmidt's or Schmalhausen's, but his basic intuitions and their experimental support were published beginning in 1942 (Waddington, 1942, 1952, 1953a,b).

The Material Basis of Evolution was a vehement attack on the nascent "modern" evolutionary synthesis, based on the view that neo-Darwinism provided a parochial explanation of evolution limited to microevolutionary phenomena. For Goldschmidt, the real challenge was to explain the evolution of new species and especially new body plans; these, he felt, were clearly beyond the limits of the theoretical framework being shaped by the synthesists. According to

Darwin, local varieties were the raw material from which new species emerge, and Goldschmidt plunged himself into empirically demonstrating this point. The first part of *Material Basis* is a long, detailed account of years of experiments with local races of moths, begun in the hopes of observing or artificially generating the actual birth of a new species. But, after more than 25 years of experiments, he concluded that new species do not simply arise as extensions of preexisting varieties.

The controversial second part of *Material Basis* discusses the evidence for and the possible mechanisms behind macroevolution. This part is remembered (like Goldschmidt himself) chiefly for the reference to the infamous "hopeful monster," a creature characterized by a systemic mutation and scorned by others as having an infinitesimal probability of changing the course of evolution. Goldschmidt's view of macroevolution, however, was much more comprehensive than this caricature. He begins his treatment of macroevolution by attacking the concept of the gene,[3] arguing that given the complexity of phenotypes, only major rearrangements of the already existing pieces were likely to bring workable alternative solutions to the morphology of an organism. In his view, this could not possibly be achieved by gradual modifications of the genetic material. He pointed out plentiful experimental evidence for the dramatic phenotypic effects of macromutations, for example, the contrast of the normal diploid tobacco plant (*Nicotiana tabacum*) with ten different trisomic lines originated by chromosomal mutation (Figure 2.2). The range of morphologies and architectures was staggering, and all the plants were healthy and capable of surviving under natural conditions!

Goldschmidt reviewed examples of environmentally induced phenotypes mimicking known mutations, for which he coined the term *phenocopy*. In his view, phenocopies (Figure 2.3) are a phenomenon of general biological importance because they provide hints about gene action. He thought that phenocopies result from environmental conditions (such as high or low temperature) that slow down or accelerate chemical reactions in an organism during development, thereby altering adult phenotypes. If a mutation and an environmental change produce the same phenotype, it is logical to conclude that they operate by similar effect—the affected genes working slower in one case and faster in the other. Goldschmidt also recognized the role of the environment in producing divergent phenotypes. For example, he elicited dramatic phenotypic differences in *Lymantria dispar* between full sib caterpillars of equal age and developmental stage, raised under normal, optimal, and suboptimal conditions.

Goldschmidt then proceeded to consider how the internal environment affects the creation of phenotypes, discussing in particular what was then

---

[3] A rather unusual thing for a geneticist to do!

Figure 2.2   Normal diploid *Nicotiana tabacum* (upper left) and several trisomic variants. Notice the range of morphologies and architectures. (From Goldschmidt, 1940, *The Material Basis of Evolution*. © Yale University Press.)

known about hormonal action. He presented the example of the developmental effects of the application of thyroxin to mudskippers (*Periophthalmus variabilis*, Figure 2.4) and the consequences for their morphology and behavior. Hormonally treated fish developed longer appendages, which they actually used as levers to move around out of the water! Goldschmidt concluded that the regulatory action of both external and internal environments can produce astonishingly different phenotypes within genetically similar backgrounds.

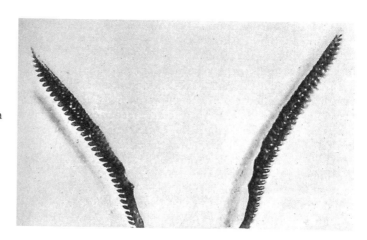

Figure 2.3 An example of phenocopy from Goldschmidt's book. On the left, an unusual lamellated antenna of a mutant *Lymantria dispar* moth. On the right, an identical phenotype obtained in a normal individual by exposure to high temperature. (From Goldschmidt, 1940, *The Material Basis of Evolution.* © Yale University Press.)

Finally, Goldschmidt discussed the effect of mutations on early development. He used examples of homeosis (fashionable then, as now) to explain the production of radically altered phenotypes by developmental mutants. The emphasis on action early in development was based on de Beer's idea that characters are more tightly related at the beginning of the ontogenetic trajectory (de Beer,

Figure 2.4 (Left, top) Normal mudskipper (*Periophthalmus variabilis*). (Left, bottom) Phenotype following thyroxin application. Notice the longer appendages in the treated animal. (Above) Thyroxin treatment induces novel behaviors, such as the use of appendages for terrestrial locomotion. (From Goldschmidt, 1940, *The Material Basis of Evolution.* © Yale University Press.)

1930)—mutation and selection acting then would be more effective in reshaping the entire phenotype.[4] He emphasized the importance of change in development as an impetus for evolution, was clearly aware of the potential importance of environment for modifying phenotypes, and espoused a view of chromosomal mutations with major phenotypic effects. The essence of these ideas is somewhat analogous to the concept of the developmental reaction norm. Our view of the importance of the evolution of regulatory genes is different, of course; but in each case, what evolves is the control system and not just the structural units of a genotype.

Schmalhausen (1949) published his main contribution to evolutionary biology under the title *Factors of Evolution: The Theory of Stabilizing Selection*. His basic idea is simple and powerful—evolution proceeds by altering the developmental systems of living organisms, changing the norm of reaction to first cope with, and then anticipate, environmental stimuli. Schmalhausen began by defining and discussing the concept of norm of reaction. He noted the manifold effects of mutation as well as the similarity of mutational and environmentally induced phenotypes (here addressing the idea of phenocopy, so clearly expressed by Goldschmidt). The emphasis on development is evident from the beginning of *Factors*: one of his first examples is the temperature-dependence of pigment formation in the hairs of Himalayan rabbits (Figure 2.5). Those results showed that disparate body parts responded to different temperature thresholds, demonstrating the developmental specificity of the interaction between the genotype and environment.

Figure 2.5 Pigment formation in the Himalayan rabbit depends on crossing temperature thresholds that are specific for different parts of the body. This indicates flexibility of the developmental system in response to environmental influences. (From I. I. Schmalhausen, 1949, *Factors of Evolution*. Reprinted 1986, The University of Chicago Press.)

---

[4] This point has been reemphasized by West-Eberhard (1989) and experimentally supported by Atchley (1984); but there is no dearth of examples of alteration of later developmental stages (see Chapter 5).

Other cases presented by Schmalhausen in *Factors* illustrate a variety of the features of phenotypic plasticity. He pointed out that variations in critical aspects of the natural environment of organisms may trigger complex morphological responses that are not easily interpreted from an adaptive standpoint. For example, the posterior segments in the crustacean *Artemia salina* are substantially modified in response to differences in the concentration of salt in the water. The number of bristles increases with salt concentration, but then decreases sharply as concentration continues to rise (it remains unclear why having more bristles at intermediate salinity would confer an advantage of any type). The pennant/vestigial system in *Drosophila melanogaster* is an example of how mutations at a single locus can produce a bewildering array of wing sizes and shapes. Only three alleles are involved, *wild type, pennant,* and *vestigial,* but the reaction norms of each allele to temperature are quite different (Figure 2.6).[5]

Other cases of developmental plasticity cited by Schmalhausen are clearly adaptive. The shape of the leaves of semi-aquatic plants is directly connected to their ability to exchange gas under water or in the air (a feature largely dependent on the total surface available for gas exchange). The case of *Sagittaria sagittifolia* is prototypical. Terrestrial populations of this species show the normal arrow-shaped (i.e., sagittate) leaves (Figure 2.7). Entirely aquatic populations, on the other hand, have a developmental system that produces elongated linear leaves. Populations subjected to periodical flooding, or that live in more variable environments, display adaptive plasticity in the ability to produce both types of leaves, one form under water, the other above (heterophylly).

Schmalhausen also introduced two essential concepts in *Factors of Evolution.* The first was the distinction between a labile and a stable organism (i.e., a plastic and a non-plastic one). The second was the idea of a "normal phenotype" (i.e., the wild type reaction norm) that is the result of past directional selection and is maintained by selection against extreme phenotypes. Schmalhausen referred to this maintenance of the normal phenotype as "stabilizing selection." It is important to note that this conception, still employed by modern evolutionary biologists, is just one of two ways in which Schmalhausen used the term (Gottlieb, 1992).

His second, expanded use of the phrase "stabilizing selection" refers to a more elaborate process that is akin to the Baldwin effect and to Waddington's "genetic assimilation." In this case, a new norm of reaction initially appears when environmental changes reveal a previously unexposed portion of the reaction norm. This is followed by selection for mutations that can improve the norm of reaction in the direction of the environmental change, and then by stabilizing selec-

---

[5] Even the dominance relationships between *vestigial* and *pennant* are altered by the environment—further linking basic genetic phenomena such as dominance, pleiotropy, and epistasis to phenotypic plasticity.

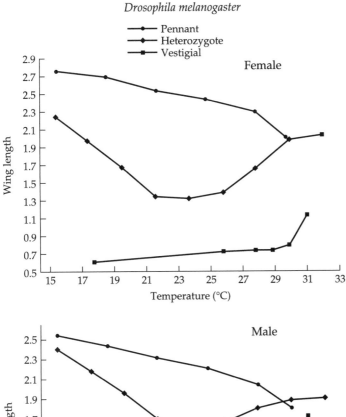

Figure 2.6   The *pennant* and *vestigial* alleles at a single locus in *Drosophila melanogaster* affect wing size and shape. They exhibit dramatically different reaction norms to temperature; however, they are also affected by homo/heterozygosity and sex. (From data in Schmalhausen, 1949.)

tion on the new norm of reaction. A scenario of this type can be created to describe the loss of heterophylly in aquatic plants. For example, if the causes of submergence become increasingly rare, then the developmental system may be selected for promptly generating sagittate but not linear leaves. If the process continues long enough (and the environment does not revert), the plants may

Figure 2.7   Heterophylly in *Sagittaria sagittifolia*. Terrestrial; forms produce saggitate leaves only; aquatic forms produce linear leaves only; but populations living in semi-aquatic or variable environments are capable of both developmental programs. (From I. I. Schmalhausen, 1949, *Factors of Evolution*. Reprinted 1986, The University of Chicago Press.)

lose the ability to develop linear leaves altogether; eventually, even if they *are* exposed to water, new mutations and new selection are necessary before the old norm of reaction can be reacquired. Schmalhausen argued that such a sequence of events implies a shift from a reaction to the environment due to differential allelic sensitivity (a simple result of the biochemistry or physiology of the organism), to a more complex regulatory system capable of not only changing with environmental conditions, but of *anticipating* environmental demands.

Overall, Schmalhausen employs a developmental perspective throughout his writing, and exhibits an implicit understanding of regulatory genetic control. These perspectives, combined with his explicit view of phenotypes as manifestations of reaction norms, bring Schmalhausen closer than Goldschmidt to our concept of the DRN. Both Goldschmidt and Schmalhausen, despite their differences, believed that a large fraction of biological phenomena was unexplainable by the neo-Darwinian synthesis. Their contributions, however, have remained largely ignored or, regretfully, ridiculed. Goldschmidt has been ensconced in the annals of biological history as the originator of the idea of the hopeful monster, an idea that in reality represented only a minor component of his treatment of macroevolution.[6] Schmalhausen's career was destroyed by the rise of Lysenko in

---

[6]He did make a conceptual mark beyond the hopeful monster, however. Mayr's 1942 book, *Systematics and the Origin of Species*, was originally conceived as a direct refutation of Goldschmidt's theses.

the Soviet Union, which rendered him incapable of further elaborating or spreading his ideas. Even though his book was championed by his countryman Dobzhansky, and was known and respected by the synthesists, its perspective was in too many ways different to fit into the neo-Darwinian paradigm. Both authors have enjoyed a revival with the reprinting of their books in the 1980s,[7] and we expect that this time around their ideas will not be forgotten for another half century.

THE EMPIRICISTS: WADDINGTON AND CLAUSEN, KECK, AND HIESEY

There are men of theory, and there are men of action. To be sure, Goldschmidt did experimental work for a long time, but his efforts there had more negative than positive results. His data failed to demonstrate the Darwinian tenet that varieties become species, and he also could not produce a systematic set of experiments to demonstrate a possible alternative. In a similar manner, Schmalhausen's reasoning and numerous examples provided a good theoretical foundation for the importance of reaction norms in nature, but there was no consistent body of evidence about how phenotypes evolve in natural populations. These voids were filled by other giants of the "alternative synthesis," Conrad H. Waddington, and the original "dream team": Jens Clausen, David D. Keck, and William M. Hiesey.

Waddington believed very strongly in the link between evolution and development, and is responsible for coining a number of the terms we currently use to describe the interface between them; for example, canalization, epigenetics, and epigenetic landscape (Hall, 1992b). Central to Waddington's perception of development was the idea of canalization. He described it as the tendency of a genotype to follow the same developmental path even in the face of internal or external perturbations, and he clearly considered it to be a consequence of natural selection. From both its definition and the examples portrayed to support it (e.g., environmentally cued metamorphosis in amphibians, heterophylly in plants), the concept of canalizing selection is very similar to Schmalhausen's idea of stabilizing selection on the new norm. (Schmalhausen acknowledged the profound similarities between his thinking and Waddington's.) The process described by Schmalhausen, of change from the old norm to the new one through an environmentally induced response, mirrors Waddington's notion of genetic assimilation. This concept was originally defined as the avenue by which a phenotypic character that is initially produced in response to some

---

[7] Controversy, however, still follows Goldschmidt. Although Gould praised him in his forward to the reissue of *Material Basis*, Templeton (1982) emphatically disagreed; his perspective was that Goldschmidt had it *all* wrong.

environmental influence is stabilized due to natural selection, and finally is found to occur in the absence of the previously necessary external influence.

Waddington demonstrated the occurrence of genetic assimilation with his famous series of experiments on *Drosophila* that were subjected to heat shock during larval development (Waddington, 1952). He began with a stock of flies that produced a novel phenotype in low frequency when subjected to heat shock at an early larval stage. Selection for this heat shock-induced phenotype resulted in a much larger fraction of the population expressing the phenocopy, indicating that the plasticity of the trait was heritable (Figure 2.8).[8] The most surprising observation came after several generations of selection when the novel phenotype was expressed in the *absence* of the heat shock! This superficially "Lamarckian" result, or the "inheritance of an acquired character," was exactly what Baldwin had predicted—selection had lowered the threshold required to switch to the alternative developmental pathway, so that the environmental stimulus was now unnecessary. In fact, it may be that selection in the opposite direction would be required in order to suppress the appearance of the novel phenotype. In modern terms, the old reaction norm had been altered

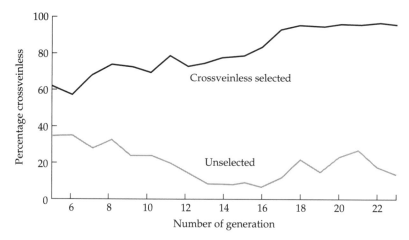

Figure 2.8 Response to selection leading to genetic assimilation in *Drosophila melanogaster*. The crossveinless phenotype initially occurs only after heat shock. The percentage of flies showing this plastic response can be increased by selection. In time the novel phenotype will be produced *even in the absence of the environmental stimulus*. (From Waddington, 1952.)

---

[8]Sweeping yet another classical genetic concept, *expressivity*, into the net of the developmental reaction norm. The level of expressivity can be altered by selecting for a change in the trait's value in a particular environment, i.e., by modifying its plastic response to that factor.

enough so that a phenotype that previously appeared only at an environmental extreme of the norm would now be the normal phenotype (i.e., a loss of plasticity). Waddington and others later generalized such findings for different traits and environmental stimuli.

Waddington's irrefutable evidence notwithstanding, his advocacy of **genetic assimilation** as a fundamental evolutionary process alienated many evolutionary biologists, who criticized his work as describing rare and unnatural phenomena. Most geneticists even today fail to appreciate that Waddington's experiments were performed under extreme conditions in order to rapidly catalyze, in the span of a few generations, what would otherwise be a very elusive, even if widespread, natural process. The criticism often raised against genetic assimilation—that it cannot be observed in nature because it is a transient stage of the evolution of new developmental systems—actually applies to *any* mechanism whose purpose is to explain the same events. This is one of the frustrating aspects of trying to bridge the gap between micro- and macroevolution (i.e., our inability to *observe* speciation events).

Waddington realized that his ideas were very close to both Baldwin's and Schmalhausen's.[9] In fact, he made a conscious attempt to differentiate among the three views (Waddington, 1961). Are there, however, any relevant differences among Baldwin, Waddington, and Schmalhausen? While the case is a bit uncertain for Baldwin (since most of his writings are couched in pre-Mendelian terminology), neither we nor Gilbert (1994) could discriminate between the ideas of Waddington and Schmalhausen, other than in their individual emphases. Waddington was the more radical one, arguing for a substantial revision of the synthesis, not just an integration of the missing developmental focus (Waddington, 1953b). Thus, like Goldschmidt before him (and unlike Schmalhausen), he caused substantial indigestion for the fathers of the neo-Darwinian paradigm.

A very different experimental approach to highlight areas of the synthesis in need of improvement was taken by Jens Clausen's group at the Carnegie Institution (including David Keck and William Hiesey, and henceforth referred to as CKH). The founder of modern ecological genetics is generally recognized to be E. B. Ford (Ford, 1931; Mayr and Provine, 1980; Reznick and Travis, 1996).[10] Although we have no argument concerning the magnitude of Ford's contribution to contemporary evolutionary biology, we wish to redress this typically animal-centered view of history. Modern ecological genetic perspectives in botani-

---

[9] Waddington was annoyed at Dobzhansky for citing Schmalhausen every time there was a need to say something about development. According to Gilbert (1994), Dobzhansky apologized, but offered that Schmalhausen needed "any support that we may give him," a reference to Lysenkoism . . . a nice example of the influence of personal and nationalistic feelings in science.

[10] It is clear that Weldon actually was carrying out an ecological genetics research program much earlier (1890s), but he left no apparent intellectual descendants (Provine, 1971; Depew and Weber, 1995).

cal research can be traced at least back to Turesson who, in 1922, developed the concept of the ecotype (see Clausen et al., 1940, for a historical overview of Turesson's and earlier work). The work of CKH is a successor in that tradition.

Some of CKH's experimental results have advanced to "textbook example" status, for instance, their now-famous demonstration of variation among genotypes in reaction to environmental conditions (Figure 2.9). We believe, however, that their contributions deserve far greater recognition. The true importance of

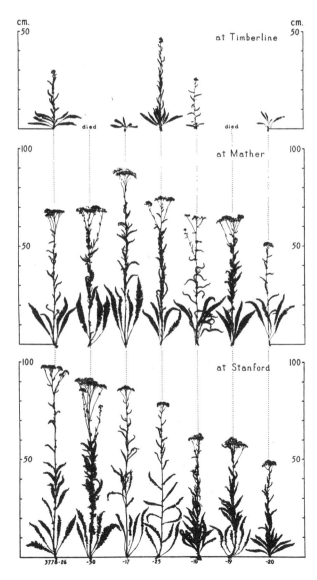

Figure 2.9 Genetic variation and differences in plasticity among seven clones of the Mather population (1400 m) of *Achillea lanulosa* in response to altitude (Timberline, 3050m; Stanford, 30m). Notice how the clones do poorly at the higher elevation, suggesting adaptation to lower elevation. Results are complementary for an alpine population (Big Horn Lake, 3350 m) subjected to the same environments (not shown). (From Clausen et al., 1948.)

their work resides in their investigation of the potential ecological and evolutionary relevance of the nature and extent of variation within and between natural populations. CKH's approach to this was based on the same fundamental assumption maintained by the neo-Darwinists—that local populations are the units of evolutionary change. However, like Baldwin more than half a century before them, they realized that variation was not limited to average differences among individuals, and that it could not be considered outside of a specific environmental context. They explicitly attempted to forge connections among ecological genetics, the effects of the environment on phenotypes, and macroevolution and speciation.

Like Goldschmidt before them, CKH were interested in the genetic bases of dramatic morphological change, and recognized the power of chromosomal alterations as a possible mechanism of evolutionary change (Figure 2.10). In contrast to Goldschmidt, however, they suggested that the nature of at least some of

Port Orford   Blodgett   Big Horn Lake

Otter Crest   Selma   Kiska

Figure 2.10 Variation among races of *Achillea lanulosa* (top row) and among races of *A. borealis* (bottom row). The effects of polyploidy are seen by comparing the middle polyploid plants (Blodgett and Selma) with the diploids on either side. (From Lawrence, 1947.)

the genetic differences between species were the same as the genetic differences between local races. They focused many of their genetic analyses on intervarietal crosses, and demonstrated that rather complex phenotypic alterations could be explained in Mendelian terms (although not necessarily by only one or two factors). For example, crosses between the prostrate maritime type of *Layia platyglossa* and the erect inland form yielded an array of intermediate forms in the F_2 generation (Figure 2.11). This focus on intraspecific variation was consistent with their view that macroevolutionary change and speciation form a continuum with ecological genetic divergence.

A superb example of the integration of their research objectives is found in the monumental *Genetic Structure of Ecological Races* (Clausen and Hiesey, 1958; this work includes an exceptional 100-page survey of ecological genetics in plants and some animal species). They made crosses among various races of *Potentilla glandulosa*, utilizing the substantial geographic variation present among populations of this California plant species. Figure 2.12 depicts typical

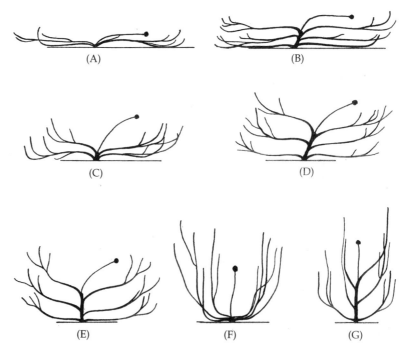

Figure 2.11 The maritime (A) and inland (G) races of *Layia platyglossa*. The variation among their F_2 progeny (B–F) indicates that a substantial number of genetic differences separate the two ecotypes. (Reprinted from Jens Clausen, *The Evolution of Plant Species*. © 1951 by Cornell University. Used by permission of the publisher.)

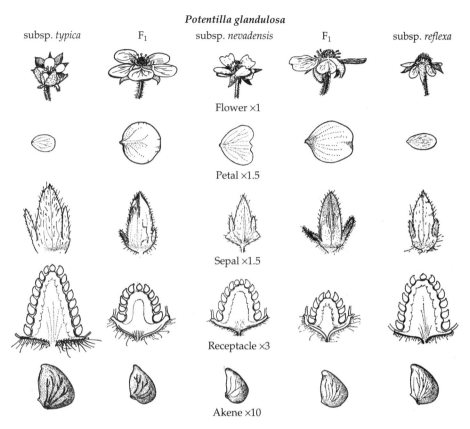

Potentilla glandulosa

subsp. *typica*   F$_1$ subsp. *nevadensis*   F$_1$ subsp. *reflexa*

Flower ×1

Petal ×1.5

Sepal ×1.5

Receptacle ×3

Akene ×10

Figure 2.12   Representative plant traits for offspring of crosses between alpine and lowland races of *Potentilla glandulosa*. Santa Barbara (15 m, ssp. *typica*) × Upper Monarch Lake (3240 m, ssp. *nevadensis*), and Timberline (3050 m, ssp. *nevadensis*) × Oak Grove (760 m, ssp. *reflexa*). (From Clausen and Hiesey, 1958.)

results of these crosses. F$_1$ offspring were grown at Stanford (30 m), and their characteristics, although usually intermediate between the parentals, occasionally transgressed the parental means. The F$_2$ progeny character means and variances often deviated substantially from mid-parental and F$_1$ values at Stanford (Figure 2.13, bottom), indicating that traits were not truly polygenic, and that there were dominance or epistasis relationships among the genes involved in trait expression.

Figure 2.13 also reveals the remarkable environmental dependence of these genetic effects. The F$_2$ offspring of the Timberline × Oak Grove cross were cloned and grown at three elevations, 30, 1400, and 3050 meters. At the two lower elevations, Mather and Stanford, leaf lengths of F$_1$ plants were more similar to the lower elevation parent (Oak Grove); while the converse was true at the

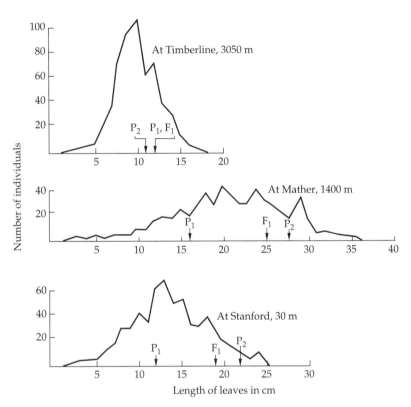

Figure 2.13 Frequency distributions (graphed lines) of the largest rosette leaves among 502 F_2 progeny cloned and grown at three elevations. Means of the parental and F_1 progeny are shown for comparison. Note the differences in the means, in the distributions, and in the degree *and* direction of transgressive segregation among the three sites. (From Clausen and Hiesey, 1958.)

Timberline station. Additionally, whereas the F_2 leaf length means were intermediate between F_1 and P_1 values at Stanford and Mather, those of F_2 plants at Timberline were smaller than all of the earlier generations. The distributions of variance also changed markedly across sites.

The authors present convincing evidence for genotype by environment interaction among the different clones examined, and even report that the broad sense genetic correlation (although they don't call it that) between leaf lengths at the different stations is highest between the two lowland stations. Clausen and Hiesey speculated about differential gene activity at different altitudes, pointing to evidence that an anthocyanin gene was expressed only at the Timberline Station (Clausen and Hiesey, 1958, p. 91). They also noted that the direction of transgression of characters was not always the same in the progeny of the two different alpine × lowland crosses.

They extended their research into the genetic differences between species to take advantage of some of the peculiarities of plants. As a case in point, they considered the complex composed of *Layia glandulosa*, an unnamed relict taxon, and their hybrid. The segregation of floral characters in the F_2 was consistent with two genes controlling the presence of rays, a few other genes involved in the number of rays, three more genes for the length of the pappus, and several genes for its color (Clausen et al., 1947). A maximum of 20 genes, most of them linked, were estimated to be involved in the overall morphological differentiation. The new taxon was an edaphic population adapted to serpentine soil, and CKH's point was that these types of differences may also characterize different genera or even higher taxonomic ranks.

CKH went on to attempt to actually quantify the degree of isolation among taxa, first focusing on intervarietal differences, and finding only partially reduced fertility among subspecies. They then considered cases of natural interspecific hybridization, not just as an approach to study the genetics of speciation, but as representative of natural situations more complex than those encompassed by Mayr's and Dobzhansky's "biological" species concept. A separate and enormous body of work in this vein (along the lines of that of other plant "biosystematists"; e.g., see Anderson, 1949; Stebbins, 1950; Grant, 1981) examines the extensive genetic compatibility spanning several species and at least four recognized sections of *Viola* (summary in Clausen, 1951), as well as extensive studies in *Mimulus* (Hiesey et al., 1971).

The major experimental work for their first book, *Experimental Studies on the Nature of Species*, I (1940) was completed between 1932 and 1937, and volumes II (1945), and III (1948), followed within the decade. However, even in retrospect, CKH are not afforded particular respect. Stebbins (1980), in his analysis of the contributions of botany in *The Evolutionary Synthesis* (Mayr and Provine, 1980), spends nearly two pages on their work. However, much of his commentary is a criticism of their publication record: "During World War Two, they thought that they would help our country by breeding better forage grasses. Their research on speciation in tarweeds was neglected and unpublished. The only publication of their work on species is a short book published by Clausen in 1951 . . . , based on data obtained between 1934 and 1941." Stebbins cites only a short paper by CKH in *The American Naturalist* from 1939, and Clausen's 1951 book; astonishingly, he does not mention any of the three Carnegie Institution books, or their insightful 1947 *American Naturalist* paper on genetics of ecological races. All these were published prior to his 1950 book.[11]

---

[11] The citation battle appears to have been launched earlier. Clausen and Hiesey's 1958 book curiously (intentionally?) makes no citation of the work of Ford or other British ecological geneticists, and the only reference to Stebbins is on the very last page of text (p. 269).

Based on their experimental evidence, CKH saw evolution as a more complex process than the simple series of branching events envisioned by modern biologists. Even though they deviated substantially from Goldschmidt in considering intervarietal differences as the actual basis of macroevolution, they also emphasized the necessity of considering complex interactions between genetics, development, and ecology in order to truly understand macroevolution. Their extension of ecological genetic techniques from intraspecific to interspecific studies is paradigmatic, leading to an appreciation of evolutionary processes that cannot be gained from adherence to the orthodox perspective. Avoiding this more tortuous path can lead us to forget that our goal is to grasp the complexity of nature, and not just to imagine a natural world simple enough to be comprehended.

CHAPTER 3

Reaction Norms and Phenotypic Plasticity

The term **reaction norm** refers to the set of phenotypes that can be produced by an individual genotype that is exposed to different environmental conditions.[1] Originally coined by the German researcher Woltereck (1909), the term was revived by Schmalhausen and was used in its current sense in his book *Factors of Evolution* (Schmalhausen, 1949; see also Dobzhansky and Spassky, 1944). Reaction norms may be plastic or non-plastic, that is, the phenotype may either change or remain fixed in response to environmental change (Figure 3.1A,B). Although phenotypic plasticity and reaction norm are often used interchangeably, plasticity always refers

---

[1] Stearns (1989a) unnecessarily restricted the term to responses to a gradient of a single environmental factor.

(A)

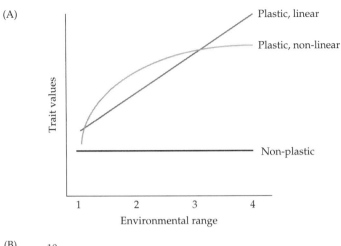

(B)

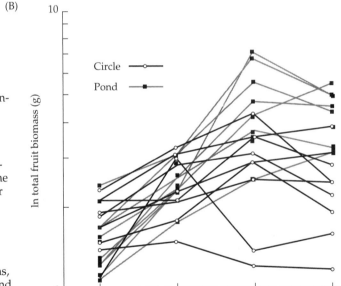

Figure 3.1 (A) Stylized representation of plastic and non-plastic reaction norms. If the environments represent a gradient, the shape of the reaction norm may be linear or non-linear. A perception of linearity may be due to the examination of a limited number of environments. (B) Reaction norms from natural populations of *Polygonum persicaria*. Note the genetic variation in response to water availability both within and between the two populations, Circle and Pond. (After Sultan and Bazzaz, 1993b.)

to a reaction norm, but a reaction norm is not necessarily plastic. Plastic responses are occasionally spectacular, sometimes producing individuals so distinct that they are classified as separate species (or even genera). Rollo and Shibata (1991) reported responses of this type in a species of terrestrial slug. Trainor (1995) described seasonal and environmentally derived variation in species of the green alga *Scenedesmus*. Typically a four-cell colony (coenobium), they also occur as unicellular forms, and there is also substantial variation in cell shape and size, and in the presence and length and number of spines. Some of these forms have in the past been described as various species and placed in at least two other genera of unicellular algae (Trainor, 1996; Figure 3.2).

Plastic responses to the environment have four attributes: amount, pattern (Bradshaw, 1965; Schlichting and Levin, 1984), rapidity (Kuiper and Kuiper, 1988), and reversibility (Slobodkin, 1968; Piersma and Lindstrom, 1997). The *amount* of plasticity refers to the magnitude of response to the environmental change (large/small). The *pattern* is the shape of response (monotonic increase/decrease or more complex curves, Figure 3.1A). *Rapidity* pertains to the speed of response, for example, fast physiological changes versus slow morphological alterations. *Reversibility* refers to the capacity for switching between alternative states, for example, a leaf's photosynthetic rate is reversible, while its shape is not. The concept of the **developmental reaction norm** embodies these attributes and a fifth element that has been largely neglected in studies of plasticity, but that is crucial to the manifestation of the other four—the *competence* of the developmental system to respond at a particular time. For example, Mozley and Thomas (1995) demonstrated that *Arabidopsis thaliana* accelerates flowering time under long-day conditions only at certain stages of development that are determined by a balance among the effects of the red/far red receptors (phytochromes A and B) and the blue light receptor. Hensley (1993) found that, although food availability affects both the age and size at metamorphosis in *Pseudacris crucifer* tadpoles, developmental rates still become fixed at a certain stage, thus locking in a particular metamorphic age. Any or all of these five elements can be modified, in principle, by selection.

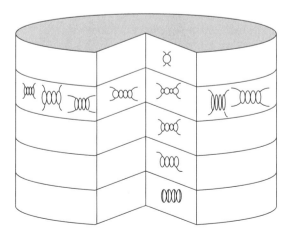

Figure 3.2   Extensive variability is possible in a single genotype of the green alga *Scenedesmus* during the course of a season (column) and within a given time period (row). (From Trainor, 1996.)

ADAPTIVE PHENOTYPIC PLASTICITY

Although Schmalhausen clearly recognized the importance of plastic reaction norms as a means of dealing with environmental heterogeneity (see Chapter 2), his approach was generally unacknowledged. The recent interest in plasticity and reaction norms can be traced to the conceptual work of Levins (1963, 1968) and Bradshaw (1965, 1973). Both of these authors recognized the importance of the ability to alter the phenotype in a context-dependent manner, and discussed the conditions that would favor its evolution. They affirmed the importance of the frequency and predictability of environmental change, and predicted that plasticity would evolve only for changes on a scale shorter than the generation time[2] (Bradshaw, 1965; Levins, 1968).

We are so familiar with the concept of phenotypic plasticity that its role in adaptation to variable environments is often assumed. Surely there is little question regarding the adaptive nature of such spectacular phenotypic responses as the cryptic matching of caterpillars of the geometrid moth *Nemoria arizonaria* to their food type (Greene, 1989); or the ability of amphibians to accelerate metamorphosis in ephemeral ponds (Newman, 1988). Nor is there much doubt about the value of phenotypic plasticity in the form of flexible animal behavior (e.g., Figure 3.3; Gordon, 1991; Jaenike and Papaj, 1992; Nager and Noordwijk, 1995). And there is a recent awareness that some parental effects may represent another adaptive form of plastic response (Schmitt et al., 1992; Carriere, 1994; Sultan, 1996; Rossiter, 1996; Lacey, 1996). However, as for any putative adaptation, it is important to *document* the fitness advantages of a particular reaction norm as compared to alternative patterns of phenotypic expression (Lacey et al., 1983; Travis, 1994; Doughty, 1995; Sultan, 1995; Dudley and Schmitt, 1996) in order to avoid falling into the well-known circular reasoning typical of the "adaptationist program" (Gould and Lewontin, 1979).

Alternative phenotypic expressions (e.g., non-plastic or different plastic responses) are often not available for direct comparison. For example, the nearly universal elongation response of plants to low light levels has been taken for granted as an adaptive response to shading by competitors. Only recently, however, has a definitive study demonstrated the fitness advantages accruing to such a response. In this elegant study, Schmitt and coworkers (1995) demonstrated the adaptive nature of shade-induced elongation in two annual plants. Two transgenic *Nicotiana* lines with blocked elongation pathways were outperformed by wild type plants under *shaded* conditions (Figure 3.4A). In a complementary experiment, a constitutively elongated mutant line of *Brassica* had reduced fitness in competition with wild type plants under *high light* intensity (Figure 3.4B).

---

[2]The evolution of adaptive plasticity does not have to be limited to within-generation responses (Scheiner, 1993a; Schlichting and Pigliucci, 1995a). For example, it may evolve when it confers increased inclusive fitness to the parent, therefore spanning at least two generations.

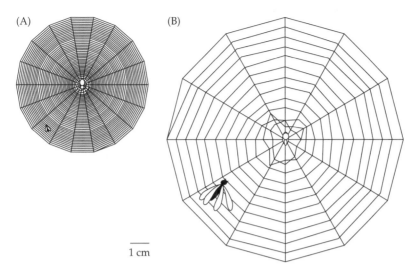

Figure 3.3 Different webs built by the spider *Parawixia bistriata* based on food availability. The web on the left, with a small mesh, is built daily at sunset, capturing mainly small dipterans; that on the right, with a large mesh, is built only during diurnal termite swarms. The amount of silk used is the same. (From Sandoval, 1994.)

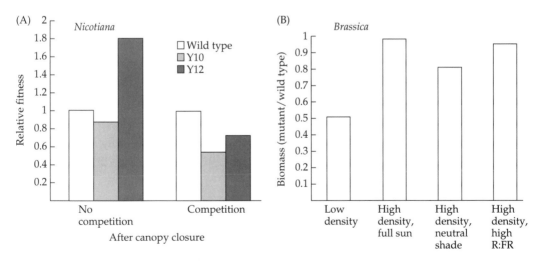

Figure 3.4 (A) Comparison of the performance of wild type and two transgenic *Nicotiana* lines with and without competition, following canopy closure. The transgenics had significantly reduced fitness relative to the wild type in the competition treatment. (B) Fitness (biomass) of the constitutively elongated *ein* (elongated internodes) mutant in *Brassica rapa* is significantly reduced relative to the wild type in low density conditions (Schmitt et al., 1995).

Induced defenses against predators represent other traits that can be manipulated to examine their adaptive nature (e.g., Harvell, 1990, 1994, 1998; Spitze, 1992; Bronmark and Miner, 1992; deMeester, 1993; Pettersson and Bronmark, 1997). Agrawal (1998) induced plant resistance to herbivores in a field experiment to evaluate the consequences of such responses for subsequent herbivory and plant fitness. Plants that were induced early in the season suffered half the herbivory by chewing herbivores and a reduction in the abundance of phloem-feeding aphids when compared with controls. Seed mass was over 60% greater for individuals that were induced.

A classic example of plasticity that has been considered to be adaptive is the phenomenon of heterophylly in many aquatic plants—submerged and emergent individuals may have markedly different leaf types; and an individual that begins growth while submerged will produce different leaves if the growing tip is exposed to air (Figure 3.5; see also Figure 2.7). The differences in leaf shape are

Submerged   Air-water Interface   Aerial

Figure 3.5   Leaf morphs in heterophyllous *Ranunculus aquatilis* (top) and *Ranunculus flabbelaris* from different environmental conditions. (From Cook, 1968.)

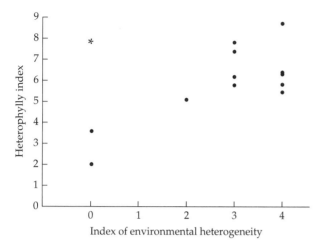

Figure 3.6 Relationship between the degree of heterophylly (aquatic vs. terrestrial leaves) and likelihood of encountering both environments for populations of *Ranunculus flammula*. The correlation is significant ($r = 0.78$; $p < 0.01$) when the Gold Lake population (*) is not included; and is $r = 0.49$ (not significant) when it is included.

only the most obvious changes; submerged leaves are often without stomata and have substantially thinner mesophyll layers than emergent leaves. Cook and Johnson (1968) examined the responses of *Ranunculus flammula* to a series of exposure treatments and found substantial variation in the heterophyllic condition in natural lake populations. They argued that populations with more pronounced responses came from habitats that were more likely to experience both submerged and emergent conditions. Cook and Johnson claimed that the Gold Lake population, which did not fit the expectation, was anomalous for being strongly heterophyllous in a constant aquatic environment (Figure 3.6). Developing consistent scenarios for the evolution of heterophylly becomes even more complicated when different species are compared. *Ranunculus flabbelaris* produces nearly continuous variation from entire to divided leaves as a function of temperature, whereas *R. aquatilis* produces distinct morphs without intermediates (Figure 3.5) in response to the presence or absence of water (Cook, 1968). These findings raise questions concerning the cues that organisms respond to (Deschamp and Cooke, 1983; Goliber and Feldman, 1990) and about the nature of the genetic machinery underlying the expressed phenotypes.

RECENT CONCEPTS OF THE REACTION NORM

Reaction norms have been depicted in various fashions. In the most familiar form, a trait (on the *y*-axis) is simply plotted against a set of environmental conditions on the *x*-axis (e.g., Figure 3.1A; Figure 3.9). Alternatively, the values of a trait in one environment can be plotted against its values in a second environment (Figure 3.7). A third option is a bivariate or multivariate plotting of traits measured in multiple environments (Figure 3.8). The selection of the graphical tech-

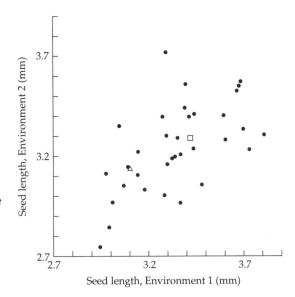

Figure 3.7 Environment by environment plot for *Crepis tectorum*. In this example, the individual values for seed length of F$_3$ families are plotted for full sun (*x*-axis) and shade (*y*-axis). If there was no plasticity, all points would fall along the diagonal. In this example the majority of families make longer seeds in full sun, and most are more plastic than the parental lines (Δ and □). (After Andersson and Shaw, 1994.)

nique will depend somewhat on which of two perspectives, or approaches, the researcher adopts—the *reaction norm* (or *polynomial*) approach and the *character state* approach (van Tienderen and Koelewijn, 1994; de Jong, 1995; Via et al., 1995).

The reaction norm approach, first advocated by Bradshaw (1965), and later by Schlichting (1986; Schlichting and Levin, 1986), considers plasticity to be a character itself and under genetic control. As such, reaction norms could be a

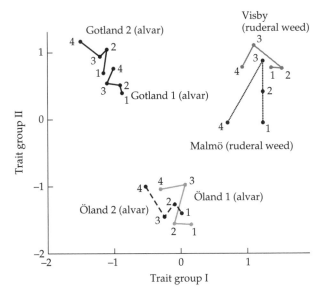

Figure 3.8 Trait group by trait group plot. In this example, the values of the principal components summarize the variation among all population treatment combinations. Six populations from two different habitat types were studied in four treatments (1–4). The vectors resulting depict the pattern of multivariate plastic response. (From Andersson, 1989.)

direct object of selection. Also taking this view were Scheiner and coworkers (Scheiner and Lyman, 1991; Gavrilets and Scheiner, 1993a) and de Jong (1990), who developed evolutionary models that focus on the consequences of selection acting directly on the parameters of reaction norms (i.e., mean, slope, and curvature—the polynomial models, Figure 3.9). Empirical support for the treatment of reaction norms as traits comes from multiple sources. There are studies that demonstrate the separate evolution of trait means and plasticities (Schlichting and Levin, 1986; Andersson, 1989; Macdonald and Chinnappa, 1989, 1990; Tucic et al., 1990; Counts, 1993; Joshi and Mueller, 1993), and experimental results that reveal that attributes of the reaction norm are responsive to selection, for example, environmental sensitivity (Jinks and Connolly, 1973; Jinks and Pooni, 1982), and yield stability in crops (Oka and Chang, 1964; Scott, 1967; Morishima and Oka, 1975; for general reviews see Falconer, 1990; Scheiner and Lyman, 1991).

The character state approach was promoted by Via and Lande (1985; Via, 1987), following Falconer (1952) and Yamada (1962; see also Burdon, 1977), and is fully formulated as a quantitative genetic model. Vectors of trait means and variances for each environment, and the covariances between trait expression in different environments (i.e., r_{ae}, the genetic correlation of a trait's values *a*cross environments), become the relevant quantities for evaluating evolutionary change. This approach was extended using the conception of reaction norms as infinite dimensional character states—for continuously varying environmental factors (e.g., temperature and humidity), the vector of means and the matrix of variances and covariances along the gradient are replaced with *functions* describing character means and variation (Gomulkiewicz and Kirkpatrick, 1992; following Kirkpatrick et al., 1990).

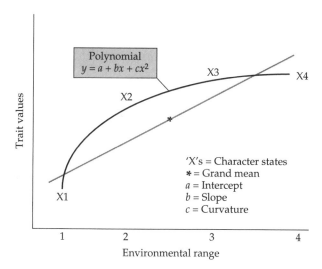

Figure 3.9 Character state and polynomial reaction norm representations. The character state representation views plasticity as the differences between pairs of means (X1–X4). The polynomial approach emphasizes the response curve, with parameters for intercept (*a*), slope (*b*), and curvature (*c*). (cf. Figure 3.1.)

A DEVELOPMENTAL PERSPECTIVE ON REACTION NORMS

This section will discuss the developmental aspects of phenotypic plasticity and its genetic basis. However, we have observed considerable confusion in the use of the terms plasticity, canalization, homeostasis, and developmental instability. Because these concepts are fundamental, and fundamentally distinct, we will first indicate the distinctions and relationships among them.

Canalization and Homeostasis

The word **canalization** was originally introduced by Waddington (1942) to describe cases in which:

> Developmental reactions . . . are adjusted so as to bring about one definite end-result regardless of minor variations in conditions during the course of the reaction.

Homeostasis (physiological homeostasis) was defined by W. B. Cannon (1932) as:

> The totality of steady states maintained in an organism through the coordination of its complex physiological processes.

and by Lerner (1954) as:

> The property of the organism to adjust itself in variable conditions, or . . . the self-regulatory mechanisms of the organism which permit it to stabilize itself in fluctuating inner and outer environments.

The term canalization has been used to designate both a process and a product (analogous to the use of "adaptation" as both a historical process and the outcome of that process; see Gould and Vrba, 1982). Canalization, when defined as the outcome of a process, has exactly the same meaning as homeostasis. When natural selection operates on the physiological mechanisms that can alter the stability of the phenotype at points along an ontogenetic trajectory, then canalization is the evolutionary process that alters organismal development (and therefore the developmental reaction norm). This leads to a consideration of the physiological processes (referred to by Cannon) that buffer the phenotype during ontogeny (i.e., developmental buffering) and that result in our perception of homeostasis or canalized phenotypes.[3]

---

[3] We mention for completeness Lerner's (1954) concept of *genetic homeostasis*. He proposed that there is a positive correlation between the level of heterozygosity and the ability of an organism to buffer inner and outer environmental changes, arguing that heterozygotes are—by definition?—characterized by a more flexible metabolism due to the greater diversity of their genetic material. (Demonstrations of such a correlation have been only partially successful; see Mitton and Grant, 1984; Mitton 1997, for reviews.) A population with higher heterozygosity would more readily absorb mutational effects. As defined, however, genetic homeostasis is a completely different phenomenon than physiological homeostasis; it is a population-based rather than an individual-based concept of buffering (and with a decidedly group-selectionist flavor).

Canalizing selection[4] is then the process by which the epigenetic system is altered to produce a more homeostatic developmental trajectory. One possible outcome of canalizing selection is the production of the same phenotype by more than one genotype (Sultan, 1987; Goldstein and Holsinger, 1992; see also Wilkins, 1997 and Chapter 9). This results from selection favoring modifiers that buffer the negative effects of new mutations; individuals thus "absorb" the effects of the mutations and the overall level of "neutral" genetic variability increases. Such redundant genetic variation can be revealed when environmental conditions change, and can be selected upon if the canalized phenotype is no longer adaptive. Note that, although selection may be able to reduce such perturbations (e.g., by controlling the error rate during DNA replication), more control implies higher metabolic costs, and genetic/epigenetic "errors" can never be eliminated completely (an inescapable consequence of the laws of thermodynamics).

From the above definitions it is apparent that both canalization (the process) and homeostasis (the result) pertain to the genetic and epigenetic components of the developmental system of a *single organism*. Canalization and homeostasis are *not* population concepts, despite their misuse in this area in the past. For example, to say that a set of reaction norms that are quite close to each other across a range of environments indicates a "zone of canalization" (Lewontin, 1974a) is somewhat misleading (Figure 3.10). Using population variance among the reac-

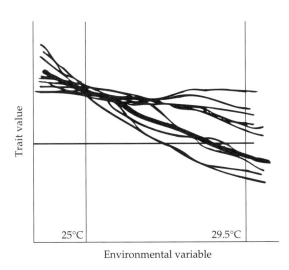

Figure 3.10 Convergence and divergence of family-level reaction norms across a temperature gradient. At 25°C the reaction norm lines are tightly bundled, while at 29.5°C there is substantial dispersion of phenotypes. "Canalization" is illusory—this diagram represents changes in variation *among* families, not along a developmental trajectory. (After Stearns and Kawecki, 1994.)

Trait value

25°C   29.5°C

Environmental variable

---

[4]In what appears to be an extra-fine distinction, Waddington (1953a) contrasted this with another form of stabilizing selection; *normalizing* selection eliminates those alleles contributing to the extremes of a phenotypic distribution.

tion norms as a measure of canalization is an extrapolation from a genotypic to a population point of view. This type of confusion, between the mechanistic description of an organism and the statistical description of a population, is at the root of the limitations of quantitative genetic models in describing evolutionary trajectories (as also partly pointed out by Lewontin, 1974a; Chapter 1 and below).

Waddington laid out the conceptual foundations of canalization in the context of his interest in the phenomenon of genetic assimilation (see Chapter 2), and from the outset distinguished environmental and genetic causes of disruption of development (Waddington, 1942). He also pioneered research into the process of canalization with a series of studies on *Drosophila* (summarized in Waddington, 1960; Waddington and Robertson, 1966). Unlike genetic assimilation, canalization studies attracted researchers carrying out artificial selection experiments whose number included some of the keener scientific minds of the 1950s and 1960s (Robertson, 1956; Falconer, 1957; Dun and Fraser, 1959; Scharloo, 1964; Rendel, 1967; reviewed in Scharloo, 1991). The hiatus that followed these initial efforts has been attributed to a lack of mechanistic information on how genes could bring about canalization (Wilkins, 1997).

Representative of those early studies are those carried out by Sondhi and Maynard Smith. They used a quantitative genetic approach to examine variation in the number and pattern of the ocelli (wild type = 3) and their associated bristles (wild type = 3 pairs) in *Drosophila subobscura* (Figure 3.11A). The base population contained a fraction of individuals with fewer than three ocelli and six bristles (Figure 3.11B). Thirteen generations of selection were performed for flies with only two ocelli, in one of two ways: (1) symmetrical selection for only the two posterior ocelli, and (2) asymmetrical selection for only the anterior and left posterior ocelli. The results were strikingly divergent—not only was the selection for symmetrical ocelli more successful, it also resulted in a more highly canalized phenotype for the number of both ocelli and bristles (contrast Figure 3.11C and D; Maynard Smith and Sondhi, 1960). In a separate experiment, selection for increased numbers of bristles resulted in the origin of a novel phenotype with a second pair of bristles in line with the anterior ocellus (Figure 3.11E). It is of interest that bristles in this position are normal in the related genus *Aulacigaster* (Sondhi, 1962).

A recent molecular genetic study of canalization in *Drosophila* examined the role of a regulatory locus, *ultrabithorax* (*UBX*). Gibson and Hogness (1996) demonstrated that increased sensitivity of the fly to ether was correlated with a loss of expression of *UBX* in the third thoracic imaginal discs. Loss of expression of *UBX* also produced phenocopies of bithorax. Furthermore, genetic variation for stability was detected, much of which could be attributed directly to polymorphism in *UBX*.

An idea closely related to canalization/homeostasis is that of **developmental stability** (Thoday, 1955). Developmental instability is typically considered to be

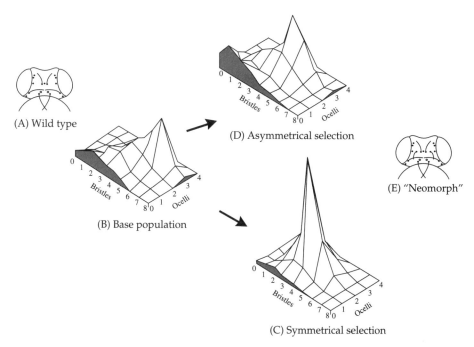

Figure 3.11 Selection experiments for canalization of bristle production in *Drosophila subobscura*. The base population (B) was subjected to selection for only two ocelli [circles on the wild type (A) and "neomorph" (E) heads], either (C) symmetrical (the two posterior ocelli) or (D) asymmetrical (anterior and left posterior). Graphs represent the distributions of individual flies with particular combinations of bristle number and ocelli. (From Maynard Smith and Sondhi, 1960, and Sondhi, 1962.)

detrimental, and represents the converse of homeostasis. The developmental instability of a genotype can be measured as the "error" around the mean in each environment (Figure 3.12). This type of error is usually included in the residual variance component in a classic quantitative genetics experiment, and is often referred to as developmental "noise." These slight variations on the intended phenotype are caused by small and localized external environmental effects, or by significant perturbations of the internal epigenetic system.

The arguments of Waddington (1960) and Lewontin (1974a) lead us to expect the degree of homeostasis to be higher (and developmental noise lower) in ranges of environmental parameters where selection has had time to act, and homeostasis to be lower (and developmental noise higher) in rarely encountered environmental zones. Of course, selection against instability (i.e., for homeostasis) can only be effective on traits for which instability has a negative effect on fitness.

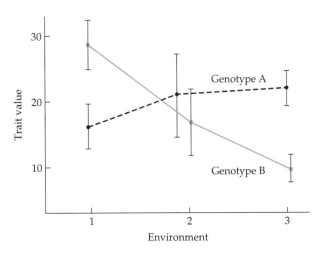

Figure 3.12   A depiction of plasticity and developmental instability of two genotypes, with clonal replicates grown in three environments. Plasticity is the variation across environments 1–3, whereas instability is represented by the error bars within each environment. Note that while B is much more plastic than A, both exhibit low instability in environment 3 and high instability in environment 2.

Developmental stability has been investigated in a variety of contexts[5] over the years (Markow, 1994), and has enjoyed a recent revival in popularity (see, for example, *Genetica* 89: vol. 2, 1993). After the pioneering work of Waddington (1942) on canalization and homeostasis, there were a variety of papers published on the relationship between heterozygosity and homeostasis/stability (Dobzhansky and Wallace, 1953; Mather, 1953; Thoday, 1955); more were prompted by Lerner's (1954) book, and another resurgence occurred in the 1980s (see reviews in Mitton and Grant, 1984; Mitton, 1997). It has been postulated that more homozygous individuals should be developmentally less stable, either due to inbreeding depression or to the lack of the beneficial effects of heterozygosity *per se.* There has been substantial support for a relationship between developmental instability and inbreeding (e.g., Lerner, 1954; Rendel, 1967; Levin, 1970; Soulé, 1979; Oostermeijer et al., 1995), although there have been no causal links between instability and diminished organism performance.

Dobzhansky and Levene (1955) found that more heterozygous lines of *Drosophila pseudoobscura* had higher, and less variable, viability across replicate cultures. Mukai et al. (1982) found heterozygotes in a natural population of *D. melanogaster* to be more homeostatic; but in a laboratory population, in which mutations had been allowed to accumulate, there was no difference between heterozygotes and homozygotes. They interpreted these results as support for the idea that natural selection has favored specific homeostasis and not heterozygosity *per se.* Wright (1977) summarized results from older experimental litera-

---

[5] And with a profusion of different measures—meristic (or continuous) variation for repeated structures; asymmetry for paired or bilateral structures; and fluctuating asymmetry (FA) for random fluctuations, in which there is no tendency for one side to deviate more often than the other (as opposed to directional asymmetry).

ture, and also concluded that there is not a tight linkage between homeostasis and heterozygosity. Although a number of studies have found a positive correlation between heterozygosity and growth rate, most of the populations investigated have a *deficiency* of heterozygotes (see discussion in Hartl and Clark, 1989). Clarke's (1993) thorough review of the literature suggests that a breakdown in genomic coadaptation is a more likely explanation for homeostasis than heterozygosity *per se*; and a meta-analysis of the various studies revealed no significant relationship between heterozygosity and fitness (Britten, 1996).

Some authors have examined developmental "noise" in traits directly related to fitness. For example, variation in the number of floral parts can directly influence the ability of pollinating insects to recognize a particular plant species, and as such stability should be a primary target of canalizing selection (Berg, 1959, 1960). There is indeed variation at both the genotypic and population level for petal number (Sakai and Shimamoto, 1965; Huether, 1968, 1969; Ellstrand, 1983; Ellstrand and Mitchell, 1988; Vlot et al., 1992; Møller and Eriksson, 1994). Sherry and Lord (1996a) found that floral traits did have lower values for fluctuating asymmetry (FA) than vegetative traits, but did not examine any fitness consequences. Møller and Eriksson (1994) detected no differences between floral and vegetative asymmetries. Evans and Marshall (1996), in a study on *Brassica campestris*, found no relationship within populations between levels of fluctuating asymmetry and organism performance.

There has been a recent spate of studies attempting to correlate fitness with measures of developmental instability in traits less directly related to organism fitness.[6] There have been numerous arguments about statistical tractability and whether or not there is heritable variation for instability (see Møller and Thornhill, 1997, and the torrent of responses to their paper). Windig (1998) found in the peacock butterfly, *Inachis io*, that FA is higher in eyespots than in pupal traits, but that it is not heritable, nor is it related to survival. We do not question that there is heritable genetic variation for factors affecting levels of instability (e.g., Batterham et al., 1996; Clarke, 1997); however, a number of studies have shown that there are not even consistent correlations among different measures of stability on the *same* organisms (Tarasjev, 1995; Brakefield and Breuker, 1996; Sherry and Lord, 1996a,b). Moreover, we have observed tendencies to confound different levels of selection (group versus individual), and to confuse anthropomorphic notions of what is good (symmetry) with the actual objects of natural selection.

It is worth re-emphasizing that developmental instability/homeostasis is a different concept than that of phenotypic plasticity—the first is a within-environment phenomenon and the second a shift in phenotype due to changes in the *macro*environment (Schmalhausen, 1949; Schlichting, 1986; Sultan, 1987).

---

[6]Much of which has been spurred by the hope of devising a magic index of the "health" of natural populations of organisms (Clarke, 1995).

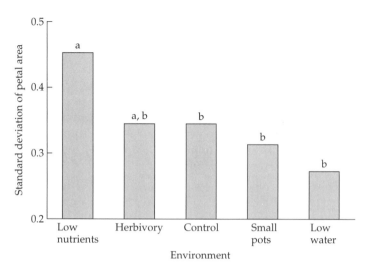

Figure 3.13 Plasticity of developmental stability in *Phlox drummondii*. Plants were grown in five environments and the petal areas of five flowers/plant recorded. Flowers in the low nutrient treatment showed significantly higher instability of flower size. Treatments not sharing letters are significantly different. (From Schlichting, unpublished data.)

Plasticity is a property of the reaction norm, developmental stability is a property of a single point along it. A tendency to instability will not necessarily translate into greater plasticity. Although there may be some commonality to their expression, the presence of genetic variability for either could permit their separate evolution. In fact there is evidence from several studies for at least a partial genetic independence of stability and plasticity (Druger, 1967, and references therein; Perkins and Jinks, 1973; Bagchi and Iyama, 1983; Scharloo, 1988). Nevertheless, a reaction norm approach can be applied to the study of macro-environmental effects on instability (Figure 3.13; see also Brakefield, 1997). Imasheva et al. (1997) found temperature-dependent differences in instability (FA) for several traits in *Drosophila melanogaster*, and in *D. buzzatii*, a high temperature-adapted species. Although the plastic responses of wing length to temperature by the two species are nearly identical, FA of wing length varies both with temperature and between species, with *D. buzzatii* having a generally lower FA than *melanogaster* (Figure 3.14).

A GENETIC PERSPECTIVE ON REACTION NORMS

There have historically been several distinct views on the relationships between the presence of phenotypic plasticity and genetic variability in natural populations:

1. Plasticity will impede (act as a buffer during) the process of natural selection.

2. Plasticity and genetic variability represent alternative solutions to the problem of environmental heterogeneity.

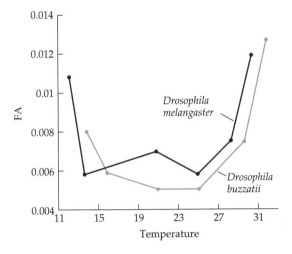

Figure 3.14 Changes in fluctuating asymmetry (FA) of wing length in response to temperature for *Drosophila melanogaster* (black line) and *D. buzzatii* (gray line). (From Imasheva et al., 1997.)

3. Plasticity is related to instability/homozygosity and is reduced by individual heterozygosity.

4. Genetic variation for reaction norms for fitness helps to maintain genetic variation in non-fitness traits.

These ideas, although by no means mutually exclusive, can lead to different predictions about the presence of genetic variation for quantitative traits. Wright (1931) was an early proponent of the first idea:

> It [plasticity] is . . . of the greatest significance as a factor of evolution in damping the effects of selection.

and this conception has been advocated more recently by Stearns (1982),[7] Schaal and Leverich (1987), Nylin (1992), and Levin (1988). According to these views, the ability of various genotypes to survive in a broad range of environments will effectively hide genetic variation from the action of natural selection. The net result will be, inevitably, a buildup of genetic variability as long as the reaction norms of different genotypes are generally congruent. Sultan (1987) has termed this phenotypic convergence (in essence, genetic redundancy for sub-portions of the reaction norms). In Wright's (1931) words:

> High development of (individual adaptability) permits the survival of genetically diverse types in the face of severe pressure.

---

[7]Stearns did, however, quickly back off on this (Stearns, 1984).

We note, along with Stearns (1984), that this perspective only makes sense for selection on reaction norms; if selection operates only on character states, then expression of traits in different environments will be convergent either by chance or because of a genetic correlation across environments.

The second idea, that plasticity and genetic variability are alternative solutions to adaptation in heterogeneous environments, was formulated perhaps most forcefully by Marshall and Jain (1968, 1979; see also Thoday, 1953; Lewontin, 1957; Levins, 1968) with the proposition that populations of organisms can be directed onto one of two evolutionary pathways to cope with environmental heterogeneity. If genetic variation is available, adaptation will be achieved via polymorphism, with different genotypes exploiting different conditions. If genetic variation is limited, the population should evolve phenotypic plasticity as the means of adaptation. The consequences are an inverse relationship between plasticity and genetic variability—populations with little variation should be plastic, those with more genetic variation should be less plastic. There have been a number of experimental examinations of this proposed relationship, with mixed results. An inverse relationship was found in studies by Marshall and Jain (1968; Jain, 1979) and Silander (1985), but not in a host of other studies (Gottlieb, 1977; Wilken, 1977; Scheiner and Goodnight, 1984; Schlichting and Levin, 1984; Macdonald and Chinnappa, 1989; Tucic et al., 1990; Houssard and Escarre, 1995; Black-Samuelsson and Andersson, 1997).

In a seminal work linking ecological and evolutionary levels of analysis, Bradshaw and Hardwick attempted to investigate the conditions of a stressful environment that would trigger the evolution of specialized ecotypes, as opposed to the development of a flexible genotype characterized by phenotypic plasticity (Bradshaw and Hardwick, 1989). They first compiled several cases of each category from the available literature. They then proposed that ecotypes are expected when environmental conditions change and then remain relatively stable (Table 3.1A). If stress occurs frequently and alternates with normal conditions, then evolution of generalists adopting phenotypic plasticity as a strategy is expected, especially if the environmental change is predictable to some extent (Table 3.1B). Notice that in some cases the same plant can evolve different specialized responses to distinct permanent stresses, as well as a plastic phenotype to cope with spatially or temporally variable ones (e.g., *Agrostis stolonifera* has developed distinct dwarf ecotypes in response to wind and a plastic pattern of root growth to cope with fluctuations in nutrient availability).

The evolutionary outcome in these situations depends on the frequency of environmental change, the temporal or spatial nature of the variation, and the type of selection (soft or hard) (Sultan, 1987 and references therein; van Tienderen, 1991, 1997; Schlichting and Pigliucci 1995a). Schlichting and Pigliucci (1995a; Pigliucci et al., 1995a) argued that framing these examples in terms of "stress versus normal" can be misleading, because from an organism's standpoint any sub-optimal environment represents some kind of "stress." A more

Table 3.1 **Examples of response to selection by genetic differentiation or plasticity**

A. Ecotypic Differentiation

Stress	Adaptation	Examples
Metal contamination	Chemical complexing	*Agrostis, Silene, Mimulus*
Low nutrients	More efficient uptake	*Festuca ovina*
Wind	Dwarfing	*Agrostis stolonifera*
Cold	Freezing control	Many species
High light	Elevated rubisco	*Solidago virgaurea*
Salt	Osmoregulation	*Enteromorpha intestinalis*
Triazine herbicides	Chloroplast membrane change	Many species

B. Plasticity

Stress	Plasticity for	Examples
Water	Stomata opening	*Sempervivum*
Predation	Phytoalexin production	Many species
Temperature	Leaf growth	*Lolium perenne*
Nutrients	Root growth	*Agrostis stolonifera*
Drought	Leaf abscission	*Theobroma cacao*
Vegetation height	Petiole length	*Trifolium repens*
Season	Leaf abscission	*Quercus*
Submergence	Leaf development	*Ranunculus flammula*
General stress	Flowering time	*Capsella bursa-pastoris*

Source: Bradshaw and Hardwick, 1989.

useful framework might be provided by assessing the frequency of occurrence of each environment. Situations that seldom occur do not exert a significant selective pressure, and the reaction norm of the organism will simply not be optimal under these conditions; we then perceive these environments as being "stressful."

The third notion is that plasticity and developmental instability are directly related. This emerges from postulated relationships between developmental instability and plasticity, and includes ideas on the buffering capacities of heterozygosity (Bradshaw, 1965; Pederson, 1968). Although the causal chain is never explicit, we reconstruct it as follows.

1. Homozygosity of deleterious recessives leads to increased developmental instability (and reduction of fitness). →

2. An increased within-environment variation translates directly into increased across-environment variation ("bad" plasticity, reducing fitness). →

3. Conversely, heterozygosity will hide these deleterious recessive alleles, leading to increasing stability and a decrease in "bad" plasticity (maintaining fitness levels). →

4. In cases where the heterozygote has two functional alleles, an additional benefit is reaped in the form of an increased breadth of tolerance (buffering of fitness).

The potential for confusion arises from the distinction between plasticity of fitness and plasticity of other traits. Organisms are selected to maintain high fitness across environments, so low plasticity of fitness is desirable (Taylor and Aarssen, 1988; Bell and Lechowicz, 1993; Lortie and Aarssen, 1996); and fitness homeostasis may even be achieved through the plasticity of other traits (e.g., morphology and physiology). Thus the predicted relationship between plasticity and developmental instability really applies only to plasticity of fitness. Additionally, link 2 has not been established (as pointed out previously), and link 3 has been shown to hold only in certain cases (e.g., Kacser and Burns, 1981; Watt, 1983).

The fourth idea, that genotype by environment interaction for fitness helps to maintain genetic variation in non-fitness traits, has a venerable history as an explanation for the maintenance of genetic variability. Levene (1953) used a population genetic model to illustrate that genetic variation could be maintained in a system where there are multiple niches and where genotypes are more or less specialized for their exploitation. Although this is not framed in a reaction norm perspective, the implication is clearly one of crossing reaction norms for fitness (e.g., Figure 3.15). In such situations, genetic variation for many other traits can be maintained. Gillespie and Turelli (1989) demonstrated a similar result from a

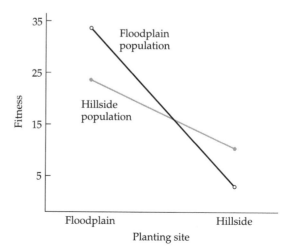

Figure 3.15 Reaction norm diagram of ecotypic variation from reciprocal transplant experiment with *Impatiens pallida* from floodplain (mesic, open circles) and hillside (dry, filled circles) sites. (From Bennington and McGraw, 1995.)

quantitative genetic perspective.[8] Prout and Savolainen (1996) contend that there is a general misconception that $G \times E$ always maintains V_G, and argue that in fact the conditions allowing this are quite restrictive.[9] However, these restrictions can be effectively overcome by organisms that choose the most suitable niche or habitat (Maynard Smith, 1962).

At this point, we have two predictions that more plasticity is associated with more genetic variation, and two that posit an inverse relationship. How can (or should?) we accommodate these divergent expectations? Lewontin (1957) discussed the alternatives of genetic variation and plasticity (he called it homeostatic ability), but in a historical context. He argued that phylogenetically "advanced" groups will have evolved phenotypic plasticity as a means of mastering their environments, thus largely eliminating the role of genetic variability in ensuring population persistence. In this context, we can see that the arguments of Jain et al. are allied with concepts of selection at higher levels of organization—population/species persistence in the face of changing environmental conditions will be related to either maintenance of genetic variability or to the individuals' ability to respond appropriately. Even this intuition does not explain why the two should be inversely related, unless there is some additional assumption, for example, a further cost to individuals that possess both genetic variability and plasticity (Scheiner and Goodnight, 1984; Schlichting, 1986).

The view of plasticity as a shield for genetic variation for trait means leads to a prediction that more plastic traits should show less evolutionary change (Stearns, 1982). Examinations of this prediction do not support it—the evolutionary divergence of traits is not necessarily related to their plasticity (Stearns, 1983; Scheiner and Goodnight, 1984; Schlichting and Levin, 1986; as well as references above demonstrating independent evolution of means and plasticities). Andersson (1989), for example, found that more plastic traits appear to evolve more rapidly. The view of plasticity as an evolutionary brake is limited by a lack of attention to the details of how plastic reaction norms are produced. Several examinations of the role of plastic variation have chosen to focus only on the amount of plastic response, represented by a mean phenotype with a "cloud" of non-directional variance (Orzack, 1985; Levin, 1988). What is overlooked is the reality that patterns of plastic response to environmental change are decidedly directional (e.g., Bradshaw, 1965; Sultan, 1993a,b,c; Pigliucci et al., 1995; Schlichting and Pigliucci, 1995b). In the next section the form of genetic control of plasticity is discussed, and the implications for the evolution of plastic reaction norms.

---

[8] Via and Lande (1987) argued that genotype by environment interaction *per se* would not maintain genetic variation. However, as pointed out by Gillespie and Turelli (1989), their model assumed that at least one genotype would have higher fitness in *all* environments.

[9] However, their own empirical example seems to indicate that at least some variation *may be* maintained due to $G \times E$. Calculations from their Table 1 indicate negative correlations between the viabilities and fecundities of the six genotypes on the 20 host plants ($r < -0.40$).

GENETIC CONTROL OF PLASTIC RESPONSES

Schlichting and Pigliucci (1993, 1995a) proposed distinguishing two forms of genetic control of plastic responses, namely, allelic sensitivity and regulatory plasticity genes. These broadly correspond to the two developmental manifestations of plasticity observed by earlier workers, that is, the graded and threshold responses of traits to environmental changes, termed by Smith-Gill (1983) phenotypic modulation and developmental conversion (Figure 3.16).[10] The key distinctions between the two forms of genetic control lie in the type and number of genes involved in the response.

Allelic sensitivity involves the response by a particular gene locus to a change in conditions. This response can be due to a change in the amount or activity of gene product, and may vary among different alleles. A classic example of this was reported by Schmalhausen for the wing reduction alleles *pennant* and *vestigial* in *Drosophila melanogaster* (Schmalhausen, 1949; see Figure 2.6). A more recent example can be seen in Figure 3.17, in which the growth rate of two strains of *E. coli*, differing for single alleles at a locus, vary in response to different limiting nutrients (Hartl and Dykhuizen, 1981). Allelic sensitivity is not unexpected as a response to temperature, given the general dependence of enzyme activity on temperature (e.g., Somero, 1978; Neyfakh and Hartl, 1993);

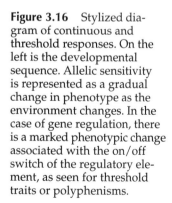
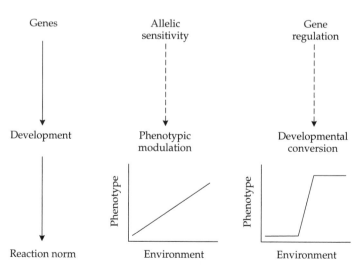

Figure 3.16   Stylized diagram of continuous and threshold responses. On the left is the developmental sequence. Allelic sensitivity is represented as a gradual change in phenotype as the environment changes. In the case of gene regulation, there is a marked phenotypic change associated with the on/off switch of the regulatory element, as seen for threshold traits or polyphenisms.

[10]These replaced Schmalhausen's mind-boggling (at least in translation) terms "dependent morphogenesis" and "autoregulatory dependent morphogenesis." His equivalent of homeostasis is "autonomous-regulatory development."

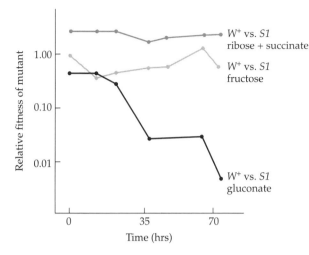

Figure 3.17 Allelic sensitivity of an *E. coli* mutant. The fitness of allele *S1* of 6PGD relative to the wild type depends on the nutrient environment. The mutant has reduced performance in gluconate, but not in fructose or ribose + succinate. (After Hartl and Dykhuizen, 1981.)

but there are also examples for such factors as pH (Dolferus and Jacobs, 1984) and alcohol concentration (Van Delden et al., 1978).

Regulatory control postulates the existence of environment-dependent gene activity controlled by the action of a regulatory switch. The activity of the regulatory pathway is seen as influencing environment-specific expression of one or many different genes (Smith, 1990). The result in terms of phenotype will be a threshold response (or an appearance of polymorphism). A classic example is the *lac* operon in *E. coli*, documented by Jacob and Monod, that turns on when lactose is present and turns off when it is absent.

Allelic sensitivity is a passive form of plastic response that can be adaptive or not, and that can result in a variety of pleiotropic effects on phenotypes. Regulatory plasticity, on the other hand, represents an active plastic response by the organism that is most likely adaptive, and that is due to the epistatic actions of plasticity genes (Pigliucci, 1996a). Several lines of reasoning can be offered to support the advantages of evolving a regulatory system (Schlichting and Pigliucci, 1995a).

1. Regulatory control results in the production of discrete phenotypes that are stable across a range of environmental conditions (Schmalhausen, 1949, p. 232). In contrast, the graded responses due to allelic sensitivity will fluctuate with changing environmental conditions.

2. Regulatory control can produce integrated phenotypes with a single switch that responds to environmental change. Integrating the responses of many allelic sensitivity loci is problematic when their numbers are large. Regulatory control permits environment-dependent expression of different groups of structural loci, which leads directly to changes in the

genetic and phenotypic correlations among traits within environments. Thus it is possible for the relationships among traits to be fine-tuned to particular environmental conditions. When tradeoffs in the abilities to grow and reproduce in different environments are expressed in the form of negative across-environment genetic correlations, then quantitative genetic models predict that there will be stringent constraints on reaction norm evolution (Via and Lande, 1985; Gomulkiewicz and Kirkpatrick, 1992). The limitation on the number of environments that an organism can adapt to may be ameliorated by the coupling and uncoupling of traits. The switching on or off of structural genes will reduce the across-environment pleiotropic relationships among traits, and thus bring about a reduction in the magnitude of negative genetic correlations across environments (r_{ae}).

3. Regulatory control allows anticipatory responses to environmental change, whereas allelic sensitivity does not. For organisms in fine-grained environments, initiating a relevant response earlier can be accomplished in two ways. First, the response can be initiated at a different environmental threshold. Second, the response can be coupled to a stimulus that is earlier and more reliable than the current one (Levins, 1968). The anticipatory nature of plastic responses has received increasing attention (Rollo and Hawryluk, 1988; Aphalo and Ballaré, 1995) as relationships between the timing of a stimulus and the subsequent response to it have been investigated in several systems. For example, winter deciduous plants respond to changes in photoperiod in anticipation of the onset of freezing (or drought) conditions. Mulkey et al. (1992) found that the perennial plants *Psychotria marginata* produce most of their leaves in two bursts of growth. Those produced at the end of the wet season (i.e., just prior to the dry season) have higher specific mass, lower stomatal conductance, and higher water use efficiency during drought periods, than leaves produced at the beginning of the wet season. "Drought leaves" continue to be formed even on irrigated plants, indicating that water availability is not the cue for their production. Some maternal effects can be placed in the class of anticipatory plastic responses, in cases where the maternal parent differentially provisions offspring based on cues that predict the progeny's habitat (Rossiter, 1996).

The mechanistic details of regulatory plasticity have provided us with a better understanding of the genetic elements necessary to build such systems. Although we tend to imagine plastic responses as a sort of environment input → phenotype output system, this black box is in fact significantly more complicated (Aphalo and Ballaré, 1995). Plasticity, in the form of developmental conversion, requires an integrated system of components (i.e., an infrastructure) to function properly. There must be a site for reception of the environmental signal, a system for the transduction and translation of the signal, and finally the sig-

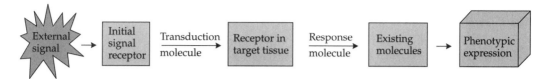

Figure 3.18   The steps necessary in a pathway from signal reception to phenotype.

nal's eventual target(s) (e.g., genes or gene products, Kuiper and Kuiper, 1988). Figured this way, there must be a minimum of three genes/gene products involved in the complete input/output pathway (Figure 3.18; Rollo, 1994), and probably many more.

An excellent example of the nature of this infrastructure in the evolution of reaction norms is found in the work on wing dimorphism in crickets. Juvenile hormone (JH) is acknowledged to play a major role in the suite of traits that produce the migratory versus non-migratory phenotypes in a number of wing polymorphic insects. Higher levels of juvenile hormone esterase (JHE) can lead to the breakdown of JH and an increased probability of macroptery (i.e., formation of wings and associated structures that are flight capable). Fairbairn and Yadlowski (1997) looked at levels of JHE in lines of the cricket *Gryllus firmus* that had been subjected to upward and downward selection for the frequency of macroptery. Although they found elevated levels of JHE in the lines selected for higher frequency of macroptery (Figure 3.19), the differences in macroptery between the selection and control lines could not be explained solely by altered mean JHE levels.[11]

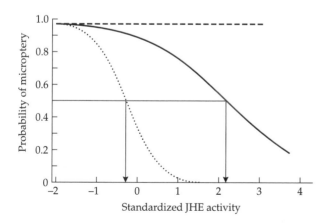

Figure 3.19   The relationship between juvenile hormone esterase (JHE) activity and the probability of producing a micropterous cricket (*Gryllus firmus*) in control (—) and lines selected for miccoptery (·····) and macroptery (---). (From Fairbairn and Yadlowski, 1997.)

---

[11] Zera and coworkers (Zera and Zhang, 1995; Zera et al., 1997b) have shown that JHE levels respond to direct selection in both juvenile and adult *G. assimilis*.

They examined the possibility that not only was the mean level of JHE being altered, but that the threshold at which it evoked the phenotypic response was altered too (Figure 3.20). Close examination of the data showed that this was indeed the case—the shapes of the curves relating JHE activity and the probability of microptery were quite different (Figure 3.19). If the observed responses to upward and downward selection were due solely to changes in JHE levels, the curves for the control and macropterous lines should be the same. The differences in the curves indicated that the response to selection also involved changes in the threshold of the phenotypic switch. The particular point in the pathway (e.g., Figure 3.18) where the threshold was altered could not be determined (i.e., altered tissue sensitivity, other hormone levels, or changed activity of the other JH denaturing enzyme, JH epoxide hydrolase). Zera et al. (1994), also working with *G. firmus*, found additional differences between wing morphs—including tradeoffs between amounts of triglycerides and other lipids in macropterous, but not micropterous, individuals[12]—which suggests that the complete picture is even more complicated than the matter of mean and threshold variation in JHE.

There are in fact a multitude of points in the pathway from nucleotide sequence to phenotype that can be acted upon to produce developmental conversion—transcription, translation, post-translation (before the protein is active

Figure 3.20 Depiction of two models for altering the frequency of macropterous individuals: (A) Shifting the mean level or activity of juvenile hormone esterase (JHE) without altering the threshold, or (B) altering the threshold point where macropterous versus micropterous phenotypes are produced, without changing mean JHE activity. S and L refer to lines selected for Short or Long wings. The vertical line in each panel represents the threshold of JHE activity necessary to induce wing formation; hatched areas are the macropterous proportion. (From Fairbairn and Yadlowski, 1997.)

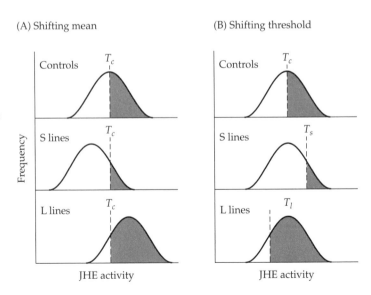

---

[12] They have also demonstrated that some phenotypically macropterous females have histolyzed flight muscles, rendering them flightless, but enhancing their fecundity (Zera et al., 1997a).

or in place), and biochemical pathways of protein activity. Environmentally exerted control can occur in two ways. The first is the direct, unmediated influence of environmental variables. This response is not actively mediated by the genotype (although it is genotype-specific), and results in Smith-Gill's phenotypic modulation. To this category belong environmental changes that affect the physico-chemical state inside cells and tissues (such as pH or temperature), and the action of toxins or the availability of nutrients (which may influence biochemical reactions or the abundance of ribosomes, tRNA, or polymerases). The second type of environmental influence is mediated by the genotype, and is classically identified as developmental conversion. Induction, attenuation, topological changes in the DNA, methylation patterns, and allosteric controls have all been shown to be capable of responding to precise environmental cues in order to closely regulate metabolic activities.

Each type of control is characterized by different reaction times and by different economic costs or consequences. For example, allosteric control on enzyme activity is much faster than induction at the DNA level, but it is also less economical. Further investigations are necessary to link the genetic bases of response to stimuli with the organism-level response. An understanding of the connection between hierarchical levels will greatly further our comprehension of the elusive epigenetic process (Chapter 8).

GENE EXPRESSION IN DIFFERENT ENVIRONMENTS

It follows from the notion of environment-dependent gene regulation that there will be differences, or the expectation of differences, in the amounts of genetic variation expressed in each environment. The fundamentals of the concept of environment-dependent genetic variability were articulated by Dragavstev in a pair of papers in 1984 (Dragavstev and Aver'yanova, 1984a,b). He argued that environmentally induced changes in gene expression (for examples see Schlichting and Pigliucci, 1995a; Pigliucci, 1996a) will necessarily change the genetic architecture of the affected traits. Such changes will reverberate pleiotropically in the genetic variances, heritabilities, and genetic correlations/covariances (Giesel et al., 1982; Gebhardt and Stearns, 1988; Clark, 1990; Stearns et al., 1991; Tucic et al., 1991; Mazer and Schick, 1991; Rawson and Hilbish, 1991; Ebert et al., 1993; Thomas and Barker, 1993; Simons and Roff, 1994; Windig, 1994; Barker and Krebs, 1995; Via and Conner, 1995). Other indirect evidence for substantial environment-specific gene function comes from analyses of the yeast genome with the observation that 70% of the yeast genes were unnecessary in a rich medium, and the suggestion that they are specific to particular environmental challenges (Oliver, 1996; Smith et al. 1996).

However, not all changes in population-level genetic parameters are necessarily due to changes in gene expression at the individual level. Several authors have determined that changes in genetic parameters are expected even without the changes brought about by regulatory control. Ward (1994) discussed the observation of higher narrow-sense heritability in new environments, and demonstrated that the parent-offspring regression coefficient will increase when the average value of the environmental contribution to phenotype increases. This result proceeds in part from his use of a metabolic flux model that produces a non-linear relationship between genotype and phenotype (as opposed to the defined linearity of $P = G + E$). Stearns, de Jong, and Newman (1991) examined a system of allelic sensitivity in which the reaction norms of two traits were linear functions of an environmental parameter. They found the genetic variances and covariances to be quadratic functions of the environment, with consequences for change in genetic correlations in different environments.

The subject of environmental effects on genetic variation is usually discussed in the context of responses to stressful and/or novel environments (Giesel et al., 1982; Holloway et al., 1990; Hoffmann and Parsons, 1991). Various hypotheses have been generated to explain why genetic variation should be either higher or lower in novel/stressful environments, with data available to support most of them (although typically not in the same study). This extensive literature will not be reviewed here. Our expectations for relative amounts of gene expression do not derive from "stressfulness," but from Fisher's fundamental theorem (tempered with the usual modern caveats concerning genetic correlations, hitchhiking, etc.): Genetic variability will be eroded in environments encountered more frequently and with stronger selection intensities. Thus we expect variation among genotypes to be highest in environments that have not (or have rarely) been encountered, or which are selectively neutral for the expressed phenotypes (Pigliucci et al., 1995; Bennington and McGraw, 1996).

PHENOTYPIC PLASTICITY: TARGET OR BY-PRODUCT OF SELECTION? WHY IT MAKES A DIFFERENCE

In recent years there has been a debate on the possibility that phenotypic plasticity is the target of natural selection and does not arise as a by-product of selection within discrete environments. This debate has become entangled with a parallel one on the genetic basis of plastic responses, and whether it is important to understand the latter in order to investigate the former (Scheiner, 1993b; Schlichting and Pigliucci, 1993, 1995a; Via, 1993a,b; van Tienderen and de Jong, 1994; van Tienderen and Koelewijn, 1994; de Jong, 1995; Via et al., 1995).

Resolution of the debate on the action of selection on plasticity and on the genetic basis of plastic responses has been dealt with at a practical level—when is one model to be favored over the other for collecting or analyzing data—and at a conceptual level—does each conception perform equally well in illuminating the process of selection on phenotypic plasticity and reaction norms. Via et al. (1995) and de Jong (1995) discussed at some length the practical aspects with the conclusion "It depends," and the curious are referred there for more detail. The conceptual aspect is of considerably more interest in our context: Does it make a difference whether the character state or the reaction norm conceptions are applied, and if so, when?

In the *character state view* of reaction norm evolution, an evolutionary force modifying the expression of characters in particular environments leads *de facto* to a change in the reaction norm of that character. In addition, this change often results in a correlated change in the expression of the character in a different environment. Such cases of presumably undirected change of other portions of the reaction norm were characterized by Via (1993a) as "by-products" of selection. The view of reaction norms as traits under their own genetic control was dismissed as unnecessary.[13] In response to Via's critique, vigorous defense of the *reaction norm approach* came from two perspectives. One group, also using the framework of quantitative genetics, demonstrated the mathematical equivalencies for quantitative genetic formulations of the character state and polynomial reaction norm approaches (van Tienderen and Koelewijn, 1994; de Jong, 1995; see also Via et al., 1995). de Jong (1995) suggested that the character state model is most suited to environment-dependent gene action, and that the reaction norm approach is used best to describe allelic sensitivity. However, although both models can incorporate regulatory effects that are (statistically) additive, neither one can describe selection on phenotypic plasticity (or for that matter any trait) that is under the control of epistatic loci (de Jong, 1995).

We argued that there is, in fact, an important *biological* distinction as well between these conceptions (Schlichting and Pigliucci, 1993, 1995a). To treat phenotypic plasticity as a character emphasizes the notion that the ability to respond appropriately to environmental change may be a direct object of natural selection. In contrast, the character state model allows for no direct selection on the reaction norm itself.

The character state approach seems a logical framework to describe the evolution of host races in various insect species—individuals experience only a single environment and specialization to one host results in the "by-product" of poor performance on other hosts (Via, 1994b). The same argument *could* be applied to cases such as the evolution of the shade avoidance response in plants—reactions to sun (or shade) might evolve in separate environments and lead to sun/shade specialization. However, the fact that a single genotype may experience both conditions opens the possibility for selection to operate on the reaction norm itself (Schlichting and Pigliucci, 1995a). In the leaf example, we know many of the relevant details of the phytochromes (light-sensitive molecules) that act as a switch mechanism to divert development to a sun or shade phenotype. Even though the developmental pathways may have evolved inde-

---

[13] From Via (1993a, p. 353): "Given the absence of detailed models . . . Bradshaw's interpretation of species differences in plasticity was reasonable. However, current models illustrate how phenotypic plasticity can evolve due to selection toward different phenotypic values in different environments. . . . Proposing separate genetic control for plasticity and trait values is no longer necessary. . . . Nonetheless, Bradshaw's views continue to be extremely influential (Schlichting 1986)."

pendently (although there is no evidence suggesting this), the coordinated switching mechanism for both pathways must have evolved in response to environmental heterogeneity (since a photoreceptor makes sense only if there is variation in light availability); and the switch mechanism itself is indicative of regulatory control. It is because of these cases of environmentally sensitive switches that we insist that a knowledge of the genetic basis of plasticity is vital for understanding the evolution of reaction norms.

The compromise position reached by Via et al. (1995) was that reaction norms may evolve either as a by-product of selection or through direct selection. However, the relative importance of these two avenues was not agreed upon. On the one hand, Via (1994a) was convinced that plasticity as by-product comprises the vast majority of cases of evolution of reaction norms. On the other hand, Scheiner (1993a) described several examples in which the reaction norm itself is the likely direct object of selection (e.g., castes in insects and family-level selection). Schlichting and Pigliucci (1995a) suggested that most examples of adaptive plasticity are directly selected reaction norms, and they outlined the environmental conditions that favor that action.

Zhivotovsky et al. (1996) examined a model for the evolution of plasticity in spatially heterogeneous environments, with the inclusion of a modifier gene that acts in an epistatic manner on the dispersability of an organism or on trait expression or optimum. They concluded that reaction norms are derived from the global evolution of plasticity (i.e., selection operates on the reaction norm). In their results, an additive genetic model does not lead to an optimum (*contra* the results of Via-Lande) because the evolution of the mean in one niche is strongly dependent on environmental structure, such as trait means and optima in other niches, niche frequencies, and selection intensities. In addition, the reaction norm may evolve even if modifier genes do not directly affect the phenotype.

There are two further elements of the current approaches to modeling reaction norm evolution that require discussion. First is the issue of the dependence of the evolutionary dynamics on the genetic basis of plasticity. Quantitative genetic models assume or imply allelic sensitivity to be the default form of genetic control (Schlichting and Pigliucci, 1993, 1995a; de Jong, 1995). The trajectories derived from additive, linear models of gene effects stand to be substantially different when non-additive effects are taken into consideration (Turelli and Barton, 1994; de Jong 1995).[14] Turelli and Barton (1994) concluded

---

[14]Statistical models *can* be constructed that ignore the actual genetics, and epistatic and non-linear effects that are statistically additive can be incorporated. What this 'weak' version of quantitative genetics gains in statistical power is offset by a loss of genetic and biological realism, unintended by the originators of the methods (Fisher, 1918; Provine, 1971; Falconer 1989; Roff, 1997). Genetic variances and covariances were meant to summarize the mechanistically additive effects of many alleles. However, if the genetics and statistical summary are incongruent, then genetic variances and covariances lose their biological interpretation (Houle, 1991; Gromko, 1995).

that "... changes in genetic variance after 10 or more generations of selection are likely to be dominated by allele frequency dynamics that depend on genetic details."

Pigliucci (1996a) examined this point with a simple genetic model of a one locus, two allele system that specifies both the mean and plasticity of a trait. He examined the predictive capabilities of the character state and reaction norm approaches and found that both were inadequate for predicting evolutionary trajectories. In particular, in the case of a negative genetic correlation across environments (r_{ae}), the outcome of selection cannot be predicted by the estimation of G and G × E components of variance, by the specific value of r_{ae}, or even by their combined information (Figure 3.21). Only when the actual genotype fitnesses and environmental frequencies are known can the outcome be forecast.

A second element that affects models of reaction norm evolution is the issue of the macroevolutionary origins of adaptive plasticity. Existing models implicitly assume that the infrastructural machinery for signal transduction is already in place. This assumption enables the quantitative genetic models to produce evolutionary change in reaction norms simply through allelic substitution. For

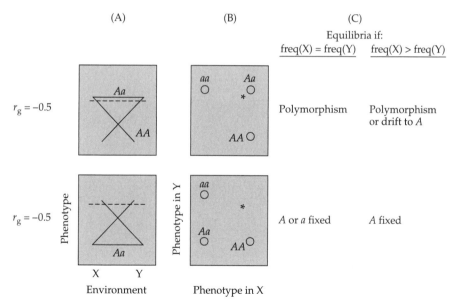

Figure 3.21   Results from a model demonstrating that different phenotypic outcomes are possible even with identical genetic correlations (r_g) and phenotypic optima. Column (A) is the reaction norm depiction of the system (dashed lines are optima); column (B) is the character state version (asterisks are the optima). The form of the equilibrium, column (C), depends on both the genetic details and the frequency of the environment. This implies the need to have information about both the genetics and ecology of a species for reliable long-term evolutionary predictions. (After Pigliucci, 1996b.)

example, a new allele might alter receptor sensitivity, or the binding capacity of an enzyme in the transduction pathway, or perhaps the conformation of the signal protein. While some such changes have clearly been important in reaction norm evolution, we regard these chiefly as tinkering. The real challenge lies in understanding how a coordinated regulatory network for translating stimulus into response can be built in the first place. This question lies outside of the scope of quantitative genetics, and requires a combination of molecular investigation along with the comparative method.

An example comes from work on the mechanisms of freezing avoidance by fishes. Chen et al. (1997b) characterized the antifreeze glycoproteins (AFGP) of two unrelated fish species, an Antarctic notothenioid (*Dissostichus*) and the Arctic cod (*Boreogadus saida*). The AFGP of the notothenioid was determined to have evolved by transformation, through enlargement and iterative tandem duplication of a portion of the trypsinogen gene. The expansion of the reaction norms of the notothenioid fish due to the origin of functionally similar, but evolutionarily unique, proteins was not caused by the allelic substitution envisioned in quantitative genetic models. And though general models of gene duplication and functional divergence have been proposed (Ohta, 1993; Clark, 1994; Hughes, 1994; Wagner, 1994; Walsh, 1995; Averof et al., 1996), these also cannot predict the evolution of the observed reaction norms.

We illustrate this limitation with a fairly simple scenario. Imagine a system with two discrete environments (A and B), with each environment evoking a distinct phenotype identifiable from early development. It may seem that in this coarse-grained environmental scheme we need only know about the trait-determining genes. But, although the environmental transformation of the phenotype may appear to be remarkably direct, the infrastructure (i.e., receptor, transducer, and activator) must be present to enable a *switch* from one developmental trajectory to the other.

Two important questions must be addressed to allow us to determine if a within-environment selection scenario is sufficient: (1) are the objects of selection the same *within **and** between* environments? and (2) will each of these selection regimes result in similar reaction norms? *Within* environment A or B, any genetic variation that leads to a trait mean closer to the optimum will be favored; that variation will most likely reside in structural or low-level regulatory genes affecting enzyme activity or protein function. *Across* environments, however, selection will focus on the genes determining a reaction norm, for example, those genes involved in processing environmental signals. Given these generally different targets of selection, similar reaction norms would be an unexpected result. In fact, it seems far-fetched to propose that the evolution of a system of appropriate plastic responses would arise from within-environment selection alone.

Alterations of reaction norms can come about due to changes in any part of the information pathway—through mutations in receptors, transducers, or targets. A developmental reaction norm view compels us to consider the mechanisms, pathways, and hierarchies involved in any transformation (G, E)→P, even for as simple a scheme as above. The devil, as usual, is in the details.

CONCEPTUAL SUMMARY

1. Reaction norms have five attributes—the *amount* of plasticity refers to the magnitude of response to the environmental change (large/small); the *pattern* is the shape of response (monotonic increase/decrease or more complex curves); *rapidity* pertains to the speed of response, for example, fast physiological changes versus slow morphological alterations; *reversibility* refers to the capacity for switching between alternative states, for example, a leaf's photosynthetic rate is reversible, while its shape is not; *competence* is the ability of the developmental system to respond to environmental stimuli only during particular time "windows" in the ontogenetic trajectory.

2. While it is difficult to demonstrate that a given plastic response is adaptive (and even more difficult to show that it arose as the result of natural selection), some clear examples of adaptive plasticity stand out as paradigmatic—shade avoidance response and heterophylly in plants and mimetic morphogenesis in insects.

3. Reaction norms can be visualized as character states, in which the expression of a trait in two or more environments is treated as several traits linked by a genetic correlation. Alternatively, one can consider the reaction norm itself as the trait. While these approaches are mathematically equivalent (and therefore neither can claim inherent "superiority"), the biological framework they refer to is radically different, with effects on the perception and action of experimental and theoretical biologists alike.

4. Plasticity has a history of fuzzy conceptual links with a variety of other terms from developmental biology and genetics. *Canalization* refers to the stability of a particular developmental trajectory in the face of random, but not persistent and predictable, environmental changes. Therefore, a reaction norm can be canalized (i.e., given certain environmental conditions, the developmental program will always yield pretty much the same phenotype), while being plastic. Canalization can be seen as a process leading to microenvironmental *homeostasis*. Canalization, plasticity, and homeostasis are not population-level phenomena; they are defined at the individual genotype level and cannot easily be extrapolated upward. *Developmental instability* is the opposite of homeostasis, but has no necessary relationship with plasticity. A given phenotype can be plastic (in response to macroenvironmental stimuli) and stable (in response to

microenvironmental noise), or vice versa. The distinction between micro- and macroenvironments is a matter of scale, and depends on the organism one is considering.

5. Theoretically, (a) plasticity can impede (act as a buffer during) the process of natural selection; (b) plasticity and genetic variability may represent alternative solutions to the problem of environmental heterogeneity; (c) plasticity may be related to instability/homozygosity and may be reduced by individual heterozygosity; (d) genetic variation for reaction norms for fitness could help to maintain genetic variation in non-fitness traits. The empirical jury on all these is still out; however, (a) is based on a misconception of the relationship between plasticity and genetic variation, (b) relates to the classical distinction between generalist and specialist genotypes, as well as to the evolution of ecotypic specialization, (c) seems to be increasingly unlikely, while (d) offers a rather novel solution to a very old problem.

6. There are two fundamental types of genetic control over plasticity—allelic sensitivity and regulatory plasticity genes. The first one translates into the developmental phenomenon known as phenotypic modulation, in which the phenotypic outcome is a continuous function of the environmental change. The second one corresponds to developmental conversion, which is characterized by the production of distinct morphs separated by thresholds. While allelic sensitivity can be adaptive and itself the target of selection, regulatory plasticity genes are virtually always likely to have been the direct target of natural selection.

7. The recent debate about plasticity as the target of selection, and the one about the existence of specific genes controlling plastic responses, is fundamental to our understanding of the evolution of phenotypes in different environments (and hardly a "matter of semantics"). In models, we can indeed treat reaction norms either as characters or as by-products of evolution in distinct environments; but the biological reality must be one or the other (of course it can be both, but not for a single character in a particular organism). In a similar manner we may be able to ignore the mechanistic basis of plastic response for short-term simulation models, but ultimately we need to figure such details out and incorporate them in stronger theories of phenotypic evolution.

CHAPTER 4

Allometry

According to Gould (1966), the concept of allometry dates
at least to Galileo's *Discorsi e dimostrazioni matematiche*
("Mathematical discourses and demonstrations," published
in 1638), in which a discussion of the biological consequences
of changes in the ratio between area and volume is presented. The modern study of allometry originated in the nineteenth century with C. Bergmann and H. Spencer, and
reached an early peak with the complex "multivariate"
descriptions of changes in shape by D'Arcy Thompson
(see Chapter 6) in the early part of the twentieth century.

Contemporary thinking about allometry was developed
by Huxley (1924, 1932) in reference to the proportional
growth of one character relative to a second character or to

some measure of overall organ/body size. In Huxley's view, allometry referred specifically to phenomena in which the rate of growth of one feature does not equal the rate of growth of a second feature, resulting in a change of shape. This is in contrast to isometry, which is the special case where two traits increase in size at exactly the same rate (Figure 4.1).

The general mathematical form of allometric relationships is summarized in the classic equation

$$y = bx^\alpha \qquad \text{(Equation 4.1)}$$

where y and x are two given characters, b is the value that y takes when $x = 1$, and α is the rate of increase of y with a unitary increment of x. In the simple case of a linear relationship, b is the intercept, and α is the slope of the line. When only two traits are compared (the bivariate case), α is called the "allometric coefficient." As Gould (1966) noted, $\alpha = 1$ is the special case for isometry for a *linear relationship* between the two variables. If the relationship is not linear, then isometry is obtained for different values of α (i.e., for $\alpha = 2$ when y increases as the square of x, or $\alpha = 3$ when y increases as the cube of x). A logarithmic transformation of the allometric function is quite common

$$\log y = \alpha^x (\log x) + \log b \qquad \text{(Equation 4.2)}$$

The log-transformation is often used because the relationships between many variables studied in biology (e.g., lengths, widths, weights) can be conveniently reduced to linear functions by semi-log or log-log plots.

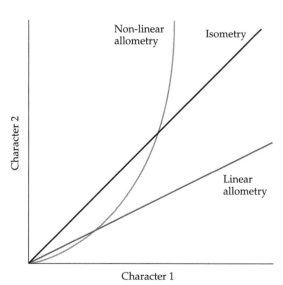

Figure 4.1 Isometric, linear allometric, and non-linear (or curvilinear) allometric relationships between two traits.

The basic allometric equation was used for the first time at the end of the nineteenth century to study relationships between brain weight and body weight; it was only later adopted and generalized by Huxley in his now classic work (1932). Huxley's synthesis, however, has been attacked from two opposing standpoints. On one hand, D'Arcy Thompson (1942) suggested that simple linear equations of the form

$$y = b + ax \qquad \text{(Equation 4.3)}$$

are more than adequate for the majority of real biological cases, suggesting that Huxley's power function is unnecessarily cumbersome. However, others have sought more complex mathematical functions in a quest to "explain" higher proportions of the observed variance. Perhaps the most famous and widely used variation of this type is the power function proposed by von Bertalanffy (1964)

$$y = a_0 + a_1 x + a_2 x^2 + \ldots \qquad \text{(Equation 4.4)}$$

Although these more complex approaches give an improved statistical fit, they do not necessarily augment biological understanding, since it is often difficult to interpret the cubic or higher terms in these equations within a biological frame of reference. Gould credits Scholl (1954) for suggesting that "a reasonably simple functional expression involving the minimum of non-interpretable parameters" should be adopted—the allometric version of Occam's razor.[1]

Jolicoeur (1989) classified examples such as those above, in which the two variables covary following some sort of power law, as "simple" allometry. "Complex" allometries do not follow *any* form of power function, and Jolicoeur proposed the use of a variation of the Gompertz growth curve to describe them:

$$X_2 = Ae^{\left\{ -C \cdot \left[\ln\left(X_{1\max} / X_1 \right) \right]^D \right\}} \qquad \text{(Equation 4.5)}$$

where X_i are the two allometrically related variables, $X_{1\max}$ is the maximum observed value of the predictor variable, e is the base of natural logarithms, A is the upper asymptote reached by the growth curve, and C is the allometry exponent. D is the time-scale factor (also known as the *complex allometry exponent*) and reflects the curvature in log-log space (curved when $D \neq 1$; straight when $D = 1$). Hence, assessment of D offers a direct test of the hypothesis of simple allometry.

---

[1] The English Scholastic philosopher William of Ockham (1285–1349) rejected the reality of universal concepts. He argued that explanations for unknown phenomena should be framed first in terms of what is already known, and that the simpler of two competing theories is to be preferred. Under the rubric of parsimony, this principle is the basis of most modern cladistic techniques for phylogenetic analyses.

ON SIZE AND SHAPE

One of the controversial dichotomies of quantitative evolutionary biology is the crucial distinction between *size* and *shape* in allometry. There is just enough intuitive difference between the two terms to generate both misconception and acrimonious discussion (e.g., Humphries et al., 1981; Somers, 1986, 1989; Bookstein, 1989; Sundberg, 1989). It is easy enough to imagine larger or smaller versions of the same organism, letting size change while keeping shape constant. However, upon reflection it is clear that enlarging *all* of the parts of an organism by the same quantity will produce something with a radically different form. This results from the different scaling of linear, area, and volume measures. For example, the doubling of an elephant's length necessitates squaring the diameter of its legs and cubing its lung capacity in order to maintain the same stable mechanical relationships. The result of this expansion will look less and less like an elephant, because its different parts are increasing at different rates. In other words, size and shape are only imperfectly correlated, and their correlation is determined by the allometric relationships of the various parts of the body.

Figure 4.2 shows several possible functions that relate traits A and B. Each linear function corresponds to a different "family" of shapes, and the curved line G indicates that a constant size can be maintained across different shapes. The only way to maintain a constant shape as size changes is to alter both A and B by the same amount, that is, along the 45° line, which is the case of isometric character relationships. Although several statistical techniques have been proposed for "eliminating" the size component, thus leaving the "pure" shape

Figure 4.2 Geometric relationships between size and shape as a result of allometry between traits A and B. The G curve indicates the combination of characters that will maintain constant size, although shape will change substantially. Lines 1 and 3 represent allometric relationships, when shape will change simply as a result of change in size. Isometry, line 2, maintains shape even with changes in size. (After Bookstein, 1989.)

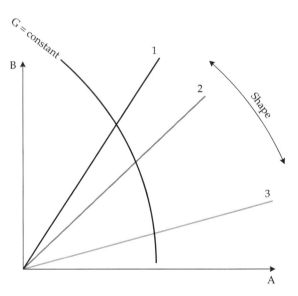

information, Figure 4.2 demonstrates intuitively why this is simply not possible—the allometric lines *also* contain information related to size, so that factoring out the special case of isometry does not eliminate size effects. In fact, wherever there is allometry, size and shape are part of the same biological phenomenon of proportional covariation between two (or more, in the case of multivariate analyses) variables. To try to separate the two and to attribute biological value to one and "noise" status to the other simply does not reflect reality.

CLASSIC AND RECENT STUDIES ON ALLOMETRY

The literature on allometry has a long and prestigious history that we cannot even begin to summarize here. However, a brief statement of some of the standard examples will help to elucidate the power of the concept of allometry to explain biological phenomena. Perhaps the most famous case of allometry is that concerning the recently extinct "Irish elk," a pan-Eurasian giant deer. The seemingly outsized antlers associated with this animal instigated rampant speculation: did they belong to an even larger organism, or were they perhaps directly responsible for its extinction?[2] Gould (1974) elegantly demonstrated that the elk's antlers are in fact not implausible at all. When he plotted the maximum length of the antlers against a measure of the size of the animal (height at the shoulder), he found that the Irish elk and other cervid (deer) species fall along the same allometric line (Figure 4.3). Gould also obtained an allometric relationship using only data from the Irish Elk itself, suggesting that the large antlers could result from selection on body size alone.

A second classic example of allometry is the well known relationship between body size and brain size (Figure 4.4; Futuyma, 1986). That larger organisms have larger brains is not surprising, but two other characteristics of these curves are biologically relevant. First, in log-log space, endotherms and ectotherms have the same slope, which suggests similarities between the physiological bases underlying the brain/body relationship. The intercept for endotherms, however, is much higher than that for ectotherms (by an order of magnitude), implying that the differences in metabolism between the two groups either require or permit a much larger brain in endotherms for similar body size.

---

[2]Such a suggestion has been made for "abnormal" morphologies in other groups of animals. As Gould (1977) pointed out, such theories ignore the fundamental mechanics of natural selection by assuming some sort of overriding directional force in evolution (not an uncommon view among earlier paleontologists, and re-emerging recently among some complexity theorists). If changes in the shape of an organism are immediately disadvantageous, they will be eliminated from the population; thus there is no possibility of their becoming fixed and causing an extinction, although it is not inconceivable that an environmental change could trigger such an event.

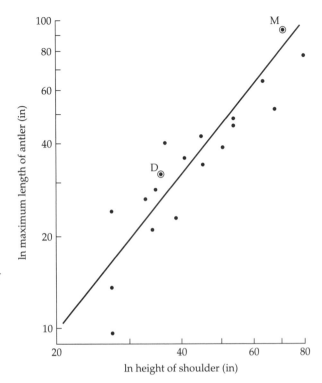

Figure 4.3 The interspecific allometry for the Irish elk (*Megaloceros*, point M) falls very near the allometric line defined by its relatives. The related, but much smaller, fallow deer (*Dama*; D) actually falls farther from the allometric line, and is thus arguably more "anomalous." (From Gould, 1974.)

The importance of allometry in macroevolution and adaptation is perhaps best illustrated by Darwin's own favorite example, the geospizine finches of the Galápagos. The well documented differentiation in bill sizes and shapes, related to feeding habit and diet (e.g., crushing versus probing bills; plant versus animal food), can be examined from an allometric perspective. Massive beaks appear to have evolved as a result of a positive allometric relationship between beak length, beak depth, and head size; furthermore, there is good contemporary evidence that beak size is a primary target of selection. Periods of drought are correlated with scarce food resources, especially in the smaller size classes of finches, and large birds (with large beaks) are more likely to survive (Gibbs and Grant, 1987). Allometry between beak depth and length also leads to a change in shape as a consequence of altered size (McKinney and McNamara, 1991).

There is no doubt that evolutionary divergence in the relationships among traits is widely recognized. To cite just one instance, Bradshaw and coworkers (Hard et al., 1993; Lair et al., 1997) have examined the genetic architecture of the pitcher-plant mosquito, examining the two traits development time and critical photoperiod for diapause. These appear to have evolved semi-independently under directional and stabilizing selection during northward expansion. Five of

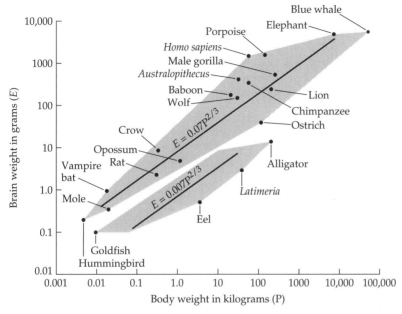

Figure 4.4 Body/brain size allometries for endotherms (upper line) and ectotherms (lower line). Notice that, although the two curves are parallel, endotherm brains are an order of magnitude larger for the same body weight. (From Futuyma, 1986.)

six populations examined had the "expected" positive correlations between these traits; one, however, has evolved a negative relationship between them.

Allometric relationships can also be remarkably resistant to significant alterations in the genetics and physiology of an organism. For example, despite threefold evolutionary changes in body size, species of *Onthophagus* beetles have maintained similar allometric relationships between horn length and prothorax width (Figure 4.5; Emlen, 1996). In another example, Shea (cited in McKinney and McNamara, 1991) compared the growth curves of dwarf and giant mice that exhibit, respectively, underproduction (due to mutation) and over-production (due to a transgenic implant) of a pituitary growth hormone. It is not surprising that the giant mice grew much faster than their dwarf counterparts (Figure 4.6, left). The femur/humerus allometry, however, remained essentially identical (Figure 4.6, right), suggesting that the physiological alterations that produce the different bone growth rates have not affected their allometric relationship.

Experiments designed to select directly on allometric relationships, however, have generally been successful, which indicates the presence of additive genetic covariation between traits. In the horned scarab beetle, there is disrup-

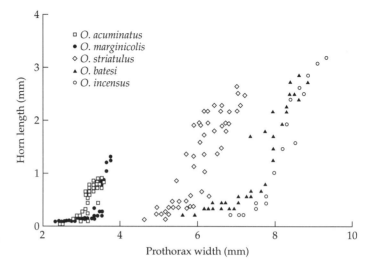

Figure 4.5 Allometry of horn length among species of *Onthophagus* beetles. Two small, one intermediate, and two large species exhibit the same S-shaped relationship during growth. (From Emlen, 1996.)

tive selection for horn size because of different mating behaviors of large and small horned males. Emlen (1996) successfully selected lines of the beetle that had significantly longer or shorter horns for a given body size (due to initiation of horn growth at an earlier or later age). Note in Figure 4.7 that although selection was effective at changing the onset of horn growth, the general sigmoidal relationship between horn and body size was not altered.

In the stalk-eyed fly, *Cyrtodiopsis dalmanni*, there is sexual dimorphism for the allometry of eye span to body length—in males the slopes are greater than one,

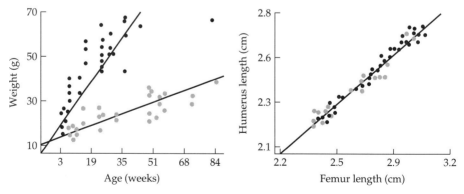

Figure 4.6 Allometric relationships for dwarf mice (gray circles) and giant mice (black circles). The divergence of the allometric relationship depicted on the left (age/weight) stands in contrast to the identical allometry displayed by the same two groups in the right diagram (femur/humerus). (After McKinney and McNamara, 1991.)

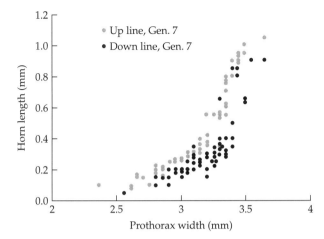

Figure 4.7 Upward and downward selection on the allometric relationship between horn size and body size in the scarab beetle (*Onthophagus acuminatus*). The curve is shifted but retains the same shape. (From Emlen, 1996.)

and in females the slopes are less than one. Wilkinson (1993) successfully selected males for both increased and decreased ratios, and found that most of the change came from increases or decreases in eye span while body length changed little. Weber (1990) selected *Drosophila melanogaster* for changes in the relationships between wing dimensions by choosing flies that had increased or decreased angles between veins, relative to a base population. The strong direct responses (Figure 4.8) were due to changes in wing contours and in movement of the location of vein intersections along the wing margins. Correlated responses to selection turned out to be specific to the particular pair of veins selected. Weber concluded that there was as much heritable variation for these allometric relationships as is typically found for quantitative traits, and that the lack of strong genetic correlations was not consistent with a view of the entire wing as an integrated unit. Guerra et al. (1997), however, concluded that correlated responses to selection for shortened wing veins in *D. melanogaster* were consistent with a view of the wing as an integrated unit.

Can some logical order be applied to the vast literature of allometric studies? Cock (1966) recognized the need to differentiate among the several phenomena that are investigated under the rubric of allometry, and distinguished between static, ontogenetic, and evolutionary analyses (Figure 4.9; Klingenberg and Zimmermann, 1992). *Static allometry* refers to patterns of character covariation among individual organisms at a particular ontogenetic stage. *Ontogenetic allometry* applies to character covariation among individuals between developmental stages, or along a growth trajectory (longitudinal studies). *Evolutionary allometry* describes the covariation among characters across related populations or species at a single ontogenetic stage.

Although Klingenberg and Zimmermann (1992) state that these three types of analyses are "tightly and reciprocally interrelated," the relationships among

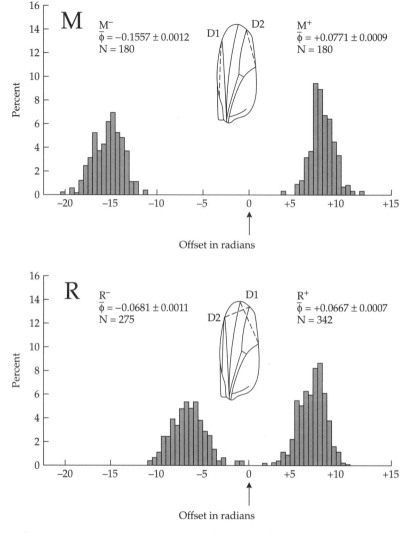

Figure 4.8 Selection for increased and decreased angles between wing veins in *Drosophila*. Shown are the frequency distributions of results for two of the five selected angles (M and R). The diagrammed wings depict the pair of veins whose angle of intersection was changed. Note in M that the response to selection was greater for decreasing angles. Arrow indicates initial population; + and – indicate the upward and downward selection lines; $\bar{\phi}$s are the final mean angles of change (± the standard error); N is the sample size. (After Weber, 1990.)

them are not necessarily straightforward. In fact, we would argue that they are not even necessarily related. This can be deduced from first principles by examining the components of variation used to derive each of the different covari-

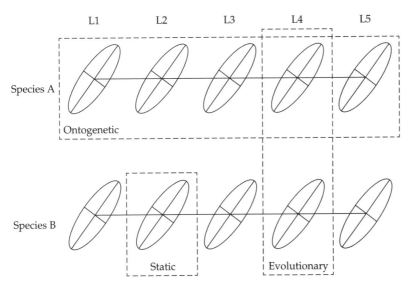

Figure 4.9 Relationships among the different types of allometry. L1–L5 are stages of ontogeny. Ellipses refer to the relationship between two traits. Static allometry is measured at one life stage in species B. Ontogenetic allometry is measured across life stages in species A. Evolutionary allometry is typically measured at the same stage in two (or more) species. (From Klingenberg and Zimmerman, 1992.)

ances (i.e., static, ontogenetic, etc.; see Table 4.1; also notice the presence of another type, *plastic allometry*, that describes changes in character covariances triggered by the environment). Stemming directly from the classic analysis of variance equation, $V_P = V_G + V_E + \ldots$ where V represents variances, the standard decomposition of phenotypic covariation is

$$\text{cov}(P)_{X,Y} = \text{cov}(G)_{X,Y} + \text{cov}(E)_{X,Y} \qquad \text{(Equation 4.6)}$$

where X and Y are two given traits and COV are phenotypic (P), genotypic (G), and environmental (E) covariances. The degree of relationship between the covariances at the different levels depends on whether individual or family observations are used. If individual terms are used at one level and family terms are used at the other, the relationship will be extremely limited. As can be seen from Table 4.1, evolutionary allometry shares no terms with the other forms in the calculation of the covariances. As an example of the distinctness of these allometries, we cite Cheverud's (1982) results on the relationships between multivariate static and ontogenetic allometries of the rhesus monkey (*Macaca mulatta*). He compared the vector angles of the first principal components derived from 48 skull traits, and found significant differences among the ontogenetic, the static phenotypic, and the static genetic principal components.

Table 4.1 **Relationships of static, ontogenetic, plastic and evolutionary covariances in allometric studies**

Allometry	Level of analysis	Covariance of traits
Evolutionary	between traits X and Y among species in environment E1 at stage D1	$COV_{SPECIES} (\mathbf{X}_{E1,D1} , \mathbf{Y}_{E1,D1})$
Plastic	between X in environments E1 and E2 for families of species 1 at stage D1	$COV_{FAMILIES} (X_{S1,\mathbf{E1},D1} , X_{S1,\mathbf{E2},D1})$
Ontogenetic	between X in stages D1 and D2 for species 1 in environment E1 (families or individuals)	$COV_{FAMILIES} (X_{S1,E1,\mathbf{D1}} , X_{S1,E1,\mathbf{D2}})$ *or* $COV_{INDIVIDUALS} (X_{S1,E1,\mathbf{D1}} , X_{S1,E1,\mathbf{D2}})$
Static	between traits X and Y at stage D1 for species S1 in environment E1 (families or individuals)	$COV_{FAMILIES} (\mathbf{X}_{S1,E1,D1} , \mathbf{Y}_{S1,E1,D1})$ *or* $COV_{INDIVIDUALS} (\mathbf{X}_{S1,E1,D1} , \mathbf{Y}_{S1,E1,D1})$

Note: Imagine an analysis of character variation that includes five families (F_1 to F_5) of four species (S_1 to S_4) grown in two environments (E_1 and E_2). Characters X and Y are measured at two developmental stages (D_1 and D_2). The focal contrasts for each type of allometry are indicated in boldface.

STATIC ALLOMETRY   Innumerable studies have been carried out examining patterns of covariance or correlation between pairs of traits, most often on mature individuals. As an example of static and not-so-static allometry, Tomkins and Simmons (1996) examined the relationship between forceps and elytra in earwig species with dimorphic (brachylabic versus macrolabic) forceps. They observed static allometry in some species—in *Oreasiobias stolickzae*, although the intercept changed, the slope of the relationship was the same for the two morphs (Figure 4.10A); in other species they found a size-related shift in the relationship between the traits—the morphs of *Eluanon bipartitus* differed in both intercept and slope (Figure 4.10B). Ward (1997) used allometry as a tool to argue against a proposal that ant soldiers in *Pheidole* are modified queens. His plots of tibia size versus pronotum width or eye length instead supported a relationship between soldiers and minor workers.

ONTOGENETIC ALLOMETRY   Plant and animal breeders have used juvenile/adult correlations to trace relationships between desirable adult features and earlier juvenile states (e.g., Rehfeldt, 1992; Vargas-Hernandez and Adams, 1992; Danjon, 1995; see also the discussion in Chapter 9 on "windows" of selection opportunity). Evolutionists are interested not only in such positive correlations between stages, but also in uncorrelated and negatively correlated stages, as

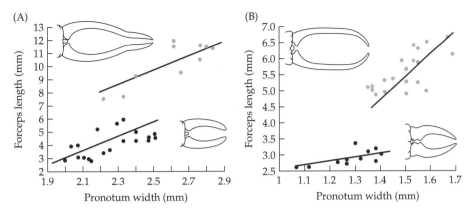

Figure 4.10   Variation in allometry of two traits in two species of earwigs (Insecta: Dermaptera) with dimorphic forceps (gray circles = macrolabic; black circles = brachylabic). (A) *Oreasiobias stolickzae* with equal slopes and different intercepts. Brachylabic: $y = 1.85x - 0.75$. Macrolabic: $y = 1.82x - 0.44$. (B) *Eluanon bipartitus* with different slopes *and* intercepts. Brachylabic: $y = 0.67x + 0.77$. Macrolabic: $y = 1.17x + 0.32$. (From Tomkins and Simmons, 1996.)

these provide insight on the genetic architecture of traits (Cheverud et al., 1983). Buis et al. (1978) examined the correlations of vegetative traits during ontogeny in *Mirabilis*, and documented a general decrease in correlation as the interval between measurements increased (Figure 4.11). Klingenberg et al. (1996) found a similar result for the correlation of size features among the instars of waterstriders, but when corrected for size effects, growth increments within instars were relatively uncorrelated (Klingenberg, 1996). For organisms with partial or complete metamorphosis between juvenile and adult stages, the question may be one of whether there are any coupled characters (Moran, 1994; Zera et al., 1997). Blouin (1992) found that the genetic correlation between larval and juvenile growth rates was near zero in the tree frog *Hyla cinerea*; and a study of heat shock in *Drosophila* showed that the degrees of resistance of larvae and adults were uncoupled (Loeschcke and Krebs, 1996).

There have been many studies that document the effects of age on allometric relationships (e.g., Atchley and Rutledge, 1980; Birot and Christophe, 1983; Atchley, 1984; Zelditch et al., 1992; Fink and Zelditch, 1996; Fiorella and German, 1997). A representative example is the study by Cane (1993) that examined the relative growth in 17 morphological traits of the common tern from embryo to adult. Although the overall allometric relations were curvilinear, separate linear phases could be identified during ontogeny (Figure 4.12). The patterns of growth were observed to change markedly during development; for example, the hindlimb bones grew somewhat faster during the embryonic stage, but the

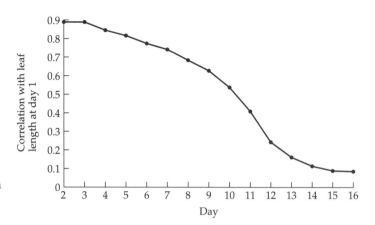

Figure 4.11 Leaf length in *Mirabilis jalapa* measured at daily intervals and correlated with leaf length measured on day 1. The general decrease in the correlation suggests that selection later in ontogeny will have decreasing effects on earlier expression of the same trait. (From Buis et al., 1978.)

forelimb bones showed a substantial posthatching growth spurt (Figure 4.12). In addition, the final relationship between traits in adult birds was quite different from those at earlier stages of development.

EVOLUTIONARY ALLOMETRY Examination of the allometric relationships among related taxa can be quite revealing (e.g., Andersson, 1996). Such comparisons have been used to produce broad evolutionary conjectures about the functionality of, or constraints on, character relationships (Lloyd, 1987; Stearns, 1992). However, because evolutionary allometry is statistically unrelated to the other

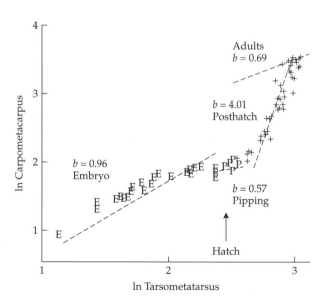

Figure 4.12 Changes in allometric relationships in the common tern (*Sterna hirundo*) between representative bones of the forelimb (carpometacarpus) and hindlimb (tarsometatarsus) during ontogeny. Although the general shape of the curve is sigmoidal, there are linear subsets of points corresponding to distinct growth phases, indicating that the bones do not share the same growth trajectories throughout ontogeny. (From Cane, 1993.)

forms (Table 4.1), a measure of circumspection is advised with regard to the interpretation of such relationships. As an example, consider the striking correlation between the number of inflorescences and the capsule number per inflorescence for 30 species of the genus *Plantago* (Figure 4.13; Primack, 1978; Primack and Antonovics, 1981). The negative relationship observed across species ($r = -0.53$) masks the positive relationship observed within most species—26 species have correlations greater than zero, 10 significantly so (Primack, 1978). Does this interspecific pattern represent the action of selection? Armbruster and Schwaegerle (1996) correctly point out that similarity of covariance patterns between taxa may be derived from any of the standard evolutionary forces; selection is just one possibility along with pleiotropy, linkage, epistasis, and genetic drift. In addition, instances of similar genetic covariances may represent cases of either synapomorphy or homoplasy.

These basic types of allometry can be linked in all possible combinations to reveal additional information about character relationships; for example, static allometry between two traits can be compared at different ontogenetic stages (see Chapter 5), or in different environments (see below).

ALLOMETRY AND THE DRN

Although most studies of allometry have focused on the morphological traits of organisms, we expand this to encompass the relationships among *any* phenotypic characteristics, as can be seen from some of the examples above (we also advocate an increased emphasis on multivariate, as opposed to bivariate, approaches). For instance, one could study the association between a morphological trait and the display of a particular behavior—the intensity of the behavior might increase directly with the trait, or it might remain unchanged until the

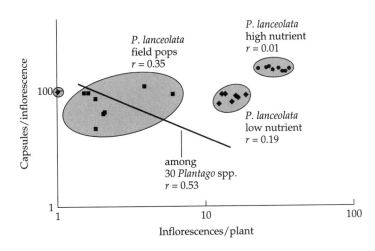

Figure 4.13 Evolutionary allometry in *Plantago*. The negative correlation among 30 species (solid line) differs from the low positive correlations measured among populations of *P. lanceolata* (ellipses). (After Primack, 1978; and Primack and Antonovics, 1981.)

trait reached some threshold value. This broadened view of "allometry" as the study of the covariation of components of the phenotype is derived from the developmental reaction norm perspective, in which we recognize the importance of the integration of an organism's many traits through ontogeny, and as environmental conditions change.

From an evolutionary viewpoint, our interest is in what causes characters to be coupled. In order to understand the mechanistic basis of trait covariation, we can apprehend the usual suspects that affect the developmental reaction norm: pleiotropy, which is the manifold effects of a single gene locus on a variety of traits; linkage, which is the physical proximity of gene loci, with the concomitant reduction of recombination causing certain allelic effects to be correlated; and epistasis, which is the suppression or enhancement of expression of a gene due to the effects of another locus. Once we understand the genetics of allometry, even more interesting evolutionary questions arise: How can the correlations among traits be modified? How can correlations be uncoupled, and how can new correlations be forged between formerly unrelated traits? What are the evolutionary forces that oppose changes in correlation? These are fundamental questions for understanding phenotypic evolution, and we will take various approaches to addressing them in Chapters 6 through 9. First we will examine in detail the category of allometry that stems directly from a DRN perspective, that is, plastic allometry, the influence of the environment on allometric relationships.

Plasticity and Allometry

As we have seen, it is widely acknowledged that allometric relationships can be shaped by evolutionary forces, and they also change during development. Much less appreciated has been the role of the environment in altering the relationships among characters. In his review, Gould (1966) gave only perfunctory treatment to environmental effects on allometry, and cited papers that principally dealt with the effects of temperature on the slope of the allometric coefficient. Although a few studies have appeared in the literature since then (and most mention neither allometry nor plasticity), it was not until the 1980s that there were directed studies to relate allometry and plasticity.

Analogous to the static, ontogenetic, and evolutionary analyses of allometry is an environment-centered approach that we call *plastic allometry*. Plastic allometry has three manifestations: the examination of the covariation of a single character across environments (Table 4.1; Falconer, 1952; Via, 1984b), the comparisons of allometric coefficients across environments (examples below), and the correlations among plastic responses themselves (see Chapter 7).

Berntson and Bazzaz (1996), as part of a long-term research project on the possible consequences of greenhouse effects and global warming, conducted an experiment to compare the responses of two species of trees, *Acer rubrum* and *Betula papyrifera*, to enhanced concentrations of atmospheric CO_2. They found that *B. papyrifera* responded to elevated CO_2 levels with a marked increase in root production and root turnover, and that *A. rubrum* did not respond significantly. However, comparisons of the allometric plots of the two species in each of the two environments revealed no differences (Figure 4.14), suggesting that, in this case at least, the correlations among characters are more resistant to environmental changes than the character means.

A number of studies have documented changes in allometric relationships of feeding structures that are associated with changes in diet (Bernays, 1986). Thompson (1992, 1998a) demonstrated significant plasticity in the head morphology of a grasshopper in response to hard or soft food (grass versus clover), differences that were positively correlated with consumption rates. Pfennig (1992a,b) documented a reversible diet-induced switch from omnivore to carnivore for spadefoot toad tadpoles, mediated by the accelerated development of certain morphological features. For example, the buccal cavity abductor muscle grew much more rapidly in the carnivores relative to overall body size (snout/vent length)—the mean ratio was 0.19 for carnivores and 0.11 for omnivores.[3]

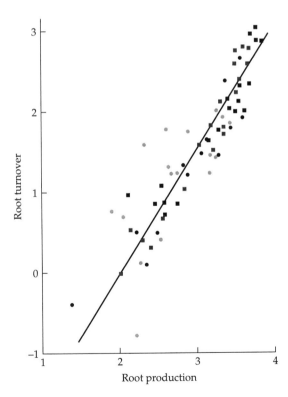

Figure 4.14 Allometry of root production and root turnover is invariant between the two species *Betula papyrifera* (squares) and *Acer rubrum* (circles) in normal conditions (gray) and with enhanced CO_2 (black). In contrast, the individual traits display variation for plasticity to CO_2 levels (not shown). (From Berntson and Bazzaz, 1996.)

[3] An interesting result of this work was that experimental manipulation of relative densities showed that the morph at the lower frequency had a fitness advantage (Pfennig, 1992b).

Plastic allometry is in some instances clearly adaptive. Most plants that experience a variety of light environments (i.e., from shade to open habitats) possess a particular type of plastic reaction known as the shade avoidance response (Schmitt and Wulff, 1993). A plant can "sense" the presence of neighbors—and therefore impending competition—by changes in the ratio between the red and far red portion of the light spectrum. Since plants absorb the photosynthetically active red wavelengths, but not far red ones, a decreased R:FR ratio serves as a reliable cue to the presence of surrounding vegetation. The adaptive value of the shade avoidance response (mediated by the phytochrome photoreceptors) has been convincingly demonstrated both by environmental manipulations of natural populations (Dudley and Schmitt, 1996) and by comparative studies of the environment-specific maladaptive responses of mutants (Schmitt et al., 1995; Pigliucci and Schmitt, unpublished).

Environmental conditions can also *qualitatively* alter allometric relationships. Weiner and Thomas (1992) demonstrated that crowding of plants can change the shape of the allometric curve from simple to complex (recall Jolicoeur's distinction discussed above). They studied the responses of stem diameter and stem height to intraspecific competition in three species of annual plants. Similar changes in allometry for all three species were observed. For plants in isolation, allometries of stem diameter, height, and biomass were linear, but under competition they were curvilinear or discontinuous. For example, in *Tagetes patula* (Asteraceae), the curves shifted upward with competition (i.e., plants were taller for the same stem diameter), and the shapes of the allometric functions became curvilinear (Figure 4.15). The authors went a step further and investigated the relationship between the static allometry within the population and the growth curves of individuals. When

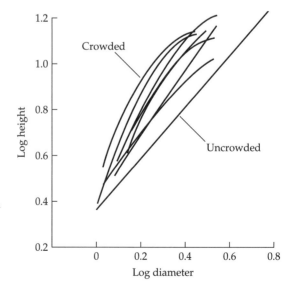

Figure 4.15 Shift of allometric relationship from linear to curvilinear in response to crowding in *Tagetes patula*. The curves for the crowded conditions represent seven sequential weekly harvests. (From Weiner and Thomas, 1992.)

conditions were uncrowded, the individual growth curves had the same shape as the allometric relationship among individuals; however, when plants experienced competition, the individual growth curves became steeper, but not curvilinear. The authors proposed that the difference between the "crowded" curves derived by the two methods can be explained if crowded plants switched to the higher growth slope at different initial sizes. They concluded that allometries measured as statistical summaries of differently sized individuals are not necessarily reflections of the growth trajectories of individual plants!

One of the problems of plastic allometry is that there are so few studies that virtually none have investigated the relationship between plasticity in allometric coefficients and reproductive or life-time fitness. As Winn (in press) pointed out, many instances of phenotypic plasticity are "plausibly" adaptive. Few, however, have been rigorously examined to see if the plastic response is congruent with a fitness advantage.

Plasticity of Character Correlations

Another way of studying plasticity in allometry is to look at environmental effects on phenotypic or genetic correlations among traits. Since correlations are simply standardized regressions, and therefore allometric coefficients, the two approaches are really dealing with exactly the same biological phenomenon. Nevertheless, they evolved quite independently in the literature, with morphologists and developmental biologists interested in allometry *sensu stricto* and quantitative geneticists involved with correlations. However, we wish to emphasize that—historical and mathematical quibbles aside—they really are one and the same.

Several authors have pointed out that both phenotypic and genetic correlations can sometimes be dramatically affected by alterations in environmental conditions (e.g., Lechowicz and Blais, 1988; Schlichting, 1989a,b; Clark, 1990; Pigliucci et al., 1995; Schlichting and Pigliucci, 1995b), and that there is genetic variation for such lability, both within and among populations and species.

Kawano and Hara (1995) illustrated the plasticity of character correlations in an experiment on rice that examined the effects of density on correlations among a variety of traits (Figure 4.16). As density declined over three orders of magnitude, the number of significant correlations also declined. Lechowicz and Blais (1988) studied the contribution of different traits to reproductive success in *Xanthium strumarium* (Asteraceae) under different levels of water and nutrient availability. Although they found a general trend of increased reproductive success with increased resource availability, the relative contribution of different traits to fitness varied with the particular environment. These changes were attributable in part to the plasticity of each single character, and in part to the plasticity of trait correlations with fitness. Plasticity of genetic correlations has also been well documented (Giesel et al., 1982; Windig, 1994a; Yampolsky and Ebert, 1994; Chapter 3), indicating that correlated responses to selection will also be environment-dependent.

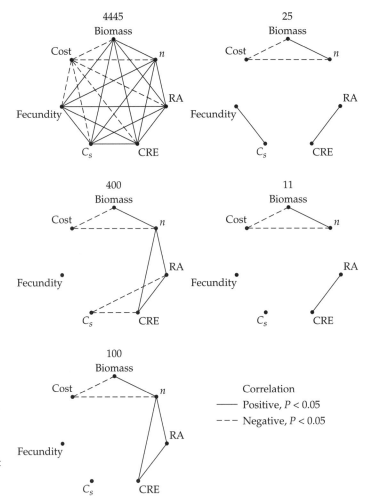

Figure 4.16 Diagram of the plasticity of character correlations in response to density (11 to 4445/m²) in rice. Solid lines represent positive correlations, dashed lines represent negative correlations. (From Kawano and Hara, 1995.)

As we remarked before, the "flat" representation of phenotypic integration resulting from the study of correlation matrices might not be representative of complex, hierarchical relationships among traits. Path analysis is one way to specify connections not only within but between hierarchical levels (Maddox and Antonovics, 1983; Crespi and Bookstein, 1989; Kingsolver and Schemske, 1991; Jordan, 1992; Mitchell, 1992; Bergelson, 1994). Complications arise, however, since the number of configurations in which characters can be combined increases exponentially with additional characters; few of these configurations will make much biological sense. In this respect, path analysis requires much more attention and thoughtful decision-making on the part of the user than most "plug-and-play" multivariate statistical methods. One intuitive way to construct a biologically plausible scenario of relationships among several characters is to follow the ontogeny of the organism under study and to connect characters expressed at the same time, or to limit paths to those from early traits to late ones, but not vice versa. This approach can lead to a restricted number of testable models when coupled with reasonable supplementary hypotheses based on knowledge of the underlying physiology of the characters considered.

When some degree of this complexity is taken into account, we still see that both genetic and environmental variability have a role in character relationships. Pigliucci et al. (1995) constructed a path model connecting eight traits in *Arabidopsis thaliana* exposed to different levels of light, nutrient, and water availability. Figure 4.17 shows the basic path model in high and low light, in which three ontogenetic phases are assumed to directly or indirectly affect reproductive fitness: vegetative growth, transition from vegetative to reproductive status, and adult reproduction. Early- and late-flowering populations were contrasted, because flowering phenology is a crucial ecological character that distinguishes local selective regimes. The changes in some of the path coefficients when plants are grown in low light can be seen in Figure 4.17. For instance, under high light the path from flowering time to number of branches is negative and significant in the early-flowering ecotypes, but it is positive and significant in the same populations under low light. It is of interest that these two traits are

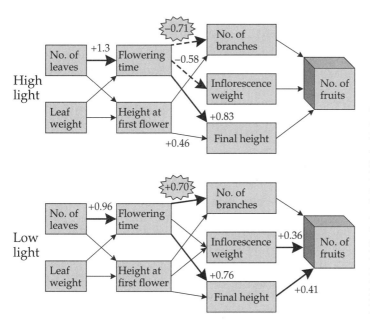

Figure 4.17   Path diagrams of interrelationships among ontogenetic traits in early-flowering populations of *Arabidopsis thaliana* in treatments of high and low light. Numbers are path coefficients (i.e., standardized partial regression). Bold lines are significant coefficients; dashed lines are negative and solid, positive. Note especially the environmentally triggered change in sign for the path from flowering time to number of branches. (From Pigliucci et al., 1995.)

uncorrelated in both environments in the late-flowering ecotypes (not shown in the figure).

A reasonable and testable explanation might be that the late-flowering plants gather enough energy even under low light conditions simply because they are already delaying sexual reproduction. The decision of when to flower, then, loses its direct consequences for plant architecture during the reproductive stage. A flat correlation analysis would have yielded a simplified and distorted scenario, because it would not have taken into account the fact that some traits (flowering time) have indirect effects on fitness, while others (branching) have a direct effect on fruit production.

How can we explain an environmentally induced change in genetic correlations? A purely mechanistic scenario has been proposed by Stearns et al. (1991) within the framework of quantitative genetics. They considered a simplified model in which the additive genetic variances (and covariance) of two traits expressed in a series of environments are determined by one locus. If the alleles at this locus have linear effects, that is, they respond to the environmental change with a linear reaction norm, then the additive genetic variance of a trait x is

$$V_{A(x)} = 2pq\alpha_x^2(E) \qquad \text{(Equation 4.7)}$$

and the genetic covariance between the two traits being

$$\text{COV}_{A(1,2)} = 2pq\alpha_1(E)\alpha_2(E) \qquad \text{(Equation 4.8)}$$

where E is the environmental effect, α is the average effect of a gene substitution, and p and q are the allele frequencies. As seen in Figure 4.18, the variances are a quadratic function of the environment, and the covariance can be negative, zero, or positive.

Stearns et al. reviewed several examples of environmental effects on covariances from the literature; however, a particularly striking one is seen in more recent work by Windig (1994a,b). Windig examined the genetic correlation between the appearance of a particular seasonal phenotype of the tropical butterfly *Bicyclus anynana* and its developmental time. The colorations of the wings of the two seasonal forms differ in overall contrast, in the presence of a black ring and an outer ring, in the degree of marbling, and in the presence of a white band. All these differences are summarized by a single principal component. When the genetic correlation between the score on this component and development time was calculated at four different temperatures, the correlation changed from positive at high temperatures to zero and slightly negative at low temperatures. When the reaction norms of extreme genotypes for the two traits were plotted, it was seen that the inversion in the correlation pattern was attributable to a crossing of the bivariate reaction norms around 20°C (Figure 4.19).

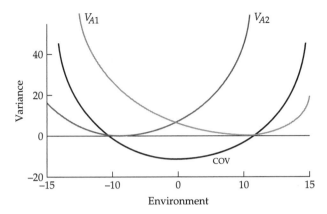

Figure 4.18 Model of environmentally induced changes in genetic correlations. The diagram shows the modification of genetic variances (V) and covariance (COV) between two traits affected by a single environmentally sensitive locus. The covariance between traits becomes negative for some environmental conditions. (From Stearns et al., 1991.)

Phenotypic plasticity, therefore, has a direct effect on the sign and magnitude of genetic correlations.

An empirical approach to the question of genetic control of integration has been taken by Pigliucci et al. (unpublished) in a study that utilized EMS-mutagenesis in *Arabidopsis thaliana*. They grew mutagenized offspring of the Landsberg line under three levels of nutrient availability, and compared the multivariate reaction norms of mutants and wild type for several morphological and life history characters related to sexual reproduction. When compared to the control line (itself selected under controlled conditions mostly for a fast life cycle), they found that several of the mutants were able to increase their reproductive fitness following one of two strategies: (1) delaying flowering to the end

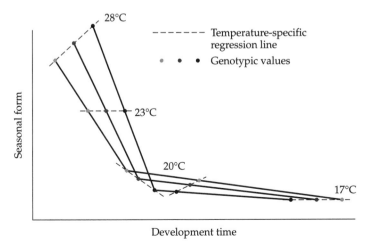

Figure 4.19 Graph of the reaction norms to temperature of *Bicyclus anynana*, representing two extreme and one intermediate genotypes. The crossing of the norms determines a switch in sign of the genetic correlation between seasonal form and development time. (From Windig, 1994a.)

of a prolonged vegetative growth, or (2) rapidly increasing growth and branching during the reproductive phase. No mutant line combined the two strategies, which may suggest the existence of a more fundamental genetic constraint that prohibits the simultaneous coupling of a longer vegetative phase with a more productive sexual phase (at least within the Landsberg line).

Stearns et al.'s (1991) theoretical discussion, and the empirical data of Windig (1994a,b) and Pigliucci et al. (unpublished) represent a new avenue of research into the genetic basis and environmental responsiveness of phenotypic integration. These works also demonstrate the importance of distinguishing *variability* from *variation* (*sensu* Wagner, 1995). *Variability* is the pre-selection variance attributable to mutation, recombination, and other purely genetic mechanisms. *Variation* is the variance observable in natural populations, following at least some screening by natural selection, and thus reflects only a subset of the *variability* of a population. Koufopanou and Bell (1991) suggested that studies of variability can help to explain macroevolutionary trends as the result of the joint effects of mutation pressures and selective forces. They also highlighted the importance of variability for understanding long-term evolutionary trajectories (see Chapter 10). The **G** matrix, which is currently used in most studies, measures only variation. Several authors have suggested that empirical research should also examine the structure of the so-called **M** (for mutation) matrix, which records variability (Houle et al., 1992; Kondrashov and Houle, 1994).

The above discussion indicates that some environmental lability of genetic correlations may be explained strictly in terms of intrinsic changes of the quantitative genetic parameters, or as the result of single gene mutations. We discussed in detail in Chapter 3, however, the extensive evidence for another genetic cause of plastic responses, namely, alterations in gene regulation. Such changes will also alter the correlation structure of a phenotype. The correlation between traits is a function of their separate variances as well as of their covariance with each other (see Equation 7.2). Any change in gene regulation that alters one or more of these three parameters can also change patterns of correlations and integration. For example, if a gene controlling one trait is turned on or off in response to an environmental stimulus, there will be a change in both its variance and in the covariance with a second trait. The environmental control of gene cascades will lead to manifold effects on the integration of various traits. Schlichting and Pigliucci (1995a) have argued that such regulatory control of entire developmental pathways is in fact a preferred route to guide the appropriate responses (plasticity) of multiple traits (phenotypic integration) to environmental change.

CONCEPTUAL SUMMARY

1. Allometry has most often been studied from a *static* perspective that examines the relationships between two morphological characters at a particular life stage. Other analyses have examined *ontogenetic* (the same trait at different stages) allometries and *evolutionary* (the same trait pair in different taxa) allometries.

2. We propose that *plastic* allometry (the same trait in two environments, or the plasticity of standard allometric coefficients) constitutes a fourth and equally important category.

3. To some extent, these four pure classes represent distinct hierarchical levels that have little or no statistical interrelationship (Table 4.1). However, these levels can be combined, for example, by comparing correlations or matrices of traits measured at two stages of development, in two species, or in two environments.

4. The **DRN** perspective leads us to consider an expanded concept of "allometry" that encompasses the covariation between any pair or suite of phenotypic traits. This suggestion arises from our perception of the importance of the integration of various facets of the phenotype to achieve coordinated changes during development.

5. The mechanistic causes of character covariation can be found in the usual genetic phenomena affecting developmental reaction norms, such as pleiotropy, linkage, and epistasis. Allometric coefficients clearly respond to selection, thereby demonstrating the existence of additive genetic variation for character correlations. On the other hand, there are also examples of stability of allometry in the face of selection on component traits.

6. Allometric relationships are altered during ontogeny and by the environment. There is increasing evidence for the presence and importance of the plasticity of character correlations. Both types of alteration will result in changes in the targets of selection and in the potential for response to selection.

7. Examples of investigations on the effects of the environment on the relationships among traits for plants and animals include some "plausibly" adaptive responses. However, few if any studies have specifically tested for the fitness advantages of plastic allometric responses.

CHAPTER 5

Ontogeny

HISTORY

The history of the study of ontogeny and of its relationships
to the evolution of life on Earth is probably as ancient as the
work of the pre-Socratic philosopher Anaximander. It was his
postulate that, since water has pre-eminence among the elements, ancestral humans must have been aquatic organisms,
which then gave rise to the modern form. As support for his
theory, Anaximander cited the fact that the early stages of
human development still happen within a liquid environment,
in the amniotic fluid. Gould (1977) presents a fascinating and
detailed history of the study of ontogeny from Anaximander
to modern times, and we will therefore only highlight the relatively recent major turns and twists of its story.

Probably the first comprehensive theory to explain differences in development among organisms was the idea of "recapitulation." This was originally proposed by the so-called "nature-philosophers" in Germany, who envisioned a classification of living things according to a sequential addition of stages, moving from simple to more complex forms. Each "superior" organism would have to go through the stages of its ancestors (recapitulating them) before "advancing" further. This idea was based on the fundamental vision of the world found in the "chain of being," in which life was seen to progress from simple, more primitive forms to increasingly more complex ones, culminating with the most perfect of all beings, *Homo sapiens*. Several nature-philosophers believed that a single explanatory principle should underlie the apparent variation of all living things, if for no other reason than to satisfy their sense of aesthetics.[1]

The principal opponent of the doctrine of recapitulation was Karl Ernst von Baer. In a fundamental book, von Baer (1828) proposed that embryonic stages of "advanced" animals are not directly comparable to adults of "lower" forms. He argued that the embryonic stage of life involves special challenges and requires particular adaptations that may be independent of the form of the adult phenotype. He proposed that development is best seen as a series of differentiation events that proceed from the simple, generalized, embryo to more and more specialized—and therefore complex—forms.

Recapitulation returned to prominence in the middle of the nineteenth century, mostly due to the work of Ernst Haeckel,[2] who in 1866 formally proposed his "biogenetic" law, that ontogeny represents the short and rapid recapitulation of phylogeny (Haeckel, 1866). This new version of recapitulation was based on a coupling of the terminal addition of stages during development with a hypothesized process of condensation (i.e., shortening of developmental stages; Ekstig, 1994). This avoided the problem of making the ontogenies of more advanced forms interminably long and provided an explanation for the apparent lack of some intermediate stages (they were there, but too "condensed" to be detectable).

The two most forceful integrators of ontogeny/embryology into evolutionary thought were C. H. Waddington and I. I. Schmalhausen (see Chapter 2). Each recognized that development was the stage on which evolutionary forces could act to bring about phenotypic change; and each had a very strong sense of the importance of the environment and epigenetic processes in mediating the mapping of genotypes to phenotypes.

As will become clear later, there are many ways to alter the ontogenetic trajectory of an organism, but in the past two centuries, the study of ontogeny from

---

[1] It was also "the most parsimonious" explanation, an argument that retains its charm even today. The French "transcendental" philosophers were also flirting with similar notions, e.g., Geoffrey Saint-Hilaire.

[2] Haeckel was also responsible for introducing the term *heterochrony* to the scientific literature.

an evolutionary standpoint has largely meant the study of heterochrony (McKinney and McNamara, 1991; McNamara, 1995). Heterochrony is loosely defined as a change in the relative timing of two developmental events that creates an adult phenotype with very different, and sometimes novel, characteristics relative to its ancestors. The modern approach to the study of heterochrony was inaugurated by the work of Gavin de Beer (de Beer, 1930). Although his system of classification of heterochronic phenomena was cumbersome, and based on the observation of patterns rather than on an understanding of underlying processes, de Beer delineated four types of heterochrony: paedogenesis, neoteny, hypermorphosis, and acceleration. Paedogenesis and neoteny both result in adults that resemble the ancestral juvenile form, and are examples of *paedomorphosis*. Hypermorphosis and acceleration, on the other hand, result in an adult form that in some sense "extrapolates" the ancestral ontogenetic trajectory, and are examples of *peramorphosis*.

The Insight of J. B. S. Haldane

Progress in science is made possible by the fact that the new generations can stand on the shoulders of giants.[3] Sometimes these giants are so perceptive that they circumscribe the fundamental concepts of an entire field, leaving new generations with the humble task of "seeing" the details. Several of the major concepts (and relevant examples) that we will discuss during this chapter were presented by J. B. S. Haldane in an article published in 1932 (!) in *The American Naturalist*.

Haldane, while attempting to devise a classification of gene action, recognized that the phenomenon of the timing of gene expression represents the intersection of the fields of genetics and developmental biology (this well before the discovery of DNA). Haldane (1932b) in fact arrived at a detailed classification of gene action with examples for each category,[4] several of which are germane to our discussion.

As an example of genetic heterochrony, Haldane discussed the case of the single locus (*L*) in *Matthiola incana* that controls the onset of the meiotic prophase. *LL* or *Ll* genotypes have normal meiosis, but the *ll* genotype has an extended chromosomal contraction phase. A common cytological consequence of this phenomenon is an increase in chromosomal mutations, entirely due to this modest shift in gene action during a critical phase of the life cycle. Haldane even went so far as to predict that so-called "maternal effects" in fact result from delayed

---

[3] *"Pigmoei gigantum humeris impositi plusquam ipsi gigantes vident"* (Pygmies placed on the shoulders of giants see more than the giants themselves). Didacus Stella in Lucan, 10, tom. ii. (Bartlett, 1901).

[4] Haldane proposed ten categories for plants, nine for animals. Haldane carefully distinguished the possibilities typical of plants from those typical of animals. Plants are somewhat more complex given the presence of both the alternation of multicellular gametophytic and sporophytic generations and of triploid endosperm (a tissue with two complements of maternal DNA and one of paternal DNA, thus distinct from both the maternal parent and the zygote).

action of genes outside the nucleus, and cited several well-characterized examples, including the now famous case of the inheritance of dextrality versus sinistrality in snails. He pointed out cases in which gene expression is dependent on the particular genetic background or is tissue-specific (such as the *C* gene in *Pisum*, necessary for anthocyanin production in flowers).[5] After making the gene/enzyme mechanistic connection, Haldane proposed that changes in the property of a cell membrane, because of an alteration in the activity of one enzyme, may affect the timing of action of several other genes, and thereby produce a cascade of heterochronic alterations. He implicated hormones as possible regulators of many simultaneous shifts in the timing of gene action.

On subjects closer to morphological heterochrony, Haldane commented on the general rule that embryonic forms resemble each other more than the corresponding adults do (in this sense following de Beer's lead). He suggested that this is the result of progressive divergence in the timing of gene action throughout development. Armed with this concept, Haldane attempted to explain some puzzling evolutionary trends, for example, the fact that complex sutures in ammonites appear first (geologically) in the adult, and in more juvenile forms only in younger rocks. This idea of a shifting in the timing of gene action could also explain other heterochronic phenomena, such as the neotenic appearance of humans as compared with their close primate relatives (an interpretation that has been questioned, e.g., McKinney and McNamara, 1991); or the possible appearance of phenotypic novelties by a combination of two heterochronies (e.g., acceleration plus neoteny)—the origin of the notochord and therefore of the phylum Chordata is one of the examples cited. Haldane does point out as a counterexample the cases of *Lepus* (Lagomorpha) and *Cavia* (Rodentia), in which the adult forms are morphologically similar, but the early embryonic stages are quite distinct.

Haldane's pioneering paper concluded with an insightful look at the ecology of heterochrony. He made the argument that *r*-selected organisms should respond with acceleration of their ontogenies, while it would be advantageous for *K*-selected organisms to develop neotenically. The 1932 paper is a masterpiece of synthesis of the then current ideas and experimental evidence, as well as a very perceptive glimpse into the future of developmental evolutionary genetics. Some of Haldane's scenarios are being addressed only now, more than 60 years after his work was first published.

A mention must be made of the work of Armen Takhtajan, the botanical equivalent of Haldane in this context. Takhtajan (1943) discussed the importance of shifts in developmental parameters for the evolution of flowering plants, as well as the possibility of insertions or deletions of developmental

---

[5]Haldane speculated that "perhaps a suitable chemical technique might detect its presence at earlier or later stages." Modern developmental genetics employs precisely such techniques.

stages. Later papers set out his general views of the relationship between phylogenetic change and modifications of developmental sequences (Takhtajan, 1972, 1976).[6] Despite the insights of Haldane and Takhtajan, and Waddington and Schmalhausen, the major catalyst for the re-emergence of developmental thinking in evolution was the publication of Stephen J. Gould's *Ontogeny and Phylogeny* in 1977.

THE DESCRIPTION OF ONTOGENETIC TRAJECTORIES: HETEROCHRONY AND SEQUENCE MODELS

The majority of studies on the evolution of development use models of heterochrony as a framework. The modern classification of heterochronic phenomena is from Alberch et al. (1979), who also introduced a simple way to quantify ontogenetic trajectories. Alberch et al. designated the endpoints of ontogeny (or any biologically defined subsection of ontogeny; Raff and Wray, 1989) as the onset (or initiation) and the offset (or termination). Any recognizable biological event, such as the switch between the vegetative and the reproductive phases in an annual plant, can represent the onset of a distinct portion of ontogeny, denoted by the symbol α. Analogously, the offset, denoted by β, is marked by a second identifiable event, such as senescence. A third variable, δ, reflects the rate of phenotypic change for the character examined during the interval α to β. The power of this system arises from the fact that these three parameters can describe virtually any change in the timing or rate of different developmental events, and that any recognizable transition during development can be used as a marking point.

The Alberch et al. approach describes six fundamental types of heterochrony; three that lead to paedomorphosis and three that lead to peramorphosis (Figure 5.1; Reilly et al. 1997). Logically, each of the three parameters α, β, and δ can increase or decrease in value while going from the ancestor to the descendant. An increase in α is called *post-displacement* and is equivalent to a delay in the start of that particular ontogenetic phase. A decrease in β is *progenesis* (hypomorphosis), which results in an earlier end to the phase. A decrease in δ is *neoteny* (deceleration), which occurs by the slowing down of the rate of growth or differentiation throughout the phase. All three types yield a paedomorphic adult phenotype. On the other hand, we could decrease the value of α (*pre-displacement*), increase that of β (*hypermorphosis*), or increase δ (*acceleration*). These would, respectively, cause an earlier onset, a delayed offset, or a faster rate of growth between the two end-points. Any of these three types of heterochrony would result in an adult phenotype with peramorphic characteristics. Of course, the six basic types represent forms of "pure" heterochrony, but typical process-

---

[6]See Guerrant (1988) for further discussion of Takhtajan's work and influence.

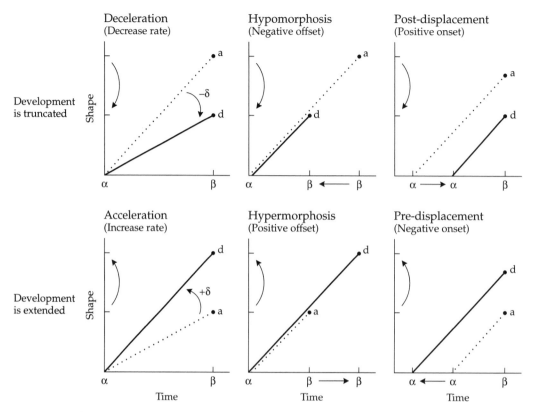

Figure 5.1 The six fundamental types of heterochrony. The diagrams depict how changing the three parameters, onset (α), offset (β), and rate of growth (δ), can alter ontogenetic trajectories. The ancestral ontogeny is designated by a, the descendant by d. (From Reilly et al., 1997.)

es of organismal evolution are likely to incorporate combinations of several of these, leading to mosaics of pera- and paedomorphic phenotypic changes that may be difficult or impossible to disentangle (Reilly et al., 1997; Rice, 1997). It is worth noting that the quantitative treatment of ontogenetic trajectories facilitates statistical or quantitative genetic approaches to examining heterochronic changes (Slatkin, 1987).

Alternatives to heterochrony models are *sequence models* (Mabee, 1993; Alberch and Blanco, 1996; Hufford, 1996). These portray the evolution of development in terms of *transformations* of ancestral ontogenetic sequences (Figure 5.2). The basic model divides ontogeny into stages and compares the descendant and ancestral taxa for changes in the timing and the sequence of events. For example, the ancestral set of ontogenetic states is A→B→C→D; additions

Ancestral ontogenetic sequence

A→B→C→D

Transformation	Descendant ontogenetic sequences		
	Initial	Middle	Terminal
Addition	X→A→B→C→D	A→B→X→C→D	A→B→C→D→X
Deletion	B→C→D	A→B——→D	A→B→C
Novel substitution	X→B→C→D	A→X→C→D	A→B→C→X
Reciprocal substitution	B→A→C→D	A→C→B→D	A→B→D→C
Repatterning	W→X→Y→Z	A→B→X→Y→Z	

Figure 5.2 Representation of various transformations in sequence models for the evolution of ontogeny. Ontogeny is divided into three phases (initial, middle, and terminal). Roman letters refer to ancestral ontogenetic states, italicized letters are states not in the ancestral ontogeny. The timing and mode of transformation in the descendant species are depicted for each of the three stages. (From Hufford, 1996.)

are insertions of new ontogenetic states into the ancestral sequence (A→B→C→D→X), and deletions are losses of ancestral stages (A→B→D). A reciprocal substitution changes the order of states (A→C→B→D), and a novel substitution replaces an ancestral stage with a new stage (A→X→C→D). There can be combinations of these different transformations; for example, Hufford's depiction of repatterning represents both additions and deletions of stages. Alberch and Blanco (1996) suggested that certain types of sequence changes, such as terminal stage changes, should be much more common, and others, such as reciprocal substitutions and additions/deletions within developmental sequences, should be extremely rare. Although the conservatism of early development is intuitive, we note that there are a number of examples of substantial changes to early ontogeny (Raff et al., 1991; Wray, 1995; Smith, 1997; Grbic and Strand, 1998; Richardson et al., 1998).

There are many ways of modifying developmental trajectories, some of which may cause heterochronic or sequence changes. At the mechanistic level, changes in induction, cell division patterns, competence, position, inhibition, and patterning all are capable of altering an ontogeny. We will next discuss heterochronic and sequence changes, and later refer to the less voluminous literature on "other" changes.

Examples of Heterochronic and Sequence Changes

The evolutionary literature overflows with examples of heterochrony described at the phenotypic, life history, or developmental levels (as opposed to more recent molecular studies of the same phenomenon), but highlights comparatively few sequence studies. Heterochrony has been cited in the explanations of numerous macroevolutionary trends (McKinney and McNamara, 1991). DiMichele et al. (1989) suggested that the origin of heterospory in plants (and thus ultimately the evolution of seed plant life histories) was catalyzed by hypomorphosis (an earlier offset) in the gametophyte generation of homosporous plants.[7] Heterochrony has also been invoked as an explanation of major evolutionary novelties such as succulence. Altesor et al. (1994) suggested that succulence in cacti evolved through a retardation of the developmental rate of woody tissues, which is an example of neoteny (deceleration). We will examine a few of the best cases in animal and plant systems to show the power (and seduction) of heterochrony as an explanation of macroevolutionary patterns.

Guerrant (1982) studied a putative ancestor/derivative species pair in the plant genus *Delphinium* (larkspurs, Ranunculaceae). The derived red-flowered *Delphinium nudicaule* displays flowers that look remarkably similar to the buds of the most common flower type in the genus, exemplified by blue-flowered *D. decorum* (Figure 5.3). A simple explanation in heterochronic terms is that the evolution of *D. nudicaule* flowers from those of a *D. decorum*-like ancestor occurred through neoteny. A reduced rate of morphological growth with no alteration of the timing of maturation (i.e., a reduction in δ, with no changes in either α or β) would indeed result in a budlike flower that is sexually mature. Several morphometric measurements confirmed a reduction in the rate of growth in flowers of *D. nudicaule* (Figure 5.4), but closer analysis demonstrated that only the externally visible morphology of the flower underwent neoteny. The nectar-produc-

(A)   (B)   (C)

Figure 5.3 Flowers of *Delphinium nudicaule* and the closely related species *D. decorum*. Notice the similarity between (A) the morphology of the mature *D. nudicaule* flowers and (B) the buds of *D. decorum*, suggesting a heterochronic shift. (C) Mature *D. decorum* flowers. (From Guerrant, 1982.)

---

[7] A plant is homosporous if all spores are morphologically identical. Heterospory refers to the production of two distinct spore types. Heterospory appears to have evolved a minimum of three times in the land plants.

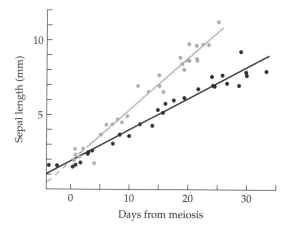

Figure 5.4 Rate of growth of the sepal in buds of *Delphinium nudicaule* (black circles) and the ancestral *D. decorum* (gray circles), showing neoteny of *D. nudicaule*. (From Guerrant, 1982.)

ing petal, for example, has evolved peramorphically by combining acceleration (+δ) and hypermorphosis (+β). This complex series of transitions might have been driven, at least in part, by selection. The generalized *Delphinium* flower with blue sepals is usually pollinated by bumblebees (as in *D. decorum*); however, *D. nudicaule* has red sepals and is pollinated by hummingbirds. It is possible that an initial mutation causing neoteny has triggered a cascade of pollinator-mediated selective episodes.[8] It is interesting to note that while a neotenic flower is more effective in attracting hummingbirds, a peramorphic nectar-producing petal is more efficient in feeding them.

A case of complex changes in both heterochrony and sequence has been described by Jones (1992, 1993) in an investigation of the developmental basis of the morphological differences in leaf shapes between two *Cucurbita*: *C. argyrosperma argyrosperma*, a cultivar, and *C. a. sororia*, its wild progenitor. Comparison of the ontogenetic series of the leaves of the two indicated that the mature leaves of *C. a. argyrosperma* resemble the juvenile leaves of the wild *C. a. sororia* (Figure 5.5). This scenario of apparently simple paedomorphosis is complicated by the fact that *argyrosperma* also displays gigantism in its leaves (presumably due to artificial selection). Later leaves of *argyrosperma* arise through a combination of conservation of early developmental features and later changes in allometric growth (Jones, 1993). Furthermore, examination of the multivariate geometry of the leaf blade revealed novel features in the cultivar (i.e., changes in sequence) that are not explainable by simple heterochronic shifts from the ancestral allometric pattern.

---

[8]Alternatively, the initial mutation may have shifted sepal color to red, attracting hummingbirds that exerted selection on floral form (C. Jones, pers. comm.).

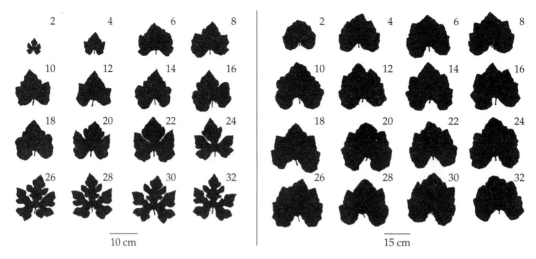

Figure 5.5 Leaf shapes of *Cucurbita argyrosperma sororia* (left) and *C. argyrosperma argyrosperma* (right) throughout development. Notice the resemblance of the adult cultivated *C. a. argyrosperma* leaves to the juvenile wild *C. a. sororia* leaves. Also note the difference in scale. Numbers refer to sequential positions on the vine. (From Jones, 1992.)

Heterochronic changes that affect life history or mating system have been described and are of obvious evolutionary importance. Various species of plants produce two types of flowers, chasmogamous and cleistogamous. Chasmogamous flowers open for pollen removal and deposition, but cleistogamous flowers remain closed and are self-fertilized. Cleistogamy is likely to have evolved as a response to maintain fecundity under environmental conditions in which pollinators are absent or scarce[9] (the downside is a decline in fitness due to increased inbreeding). Lord et al. (1989) described the role of heterochrony in the evolution of cleistogamous flowers in *Collomia grandiflora* (Polemoniaceae). They demonstrated that the alterations in anther size and shape that characterize cleistogamous flowers are due to decreased rates of cell division at a critical developmental stage (Figure 5.6). Since the chasmogamous form is certainly ancestral, this heterochronic change has supplied the plant with an alternative mating system to assure reproduction even under unfavorable conditions.

Strauss (1990) investigated heterochronic changes that affect the process of ossification in the cranium in five species from three genera of poeciliid fishes:

---

[9] The presence or percentage of cleistogamous flowers on a plant is known to be a plastic trait in many species. The plants are not responding to the absence of pollinators *per se*; instead, they make use of cues correlated with conditions in which pollinator service may be rare (e.g., low light; Le Corff, 1993). This is an example of anticipatory phenotypic plasticity (Chapter 3).

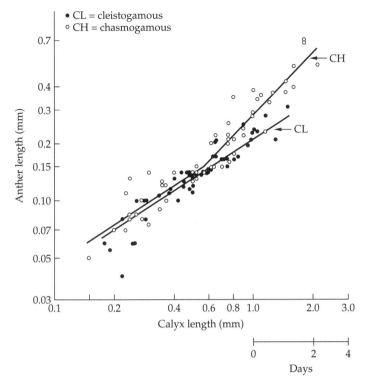

Figure 5.6 Allometric growth of anthers in chasmogamous and cleistogamous flowers of *Collomia grandiflora*. Notice the acceleration of growth in anther length relative to calyx length in chasmogamous flowers. Arrows represent the mean floral dimensions at the tetrad stage of meiosis. (From Lord et al., 1989.)

Xiphophorus, *Poecilia*, and *Poeciliopsis*. He mapped onto the phylogeny of these species the character state changes representing acceleration and retardation in patterns of ossification relative to overall body size (Figure 5.7). The branches show a mosaic pattern of both accelerations and retardations of different ossification processes that contribute to the phenotypic differentiation of the five species. Some bones tend to change in only one direction—for example, the parasphenoid accelerates twice, but never decelerates, and the frontal bone is retarded twice (along the same branches as the parasphenoid changes), but never accelerates.

A similar comparative study was carried out by Kellogg (1990) on the shape of the lemma (the lower of the two bracts situated beneath each floret) in six species of *Poa* (Poaceae). When she plotted lemma length versus a measure of developmental time (the length of the longest anther), she found two very distinct patterns (Figure 5.8). In three of the six taxa, lemma growth continued well after meiosis. In the remaining three, elongation ceased at meiosis. From the standpoint of Alberch et al.'s model, β has been altered. However, determining directionality, that is, whether this is due to hypomorphosis or hypermorpho-

Figure 5.7 Accelerations and retardations of ossification patterns of skull bones in poeciliid fishes belonging to the *Poecilia* and *Xiphophorus* lineages relative to the ancestral *Poeciliopsis occidentalis* (acceleration = shaded circles; retardation = closed circles). Both types of change occur repeatedly and independently; for example, the preopercle/opercle ossification is accelerated in *Poecilia* and retarded in *Xiphophorus*. (From Strauss, 1990.)

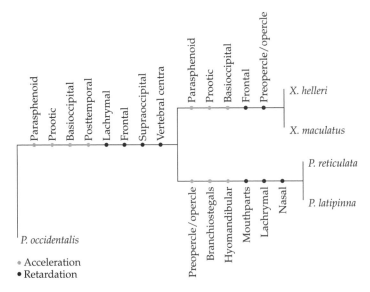

sis, is impossible in this case because of the absence of a reliable phylogeny for the group.[10]

Natural populations can harbor the raw material for microevolution by heterochrony, that is, genetic variation in the parameters that control ontogenetic trajectories. Pigliucci and Schlichting (1995) and Pigliucci et al. (1997) investigated genetic variation for α, β, and δ in two species of *Lobelia* (Lobeliaceae): the red-flowered, hummingbird-pollinated, evolutionarily derived *L. cardinalis*; and the putative ancestral form, blue-flowered, bumblebee-pollinated *L. siphilitica*. Genotypes within each species displayed ample variation in the timing of the cessation of reproductive growth and in the rates of leaf production and inflorescence elongation. In contrast, the onset of the reproductive period was fairly constant. One of the most striking findings was that the same mature phenotype can be reached by a combination of different values of β and δ. If selection targets the mature phenotype only, then genetic variability in these parameters can be maintained within populations, since the same "target" phenotype can be obtained by several effectively neutral combinations of increases/decreases in β counterbalanced by decreases/increases in δ.

That heterochrony can evolve on a short time scale has been demonstrated through artificial selection in the facultative metamorphic salamander, *Ambystoma talpoideum* (Semlitsch and Wilbur, 1989). In ponds that do not dry out, and in which the competition for food is not too intense, *A. talpoideum* is

---

[10] It is known, however, that *Poa annua* is distantly related to the other two species with continued lemma growth, implying convergent evolution of this developmental pattern.

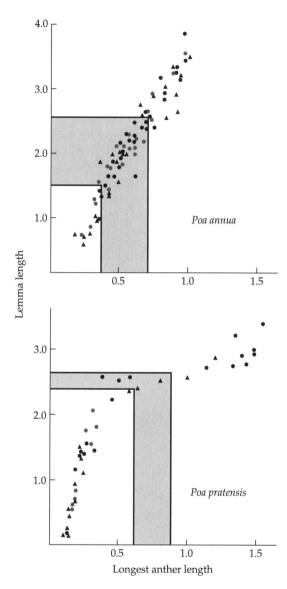

Figure 5.8 Allometric growth of lemmas in two of the six species of *Poa* studied by Kellogg (1990), revealing the existence of two distinct patterns of development. In *Poa annua* growth continues after meiosis (represented by the shaded areas), while in *Poa pratensis* the time of meiosis coincides with the cessation of lemma growth. The different symbols refer to successive florets. (From Kellogg, 1990.)

characterized by a paedomorphic phenotype that reaches sexual maturity with a larval morphology. However, when the conditions in the pond are not favorable, then metamorphosis to the terrestrial form occurs. Although this exposes the individual to a different, unpredictable, and harsh environment, it beats death by desiccation or starvation, and offers the opportunity to colonize new ponds. Semlitsch and Wilbur subjected two populations to artificial selection for paedomorphosis through four generations; the fifth generation was raised in a "common pond." Both populations responded to selection under the two environmental conditions (drying and constant ponds) by increasing the proportion of paedomorphic individuals and reducing that of metamorphic ones (thereby demonstrating genetic variation for the trait). However, the intensity of the response differed between populations, which perhaps reflects their idiosyncratic evolutionary histories (Figure 5.9).

Lammers (1990) presented an intriguing scenario for the paedomorphic evolution of leaf shape in the *Cyanea solanacea* complex (Lobeliaceae) in Hawaii. The four species of the complex exhibit a diversity of leaf shapes, from the entire leaf of *C. profuga* to the highly dissected leaf of *C. shipmanii* (both now extinct; Figure 5.10). The phylogeny and biogeography of the group are congruent. The two basal taxa *C. profuga* and *solanacea* are known only from the oldest island of Molokai; the intermediate *C. asplenifolia* is found on Maui; and *C. shipmanii* occurs on the youngest island, Hawaii. A comparison of adult and juvenile leaves of each taxon reveals a striking similarity between the adult leaf in one species and the juvenile leaf of the species immediately more basal to it (Figure 5.10), which suggests a "continuous" trend of increasing acceleration in this group over a million years.[11] However, a detailed analysis of the ontogenetic trajectories of the two extant species has not been carried out to test this hypothesis. Among the questions left unanswered is, for example, how did a novel juvenile form arise in each species? (C. Jones, pers. obs.).

Figure 5.9 Response to artificial selection for paedomorphosis (left) or metamorphosis (right) in two populations of the salamander *Ambystoma talpoideum*. White circles represent ponds that dry out; black circles are permanent ponds. E and S are two different populations. (From Semlitsch and Wilbur, 1989.)

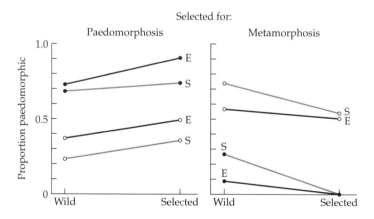

[11] Lammers suggested that this may have resulted from *r*-selection for increased fecundity (through accelerated development). This *r*-selection regime is consistent with colonization of novel or highly disturbed environments, such as the Hawaiian archipelago, which is geologically very young and active.

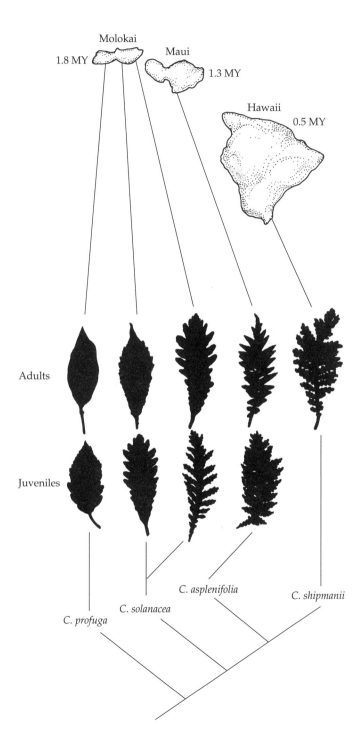

Figure 5.10 Correspondence between the biogeography and phylogeny of leaf form in four species of *Cyanea* (Lobeliaceae) in the Hawaiian Islands (numbers are island ages). Note the similarity between the adult leaves of each species and the juvenile leaves of the next more basal taxon. (From Lammers, 1990.)

One of the classic examples of heterochronic change also reveals a superb instance of a sequence transformation. Echinoderm development has been under intense scrutiny for years, and has delivered more than its share of insights into ontogenetic evolution. An especially well studied system has been the development of sea urchins (e.g., Wray and Bely, 1994; Wray, 1996). The basal ancestral pattern has been identified as egg→pluteus larvae→juvenile→adult (Figure 5.11). *Heliocidaris tuberculata* and *H. erythrogramma* have nearly indistinguishable juvenile and adult forms, but *H. erythrogramma* does not have the feeding pluteus larval stage. Instead, it forms a very different, short-duration, non-feeding larva that will achieve the juvenile form several weeks prior to *H. tuberculata*. The earlier onset (and equal duration) of the juvenile stage results in isomorphism at the adult stage. The distinctive larval morphology is not explained by timing changes (but see Reilly et al. 1997). The presence of a stage not seen in the ancestral ontogeny would be classified as a novel substitution in the sequence model.

Another excellent example of combinations of heterochronic and nonheterochronic changes comes from the work of Friedman on the evolution of male and female gametophytic stages in seed plants (Friedman, 1992, 1993, in press). The gametophyte stages are of interest because, although they are small and relatively short-lived, there is considerable diversity in their form among seed plants. Friedman and Carmichael (Carmichael and Friedman, 1996; Friedman and Carmichael, 1996; Friedman, in press) have examined in detail the development of the female gametophyte in *Gnetum*, a member of the

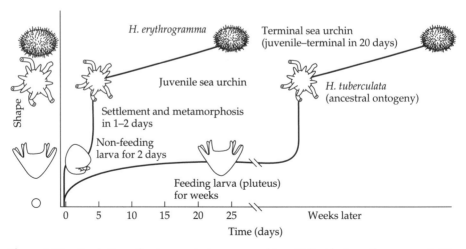

Figure 5.11 Evolution of ontogeny in two sea urchins. *Heliocidaris erythrogramma* initiates the juvenile stage earlier (pre-displacement) and has a non-feeding larva rather than the feeding larva (novel stage substitution) found in *H. tuberculata*. Both juvenile and adult forms are nearly indistinguishable between species. (From Reilly et al., 1997.)

Gnetales, the group thought to be the sister clade to the flowering plants. The female gametophyte in seed plants generally consists of three phases: free nuclear development, cellularization, and growth (Figure 5.12A). In *Gnetum*, the time of fertilization occurs much earlier with respect to somatic development (i.e., earlier β) than in other groups. Thus development is completed after a precocious sexual maturation, and nutritive tissue is formed after fertilization rather than before (Figure 5.12B). There are however, other changes to the sequence of events that are not heterochronic in nature. These include alter-

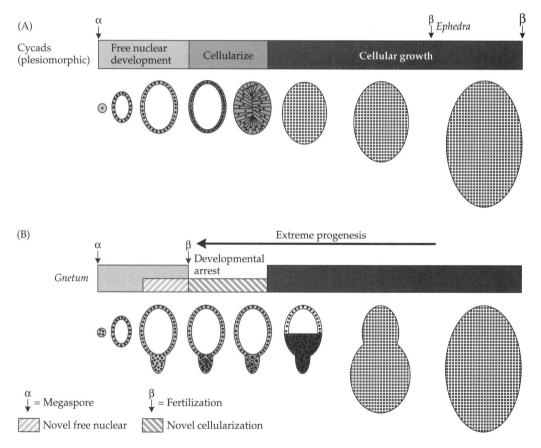

Figure 5.12 Comparison of the development of the female gametophytes of the cycads, representing an ancestral state, and *Gnetum*, representing a highly derived state. *Gnetum* differs in three key events: first, it has a markedly shorter time to fertilization; second, it produces a distinctive free nuclear zone at the chalazal end of the female gametophyte; and third, cellularization is confined to the chalazal region. (From Friedman, in press.)

ations of both the free nuclear phase and cellularization phase. The free nuclear phase is distinguished by a proliferation of free nuclei at the chalazal end (the end opposite the site of fertilization by sperm). Cells are formed only in that area, and each contains numerous nuclei that subsequently fuse to form a polyploid uninucleate cell. The rest of the gametophyte is in apparent developmental arrest. These alterations are phase-specific, without any conspicuous ramifications for later development.

These examples from critical studies indicate that there are probably few cases of pure heterochronic or sequence change. This suggests that investigation of the details should precede the "identification" of patterns or processes.

MODELS OF HETEROCHRONY

The profusion of empirical data concerning heterochrony stands in contrast to the paucity of models that incorporate heterochrony into mathematical frameworks currently in use by evolutionary theory. We briefly discuss two exceptions and point out their insights as well as their limitations.

Slatkin (1987) proposed a quantitative genetic model of heterochrony based on the Alberch et al. (1979) formalism.[12] In it, he considered a sequence of bifurcating events. At each bifurcation, several traits develop from the same tissue until the next bifurcation. Slatkin examined directional and stabilizing selection on two or three traits and the expression of mutational variance. The major conclusions of Slatkin's study were that: (1) the response of heterochrony to selection on adult traits depends on growth rates at times different from the selection event itself; (2) the intensity of selection on growth rates at particular stages of development depends on the length of those stages; (3) mutations that affect only the means of heterochronic parameters will affect both means and variances of adult traits; (4) developmental processes constrain the available range of changes in the allometric relationships among traits; and furthermore, (5) it is not possible to estimate means, variances, and covariances of developmental parameters from data on adult traits only, which highlights the importance of gathering stage-specific data throughout an ontogenetic trajectory (e.g., Pigliucci and Schlichting, 1995).

Slatkin's model is very general, and represents a first attempt to incorporate heterochrony into quantitative genetics.[13] However, the generality of the attempt leads to somewhat trivial expectations and to difficulties in constructing specific quantitative tests of the model's predictions. In fact, in the time since Slatkin's

---

[12] The Alberch et al. "model" was not cast in terms of either quantitative genetics or optimization theory, the two leading frameworks of modern evolutionary theory. It was rather a heuristic approach to classifying heterochronies.

[13] One of the perhaps necessary, but certainly unrealistic, features of the model is that it does not consider dominance, epistasis, or genotype by environment effects, all potentially fundamental features of real heterochronic alterations.

attempt, no empirical (or further theoretical) work along these lines has appeared.

A very different approach to modeling heterochrony is that of optimality or game theory. An example is represented by the work of Whiteman (1994) on the evolution of facultative paedomorphosis in salamanders. He proposed three different scenarios for the maintenance of this form of developmental plasticity (Figure 5.13). The first, called "paedomorph advantage," simply predicts that whenever permanent aquatic habitats are surrounded by harsh terrestrial conditions, paedomorphs will have a higher fitness (because they will be larger) than metamorphs. Metamorphosis is maintained through the occasional intense selection provided by drought years that force the animals out of the pond. In the second scenario, called "best of a bad lot," paedomorphs will have higher fitness only with respect to those metamorphs that delay reproductive maturity. Larvae that reach a certain size threshold will metamorphose; those that cannot achieve the threshold mature as paedomorphic individuals. The third situation combines the previous two and creates the scenario in which two types of paedomorphs are produced ("dimorphic paedomorphs"). For this to occur, there must be three size classes of larvae: the largest become paedomorphic and outcompete every other size class; medium-sized larvae go through metamorphosis to escape competition; the smallest larvae cannot achieve the threshold metamorph size and are forced to reach sexual maturity as paedomorphs. Figure 5.13 shows the predicted frequency distributions of individuals of each type. It is clear that co-occurrence of metamorphic and paedomorphic individuals can be the norm, dependent on the variability in body size within a given population. The results of this study suggest that selection (for body size) can favor developmental plasticity based on heterochrony through more than one mechanism. Although preliminary data suggested support for both the paedomorph advantage and the best of a bad lot scenarios, detailed studies in one location indicate that, as we have come to

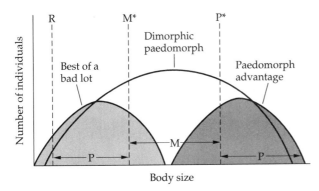

Figure 5.13 Distributions of body sizes for three mechanisms that possibly maintain facultative paedomorphosis in amphibians. R, M*, and P* represent three size thresholds. An individual larger than R but smaller than M* becomes a paedomorph (P) through the best of a bad lot mechanism; organisms larger than M* but smaller than P* metamorphose (M) according to the dimorphic paedomorph model; individuals larger than P* are paedomorphs because of the paedomorph advantage mechanism. See text for further details. (From Whiteman, 1994.)

expect, reality is even more complicated: selection appears to favor paedomorphic males and metamorphic females (Whiteman, 1997).

Optimality modeling of this sort can be tested against empirical data, since the alternative hypotheses are couched directly in terms of the particular experimental system under study. However, this approach suffers to some extent from the opposite problem of Slatkin's: Once one tests hypotheses for a specific system, what fraction of these conclusions can be generalized and incorporated into evolutionary theory at large? This tradeoff between the generality and testability of models plagues modern evolutionary biology.

THE MECHANISTIC BASES OF ONTOGENETIC CHANGE

Many of the classical examples of heterochrony represent studies conducted at the morphological level, in which the timing or rate of visible developmental events is described in terms of paedomorphosis or peramorphosis. However, an understanding of these biological phenomena would not be complete without investigation of their mechanistic basis and of the relevant molecular aspects. As we shall see, this can be done for heterochrony, but it typically raises more questions than it answers and introduces more complexity than simplicity (neither, of course, is a good reason *not* to do it!).

A good starting point is the list compiled by Raff and Wray (1989) of developmental mechanisms that underlie heterochrony. They divided these mechanisms into those that directly involve timing and those that do not. Mechanisms that *directly involve timing* include hypermorphosis by late expression of duplicated genes or by an alteration in the timing of mitosis in order to lengthen the growth phase. Pre-displacement and acceleration can be obtained by excision of genes expressed in early development or by early initiation of adult genes. Persistence of a protein normally shut down during late development can explain cases of neoteny. Progenesis may be obtained by again changing the timing of mitosis, in this case in order to shorten the growth phase. The examples of mechanisms that *do not directly involve timing* include the failure to activate connections between hormone-producing glands or, in the case of neoteny, the inability of target tissues to respond to stimuli. Acceleration may result from altered volume ratios between the nucleus and the cytoplasm. Pre-displacement can be caused by alterations in the plane of cleavage. Hypermorphosis can result from higher concentrations of certain proteins. This classification does not lead us to a one-to-one correspondence between mechanisms and effects. In fact, the main message seems to be that developmental events that can be classified as heterochronic at one level may be the result of heterochronic or non-heterochronic events at different levels (Raff and Wray, 1989).

Let us consider some examples in detail to gain an insight into the complexities of the genetic basis of developmental changes. Collazo (1994) analyzed the patterns of temporal expression of fibronectin during gastrulation in several vertebrates, focusing on the macroevolutionary separation between fishes and amphibians. A cladogram shows that a heterochronic shift in this pattern occurred in urodele amphibians relative to anuran amphibians (Figure 5.14). Urodeles are usually presumed to represent the ancestral developmental sequence among amphibians, with anurans as the derived group. While Collazo did not challenge this fundamental assumption, his work clearly shows the appearance of developmental novelties in basal groups (and, incidentally, the usefulness of outgroups in comparative studies in order to determine the direction of evolutionary change).

Chanoine and Gallien (1989), also studying urodele amphibians, examined the patterns of expression of myosin in the fast skeletal muscle in normal and neotenic individuals. They showed that the light chain appears in three forms that are expressed throughout development. The heavy chain is normally represented by two developmental forms, one expressed during the larval stage, and the other in the adult stage. Neotenic individuals, however, express both the larval and the adult form of the heavy chain, a rare instance of congruence between molecular and morphological heterochrony.

One of the organisms in which heterochrony has been studied from a mechanistic perspective is the nematode *Caenorhabditis elegans*. Ruvkun and Giusto (1989) found that the gene *lin-14* creates a temporal gradient during development, in that it is expressed early in development but not after the early larval stage (Figure 5.15). This gene product induces cells to switch from an early fate (S1) to later fates (S2 . . . S4). A gain of function mutant (*lin14-n355*) maintains the expression of *lin-14* throughout development, which causes an indefinite reiteration of the S1 fate. A loss of function mutation of the same gene (*lin14-*

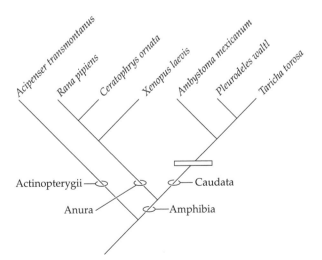

Figure 5.14   Phylogeny of anuran and urodele amphibians, with the indication of the evolutionary time of occurrence of a heterochronic shift (bar) in the expression of fibronectin. Circles identify four clades. (From Collazo, 1994.)

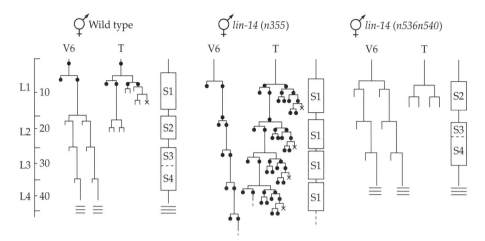

Figure 5.15 The wild type temporal gradient produced by the gene *lin-14* in *Caenorhabditis elegans* (left), and its disruption in two mutants (center and right). The gradient controls the switch between early and late cell fate. S1–S4 represent successive stages of development. Notice that one mutant is "stuck" at S1 (center), while the other mutant (right) skips that stage. (From Ruvkun and Giusto, 1989.)

n536n540) causes the *lin-14* gene never to be expressed, which translates into the absence of cell fate S1 even from the early larval stage. Later studies uncovered an ensemble of *lin* genes that are organized in a hierarchy of heterochronic switches (Liu and Ambros, 1991; Ambros and Moss, 1994). Some of these are positioned in the hierarchy so as to control a wide array of developmental events, and some are more specific, affecting the switch between the larval and the adult stage, or between specific hypodermal cell fates.

Crop plants have provided several interesting mechanistic examples of heterochronic changes. Poethig (1988) has described three semi-dominant mutations affecting shoot development in maize—*Teopod (Tp) 1, Tp2,* and *Tp3*—named after teosinte and pod corn, two of the putative ancestors of cultivated maize. All three mutations produce similar alterations of the basic phenotype (Figure 5.16). They increase the number of vegetative structures, ears, and tillers; decrease the size of the leaves, ears, tassels, and the length of the internodes; and transform reproductive structures into vegetative ones (thereby apparently acting as homeotic genes; see below). In all three cases the mutant phenotype is due to a prolonged expression of the vegetative developmental program into the reproductive phase. It is of interest that the expression of these mutations depends on the specific genetic background, as it once again illustrates the importance of the interaction between genes with major effects and "modifiers" in shaping the actual phenotype. Because their phenotypes resemble the ancestral condition for these plants,

Figure 5.16 Photograph of a wild type maize plant (on the right) and of heterozygotes for three mutations of the *Teopod* class that transform reproductive structures into vegetative ones. (From Poethig, 1988.)

the possibility is raised that the modern crop evolved by a progressive truncation of the juvenile (vegetative) phase of development.

Another illuminating study on the mechanics of heterochrony using a crop plant is the research by Wiltshire et al. (1994) on *Pisum sativum*. The authors identified nine distinct mutations that produce heterochronies ranging from progenesis to neoteny, and from hypermorphosis to acceleration. Furthermore, they showed that the onset of flowering can be shifted independently of vegetative transitions of the leaf form that are coupled with it in the wild type, uncovering the existence of two distinct developmental programs where only one was thought to exist. Moreover, Wiltshire and collaborators demonstrated that the same heterochronic pattern can be caused by genes controlling distinct physiological processes.

We note that mutations are not the only type of alteration of the DNA that can be responsible for heterochrony. There has been a series of reports through the years of apparently Lamarckian phenotypic changes (for a review, see Cullis, 1986). Sano et al. (1990) investigated the inheritance of dwarfism in rice (*Oryza sativa*), and showed that a single exposure to the chemical 5-azacytidine (azaC) immediately after germination is sufficient to induce dwarf plants. The progeny of these plants segregated into 35% dwarf and 65% tall types. Selfing of the dwarfs produced only dwarfs, while selfing of the tall plants produced only tall plants. When Sano and coworkers analyzed the genomic DNA of their plants, they found that the azaC treatment had induced 16% reduction in DNA methylation relative to the parental plants. This pattern of methylation was maintained through the M_1 and M_2 generations, explaining the observed "Lamarckian" effect.

Heterochrony: *Caveat Emptor*

Many of the above examples, as well as numerous others (e.g., Gould, 1977; McKinney and McNamara, 1991; Conway and Poethig, 1993), have led to a sense that heterochrony is capable of explaining nearly all evolutionary transitions. Raff and Wray (1989) discussed the problems associated with inferring heterochronic processes from observable patterns (see also Reilly et al., 1997) and made the following three general points:

1. Developmental sequences of events are not necessarily causally linked. Non-heterochronic changes in gene action can cause visible heterochronies. Alterations in the timing of gene action may be reflected as homeotic change or as another type of change in the developmental program that may not appear to involve heterochrony.

2. Heterochrony can potentially occur at any stage of the life cycle and in any process.[14]

3. An appropriate temporal frame of reference is essential for valid comparison of patterns or processes.

The potential uncoupling of heterochrony of pattern and process pointed to by Raff and Wray has been elegantly demonstrated by Nijhout et al. (1986), who used computer simulations of cellular automata to generate complex morphologies. They first established a series of genetic rules to describe several possible patterns of cell differentiation (Figure 5.17A). The fundamental idea was that the simulated "genes" would act like real genes and not follow a predetermined "blueprint," but sense and react to the local cellular/tissue environment. So, for example, cells of type 8 originate only from cell line I and never undergo further mitosis. By combining these rules with others concerning production and diffusion of "morphogens," Nijhout et al. produced the "wild type" ontogeny illustrated in Figure 5.17B. They then explored the effects of "single-rule" mutations on the overall ontogeny and on the resulting adult forms (Figure 5.17C), and discovered that, in numerous cases, genetic heterochronies did not necessarily lead to morphological heterochronies.

Another source of error in the inference of process from pattern is the potential for concurrent changes in more than one of the parameters α, β, and δ. Reilly et al. (1997) point out that not only can changes in multiple parameters result in paedomorphic or peramorphic morphologies, but that balanced changes among α, β, and δ can lead to *isomorphosis*[15]—heterochronic changes that do not result in a change in shape. Rice (1997) has illuminated another potential source of confusion, concerning the validity of comparisons of ancestral and descendant trajectories that assume that the same linearization of ontogeny (i.e., the calculation of slopes and intercepts) applies to both taxa. If the "rules" of development have changed, then the descendant trajectory may not be linearized by the same transformation.

---

[14]It is now feasible to investigate the mechanistic basis of heterochronic changes at different levels in a variety of organisms; this includes developmental patterns, molecular details, and genetic elements.

[15]Oxymoronic, but apt.

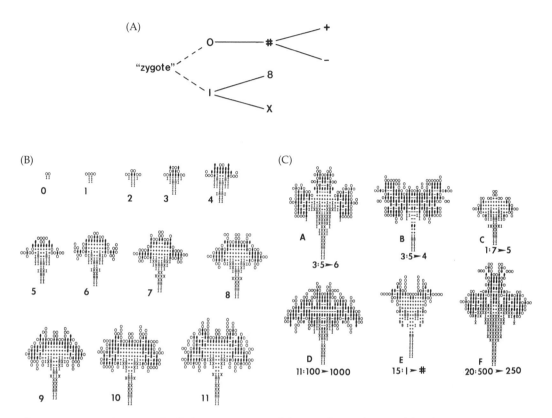

Figure 5.17   Application of cellular automata modeling to the evolution of development. (A) Cell lineage diagram depicts the rules for cell differentiation starting from a zygote. Broken lines indicate several cell divisions from the zygote before differentiation starts. (B) "Wild type" ontogeny characterized by eleven "stages" of growth and differentiation. (C) "Mutant" adult forms originated by single-rule mutation in the simulation study. Numbers refer to the specific type of mutation. For example: 3:5→6 indicates that a type 3 mutation has occurred that alters the end-point of cell division for cell type *I*, and postpones it from age 5 to age 6. This heterochronic "mutation" (corresponding to post-displacement) does not result in a juvenilized adult phenotype. (From Nijhout et al., 1986; see original for details of model.)

Given the complex scenarios that can be obtained by even simple modifications of the developmental system, we strongly support the caution urged by Raff and Wray, Reilly et al., Rice, and others.[16] We venture to suggest a general framework for the study of heterochrony. Ideally, "heterochronic" phenomena should be examined experimentally, with organisms grown under the same

---

[16] For example, Klingenberg and Spence (1993) show in their examination of waterstriders that allometric relationships of ancestor-derived species are not reliable indicators of true heterochrony.

environmental conditions, because of the potential for environmentally induced shifts in α, β, and δ (see below). Absolute time should be used as a baseline for comparison of ontogenetic trajectories, and all contrasts should be done within a clear phylogenetic context. These restrictions prohibit the use of paleontological data in studies of heterochrony. We lament the fact that many "classic" examples might therefore be questioned, but there is no getting around the lack of reliable ontogenetic and environmental data for fossil specimens (e.g., Bales, 1997).[17]

HETEROTOPY

Not all evolutionary changes of ontogenetic trajectories occur through heterochrony. In fact, it is arguable that the logical alternative mechanism, heterotopy, is even more important at a macroevolutionary scale (Zelditch and Fink, 1996). Heterotopy refers to changes in the *place* (as opposed to the *time*) of action of genes. The term has been used much more sparingly than heterochrony, and researchers do not necessarily realize that there are more examples of heterotopy in the literature than is commonly acknowledged; they are just not labeled as such. However, as we shall see, heterotopy is intrinsically much more difficult to study than heterochrony, and, furthermore, the mechanistic basis of the two phenomena may be interchangeable to some extent (i.e., alterations in the timing of gene action can cause developmental/morphological heterotopies, while alterations in the place of gene action can cause phenomena that may be classified as heterochronic at the phenotypic level).

A case of co-occurrence of heterochrony and heterotopy is described by Wray and McClay (1989). They considered seven species of echinoids for which the phylogenetic relationships were known. They analyzed the spatial and temporal patterns of expression for three genes during the early phases of development. The gene controlling the expression of the msp-130 protein was expressed in the primary mesenchyme in all seven species, but the time of its initial expression (onset) was variable, a case of "pure" heterochrony (Figure 5.18A). The expression of the meso-1 protein, on the other hand, occurred in different tissues according to taxon. Four species expressed the gene in the mesenchymal cells, but in one species only pigment cells showed gene activity. The time of expression of the gene also varied between taxa, producing a case of mixed heterochrony and heterotopy (Figure 5.18B). The third protein, sp-12, was confined to the primary mesenchyme in six species, but was present in all mesenchyme cells in the remaining taxon. The time of expression was again variable across taxa (Figure 5.18C). When these differences were mapped on the available phy-

---

[17] Raff and Wray suggested distinguishing two categories of heterochrony: *process* heterochrony, when the change in ontogenetic parameters (onset, offset, or rate) can be evaluated, and *pattern* heterochrony, for cases in which this is not possible. Paleontologists would typically document *pattern* heterochrony, but rarely *process* heterochrony.

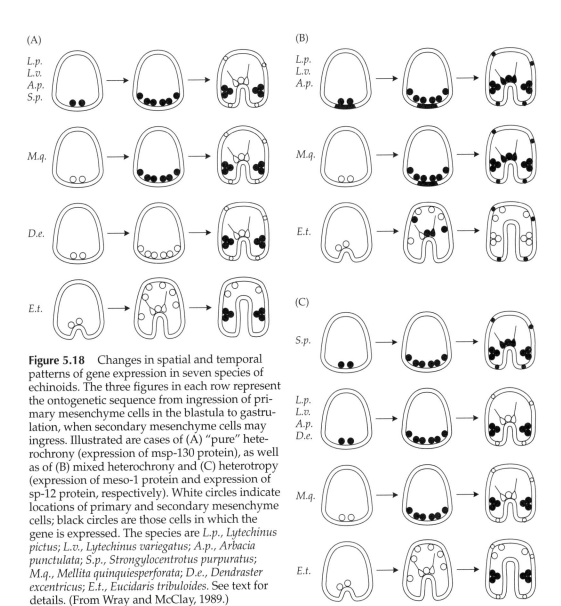

Figure 5.18 Changes in spatial and temporal patterns of gene expression in seven species of echinoids. The three figures in each row represent the ontogenetic sequence from ingression of primary mesenchyme cells in the blastula to gastrulation, when secondary mesenchyme cells may ingress. Illustrated are cases of (A) "pure" heterochrony (expression of msp-130 protein), as well as of (B) mixed heterochrony and (C) heterotropy (expression of meso-1 protein and expression of sp-12 protein, respectively). White circles indicate locations of primary and secondary mesenchyme cells; black circles are those cells in which the gene is expressed. The species are *L.p., Lytechinus pictus; L.v., Lytechinus variegatus; A.p., Arbacia punctulata; S.p., Strongylocentrotus purpuratus; M.q., Mellita quinquiesperforata; D.e., Dendraster excentricus; E.t., Eucidaris tribuloides*. See text for details. (From Wray and McClay, 1989.)

logeny, they showed a complex mosaic pattern, in which different combinations of heterochronic and heterotopic alterations occurred in distinct parts of the phylogenetic tree.

Tissue-specific gene expression (the obvious molecular basis for heterotopy) is a booming field of inquiry, and we will give a few examples of this evolutionarily important phenomenon. An interesting instance involves the control of the alcohol dehydrogenase gene, *Adh*, in *Drosophila*. Although *D. affinidis-juncta* and *D. hawaiiensis* are phylogenetically very closely related, *Adh* specificity of tissue expression is dramatically different—the gene is expressed in the larval midgut and Malpighian tubules in *D. affinidisjuncta*; in *D. hawaiiensis* it is expressed in low levels in the midgut and not at all in the tubules (Brennan et al., 1988). A series of chimeric or deletion versions of the *Adh* genes were created by arranging 5′, coding, and 3′ regions from these two species, which were then inserted in *D. melanogaster*. Only those genes containing *D. affinidisjuncta* 5′ elements resulted in Adh production in both the midgut and the Malpighian tubules (Fang et al., 1991). The expression of those genes chimeric for the coding region or 3′ segments was context-dependent.

A study that used chimeric genes in peas also revealed a complex pattern of gene expression. Simpson et al. (1986) examined the organ- and tissue-specificity of two gene families, *rbcS* (coding for the small subunit of ribulose-1,5-bisphosphate carboxylase, otherwise known as Rubisco, a fundamental enzyme in carbon fixation) and *LHCP* (light-harvesting chlorophyll a/b-binding proteins). Although both gene families are intimately involved in photosynthesis, a complex array of regulatory factors produced differing patterns of environment-dependent and tissue-dependent induction.

A phenotypic hallmark of heterotopy is referred to in the classical literature as *homeosis*. The appearance of a given morphological feature in the "wrong" place during development is a clear indication that genes are being expressed in unusual tissues or cell types (though it could also be the result of gene expression at unusual times, thereby unifying, or confounding, the molecular basis of heterotopy and heterochrony). We will not review the vast literature on homeosis here, but a few examples will illustrate the relationships between homeosis, heterotopy, and heterochrony.

Schultz and Haughn (1991) investigated the transition between vegetative and reproductive phases of development in *Arabidopsis thaliana*. In this plant, basal leaves and bracts develop first. These leaves subtend quiescent meristems that can later form basal or lateral inflorescences. In the wild type, bolting (switching to the reproductive phase) is followed closely by initiation of the first flowers. Lower flowers develop first, while the upper lateral and basal inflorescences are the first to elongate (Hempel and Feldman, 1994). Schultz and Haughn identified a mutation, *leafy*, that transforms flowers into inflorescences (Figure 5.19). Even though this is an apparent case of homeosis (at the meristem level), the roles of heterotopy versus heterochrony are not clear. *Leafy* may not be expressed in the right cell tissues (the primordia that typically produce the flowers), or it may not be activated at the right time (in the beginning of the reproductive phase).

Homeotic mutations, however produced, are potentially responsible for major evolutionary transitions. Carroll et al. (1995) discussed the evolution of insect wing number in terms of mutations affecting homeotic genes (i.e., genes capable of "transforming" other body segments into wing-bearing segments).

Figure 5.19 Comparison of (A) wild type and (B) *leafy* mutant of *Arabidopsis thaliana*. Circles indicate flowers, arrowheads signify inflorescences. Flowers are transformed into inflorescences in the mutant. (From Schultz and Haughn, 1991.)

The wings in pterygote (winged) insects originated from a segment of the leg, and fossil evidence suggests that early winged insects had wings on all thoracic and abdominal segments (Figure 5.20). The interesting finding by Carroll et al. is that a series of homeotic genes *repress* wing formation instead of promoting it. They obtained a series of *Drosophila* embryos with an increasing proportion of the homeotic genes knocked out. The phenotypic effect was the appearance of progressively more wing primordia. Thus, a clear example of heterotopy (genes that are not expressed in some of the segments) can potentially explain a major trend in animal evolution.

In conclusion, although heterotopy is studied mostly at the molecular level and is rarely included in an explicitly evolutionary perspective, it is still expected to be a major mechanism for the alteration of developmental patterns that may profoundly affect adult phenotypes. In principle it is distinguishable from heterochrony, but we suggest that there are clear examples in which such distinctions are impossible, or at least not useful, to make at the most fundamental mechanistic level. Nevertheless, it is important to realize that changes in the timing of and/or the site of the action of genes (both of which are kinds of *regulato-*

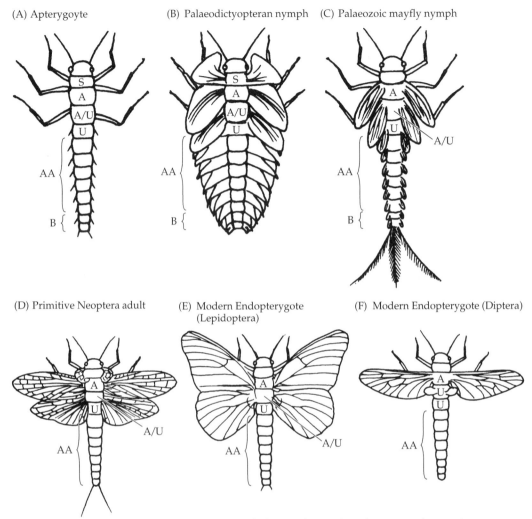

(A) Apterygoyte   (B) Palaeodictyopteran nymph   (C) Palaeozoic mayfly nymph

(D) Primitive Neoptera adult   (E) Modern Endopterygote (Lepidoptera)   (F) Modern Endopterygote (Diptera)

Figure 5.20 Depiction of the morphologies of a number of extinct and recent insect groups showing the hypothesized control of wing production. Letters on segments refer to the action of homeobox genes identified in *Drosophila* and presumably present in all species depicted in the series. S, *Sex combs reduced*; A, *Antennapaedia*; U, *Ultrabithorax*; AA, *abdominal-A*; B, *Abdominal-B*. A/U indicates expression of both A and U. (From Carroll et al., 1995.)

ry as opposed to *structural* evolution) are keys to the understanding of both microevolutionary and macroevolutionary patterns of phenotypic change.

NON-HETEROCHRONY AND NON-HETEROTOPY

We have made the point that not everything that changes is due to heterochrony, and all the remainder is not due to heterotopy. There are a number of examples in the literature of other categories of changes in gene action that also have had profound evolutionary importance. Changes in the plane of cell division, cell competence, cell shape or wall properties, or cell migratory behavior could have huge consequences. Changes in the expression of genes that produce receptors or signaling molecules could be of considerable importance. In fact, any number of the examples attributed to heterochrony might be mechanistically determined to belong to one of these categories.

For example, the different varieties of cotton, although fundamentally similar, display an array of leaf shapes from highly divided to almost entire (Figure 5.21). Changes in the local concentration of the plant hormone auxin are responsible for similar alterations in peas. This could be construed as an example of heterotopy (the auxin gene is expressed in the "wrong" place), or of heterochrony (at the "wrong" time); it is more likely an example of change in the *intensity* of gene action, and not in the time or place of expression. A second case is exemplified by the *ein* mutant in *Brassica rapa* (Rood et al., 1990). This mutation causes an increase in the rate of shoot elongation (a typical "heterochronic" phenotype), faster development of flowers, and elongated internodes and inflorescences. As *ein*'s effects can be counteracted by the application of a growth retardant, Rood and coworkers applied plant hormones exogenously and dis-

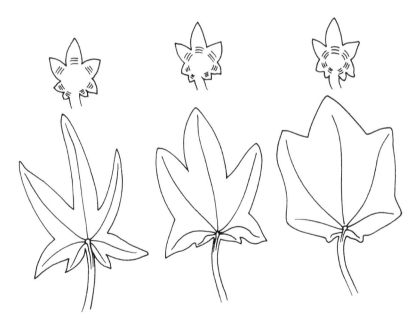

Figure 5.21 Leaf shape in three varieties of cotton. The top row shows leaf primordia and the hypothesized locations of increased intercalary growth that would lead to the actual shapes in the lower row. The range of observed shapes present in different cultivars is controlled by different alleles at the same locus. (From Sachs, 1988.)

covered that the application of gibberellin converted normal wild type plants into *ein* phenocopies. They then confirmed that the *ein* mutation results in a higher rate of production of GA_1, a fundamental precursor of bioactive gibberellin. Other examples could be detailed, but the salient points are that phenotypes are affected by internal environments, and that the responses to these conditions will depend upon the characteristics of both signals and receptors (see also the conclusion of Chapter 3).

Other classes of gene action are of course possible, including changes in the qualitative effects of protein action at the cellular or subcellular level, for example, alterations of the plane of cell division. Valentine (1997) examined the distribution of embryonic cleavage types, unquestionably a distinctive evolutionary change early in metazoan evolution, from a phylogenetic perspective. He argued that only one change from a radial ancestral condition (characteristic of deuterostomes) to spiral cleavage (characteristic of protostomes) and a further change to "idiosyncratic" cleavage is required to conform to the arrangement of recent molecular phylogenies. The shift in cleavage is hypothesized to alter the fates of blastomeres, which in turn could change the choice of which cells form mesoderm, and eventually how the coelom is formed.

Such "not-place, not-time" mutations may represent just plain structural gene alterations. Regardless, it is fundamental that evolutionary biologists recognize the existence of diverse possibilities, and especially that they recognize that there is no necessary one-to-one correspondence between the qualitative, phenotypic effects of a mutation and its mechanism of action at the molecular level. It might be argued that evolutionary biologists could safely ignore such details, as long as they know that mutations *can* cause certain effects. However, this position is not only philosophically unsatisfactory (in that we would have only a partial explanation of what is going on), it could even be misleading. For example, the frequency and directionality of some genetic/epigenetic changes is not likely to be uniform, a condition that might explain the prevalence of some macroevolutionary trends and the absence or rarity of others, with no necessary relationship to externally imposed selection pressures.

WHAT IS THIS THING CALLED HOMOLOGY?

A fundamental difficulty facing evolutionary developmental biology, and one that has generated endless controversy since the middle of the nineteenth century, concerns homology: its definition, its causes, and most of all, the evolution of homologous structures. Homology was originally defined by Owen in 1848 as the same organ in different animals having a variety of forms and functions. The complexities hidden in the concept of homology are key to fundamental and diverse evolutionary phenomena such as the origin of phenotypic novelties, the mapping of genotypes onto phenotypes, the response to selection, the maintenance or origination of genetic variation, and the evolution of development. In this section we will briefly discuss our point of view on homology and its consequences via a mosaic of the opinions of the many authors who have been recently engaged in this debate (van Valen, 1982; Roth, 1988; Wagner, 1989a,b, 1996; McKitrick, 1994; Hall, 1994, 1995; Müller and Wagner, 1996; Bolker and Raff, 1996; Sanderson and Hufford, 1996; Abouheif, 1997). As usual, the environmental component is chronically missing from such discussions, and it will be our role to reintroduce it in order to elucidate some otherwise obscure points.

The first myth to dismantle is that there must be a correspondence between homology at the phenotypic and genetic levels. The fact that two genes are homologous does not require that their phenotypic effects also be comparable (Dickinson, 1995). Conversely, a perception of phenotypic or developmental homology does not require that the genetic bases of those phenotypes are identical. Two biological examples from the development of the eye in *Drosophila melanogaster* will make the point clear. Two recessive mutations, *vermilion* and *cinnabar*, both result in orange eyes. Crosses of orange-eyed flies that are homozygous for *vermilion* and heterozygous for *cinnabar* (i.e., *vvCc*) with red-eyed (wild type) flies (wild type *vermilion*, heterozygous *cinnabar*—*VVCc*) produce only one-quarter orange-eyed progeny. While the orange eyes of the parents will be due to the effects of the alleles at the *vermilion* locus, the orange coloration in the progeny is due instead entirely to the *cinnabar* locus. In this particular case, it turns out that the two loci are part of the same biochemical pathway, and that blocking it at different levels impedes formation of the normal pigment (van Valen, 1982). Therefore, the same (apparently homologous) phenotype can be determined by two entirely distinct genetic bases.

An even more extreme case is that of the *eyeless* mutation in *D. melanogaster*. The *eyeless* allele causes complete loss of the eyes by blocking the normal developmental sequence. However, it is possible to select for the reappearance of normally functional eyes by selecting for modifiers of *eyeless* (Spofford, 1956; de Beer, 1971). The genetic basis of a highly sophisticated and evolutionarily conserved structure such as the compound eye can be dramatically altered in a few generations of artificial selection! Wagner and Misof (1993) presented a number of examples in which adult phenotypes are highly conserved despite substantial variability in the underlying developmental pathways.

Homologous structures can also give rise to non-homologous ones, as noted by Wagner (1989a). It has been surmised that insects are derived from annelid-like ancestors. In this scenario, each differentiated segment of an insect's body would be homologous to the mostly undifferentiated segments of annelids. However, the central segments of the body of the Insecta form a completely new structure, the thorax, which characterizes the whole subphylum but has no equivalents in their sister group. The thorax can therefore be considered a phenotypic novelty originating from the permanent assemblage of distinct parts, each with evolutionary homologues in related groups, even though the novel phenotype has none.

This example leads to another highly debated question—are phylogenetic and iterative homologies the same? Phylogenetic homology traces the evolution of structures through common ancestors, regardless of the specifics of their actual genetic or developmental basis; for example, the thoracic segments are derived from undifferentiated ones. Serial (iterative) homology refers to the presence of similar or identical repeated structures in a single organism. All segments of the body of a given insect are considered homologous, although differentiation into distinct and specialized structures must be determined by differences in their developmental/genetic machinery. However, there might not be much point in categorizing too rigidly, since evolutionary homologies can easily result from natural selection fixing variations around serial homologies at the individual level. Neither the differentiation of species nor the development of individuals necessarily represent discontinuous processes, and analogously, the question of whether homology is an all-or-nothing phenomenon is moot. Morphological homology can be maintained while its genetic bases change dramatically, and many features of organisms share some degree of homology (such as gene sequences). Homology must also be a matter of degree simply because evolution can proceed gradually (even though it does not have to). It is important to recognize that homology may exist at some, but not all, hierarchical levels (Hall, 1995; Abouheif, 1997), and to specify which level is being studied.

Some systematists have argued that the term *homology* should be restricted to synapomorphies; that is, to morphologies shared by a monophyletic group of species (Patterson, 1982). While this may be convenient for problems of phylogenetic reconstruction, it ignores the biology of the phenomenon, for example, by automatically excluding iterative homologies within an individual. On the other hand, the comparative approach can be used to test the evolutionary homology of particular structures (e.g., McKitrick, 1994).

The idea that homologous characters can have distinct genetic bases led Roth (1988) to propose her concept of "genetic piracy." She defined the phenomenon in these terms:

> New genes, not previously associated with the development of a particular structure, can be deputized in evolution; that is, brought in to control a previously unrelated developmental process, so that entirely different suites of genes may be responsible for the appearance of the structure in different contexts.

Roth further reasoned that the very definition of pleiotropy excludes it as a criterion on which to base homology; most mutations have pleiotropic effects that span characters that no systematist or evolutionary biologist would ever consider "homologues." For example, natural populations of *Drosophila mercatorum* are polymorphic for the abnormal abdomen phenotype, with two alleles segregating at one locus. The differences in the effects of the two alleles are temperature sensitive. The *Ab* allele causes the maintenance of juvenile abdominal cuticle in adults; therefore, it qualifies as a heterochronic mutation. Its phenotypic effects are diverse, with influences on pigmentation, segmentation, and bristle pattern (Templeton and Johnston, 1988). These three characters, however, are not considered homologous simply because they are affected by the pleiotropic results of a single allele. Roth also pointed out that so-called "homeotic" mutations in *Drosophila* not only change some structures into homologous traits (such as the transformation of an antenna into a leg), but also transform structures into others (e.g., the transition between antennae and eyes) that no biologist would claim as evolutionarily homologous! From a molecular genetic standpoint, these "unorthodox" transformations occur because all these structures share a certain number of fundamental developmental switches, but the *fact that disparate pieces of the genetic machinery have been brought under similar control does not imply that evolution followed the same path from a morphological point of view.* One of the consequences of the concept of genetic piracy is that it might provide a very powerful model for explaining the evolution of new gene functions as well as explaining the maintenance of genetic variation for regulatory loci in natural populations.

Environmental effects (phenotypic plasticity) also enter into the study of homology and offer insights into the genetic/developmental phenomena underlying homology. Goldschmidt (1940) proposed the term *phenocopy* to characterize this link—a phenotype induced by environmental effects in a wild type genotype that mimics a phenotype induced by mutation. Environmentally induced perturbations can be useful clues to the mechanistic basis of some phenotypes. For example, Gibson and Hogness (1996) found that it is possible to select for an increase in the rate of the appearance of phenocopies of the *Ultrabithorax* mutation phenotype in *Drosophila melanogaster*. The mutation causes the development of a second thoracic structure in the fly, and exposure to ether vapors also causes the same phenotype. An increase in sensitivity to ether is very strongly correlated with the loss of expression of the *Ubx* wild type gene. The implication is that the product of *Ubx* has a role in metabolizing ether, or in somehow neutralizing its biological effects.

What, then, is homology? Innumerable definitions have been proposed, discussed, rejected, and rediscovered, but given the above considerations, we lean toward van Valen's (1982) as providing the necessary balance between historical and biological contexts:

> Homology is a correspondence between two or more characteristics of organisms that is caused by continuity of information.

Information and continuity are the key words here, and they are intertwined. The classical concept of homology implicitly refers to genetic information. But some of the examples mentioned above make it clear that we can have perfectly recognizable homologies even though their genetic bases might be dramatically altered. Continuity refers to the necessity of an evolutionary link between ancestors and descendants. Even though, as pointed out by Wagner (1989a), the word *information* cannot be applied in biology with the same precise meaning that it has in mathematics, there must be transmission of information in order to generate biological form. That information can be genetic, epigenetic, external environmental, or a combination of the three. The challenge for modern biologists is to determine when and how these components interact to yield what we perceive and discuss as homologies.

ENVIRONMENTAL INFLUENCES ON DEVELOPMENTAL PHENOTYPES

Recognition of the role of alterations in the developmental program of living organisms has always been considered fundamental to our understanding of the evolution of phenotypes. To the best of our knowledge, no recent biologist has questioned the existence of an underlying genetic basis for heterochronic, heterotopic, or other types of changes occurring during epigenesis, even though the details of the genetic machinery are guessed at more often than not. On the other hand, historically, little attention has been paid to the fact that developmental trajectories can also be plastic; they can respond to environmental changes, and the intensity and pattern of these responses may have profound implications for microevolutionary and macroevolutionary trends. In this section we will discuss a few studies on the plasticity of ontogeny, and the change in perspective they can catalyze by drawing together ecology and developmental biology (in what could be termed "developmental ecology").

One modern merger of plasticity and development is represented by the work of Smits et al. (1996a,b) on trophic morphs of the haplochromine cichlid *Astatoreochromis alluaudi*. They examined differences in features of the skeleton and musculature of the fish from insectivorous and molluscivorous populations. When fish from molluscivorous populations are fed insect larvae, they are indistinguishable from the insectivorous morph, so the observed differences are due to diet. Although adult fish of similar standard lengths did not differ for skull height, the snail-eaters had an astounding 55% greater volume in the pharyngeal jaw muscles and associated skeletal elements (Figure 5.22A). Additionally, in the molluscivore there are fewer pharyngeal teeth but more molariform teeth (Figure 5.22B), and there is a greater jaw angle.

The evolutionary implications of this study are perhaps even more interesting than the ecological ones. Cichlid fishes are renowned as examples of adaptive radiations, with highly speciose assemblages characterized by dramatic alterations in their functional feeding morphology (trophic niche differentiation; Meyer, 1987, 1993; Galis and Metz, 1998). Two possible explanations are: (1) some taxa that are currently considered distinct species may in fact represent

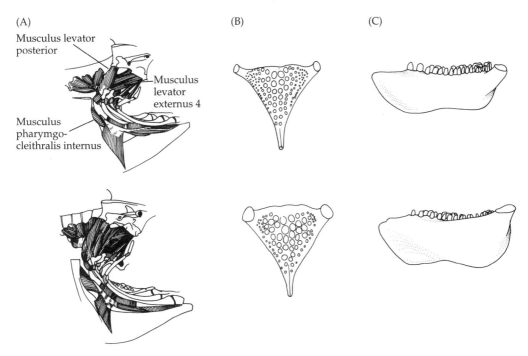

(A)

Musculus levator posterior

Musculus levator externus 4

Musculus pharymgo-cleithralis internus

(B)

(C)

Figure 5.22 Differences in jaw morphology of *Astatoreochromis alluaudi* fed on insects (top row) or snails (bottom row). (A) The pharyngeal jaw apparatus (right lateral view). (B) Lower pharyngeal element (dorsal view). (C) Lower pharyngeal element (lateral view). (From Smits et al., 1996a.)

the results of an extremely flexible reaction norm of a single species; (2) perhaps the initial morphological differentiation among some of the species of cichlids occurred *because* of plasticity (i.e., independently of genetic variation; Wimberger, 1992).[18] The divergent phenotypes could have been fixed during the course of specialization on different diets, either through the loss of plasticity by genetic drift, or by selection for canalization of the adaptive morphology in a more "constant" environment. In fact, there are other haplochromine species that exhibit little apparent plasticity, specializing instead on either insect larvae or mollusks (Smits et al., 1996a).

Another case in which plasticity could have represented the first step in a macroevolutionary change is found in the work of Strathmann et al. (1992) on the developmental plasticity of larvae in sea urchins. They exposed full siblings

[18]Plasticity as a macroevolutionary catalyst. See also West-Eberhard, 1989 and discussion in Chapter 10.

to different levels of food availability and obtained a developmental sequence of diverging morphologies (Figure 5.23). The observed morphological changes were in the direction expected on the basis of functional adaptive arguments. The authors observed an increased allocation to the larval apparatus for catching food under scarce available resources, and an allocation to juvenile structures when the food was abundant. These results mimic the evolutionary trend that represents the transition from a developmental sequence with feeding larvae to one with non-feeding larvae.

Higgins and Rankin (1996) surveyed the literature on variations in arthropod life stages. They categorized the relative plasticity or canalization of three juvenile developmental parameters (the number of instars, the intermolt interval, and the capacity for ecdysis at a different size) into six possible combinations. Despite the presumed benefits of maximal flexibility, only five of twelve species were plastic for all three; and only three of the other five syndromes were represented. The two possible syndromes that were not found—plasticity of intermolt interval with canalized number of instars and size at ecdysis; and plastici-

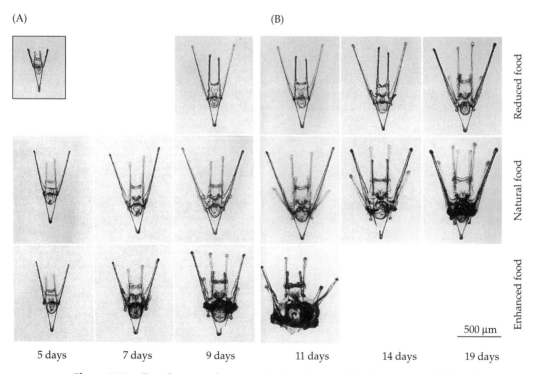

(A) (B)

Reduced food

Natural food

Enhanced food

500 μm

5 days   7 days   9 days   11 days   14 days   19 days

Figure 5.23 Developmental patterns in larval sea urchins *Paracentrus lividus* when grown with different levels of food. (A) The 2-day-old pluteus larva prior to the feeding treatments. (B) Developmental sequences from 5 to 19 days for the three treatments. Larvae with enhanced food metamorphose after 14 days. (From Strathmann et al., 1992.)

ty of size at ecdysis, but canalization of intermolt interval and instar number—
were suggested to be maladaptive because they fix either the age or the size at
maturation.

A textbook case of the plasticity of ontogenetic trajectories is that of the ecology of facultative metamorphosis in amphibians (see the discussion of the theory above). Semlitsch (1987) studied the expression of paedomorphosis and
metamorphosis in the salamander *Ambystoma talpoideum*. Paedomorphic salamanders do not metamorphose, but retain their aquatic forms and achieve sexual maturity in the pond. Metamorphosis in this species is thought to be an
adaptive plastic switch that allows the animals to leave an unfavorable pond if
the chances of making it in the terrestrial world are reasonably higher. Semlitsch
exposed *A. talpoideum* to a combination of two environmental factors, drying
ponds and constant ones, and three levels of density of conspecifics. Density
alone did not affect the propensity to metamorphose in constant ponds, with
very few animals, if any, taking that opportunity (Figure 5.24). This was surprising because one would expect that escape from the pond would be a mechanism for avoiding stiff competition at high density. It turns out that the salamanders develop an alternative way of coping with competition in constant
ponds: they become cannibalistic.

On the other hand, metamorphosis was prevalent in drying ponds when larval density was low. This is in accord with the adaptive plasticity expectation.
However, almost no salamander metamorphosed in drying ponds with a high
density of conspecifics. The lack of conformation with the expected adaptive
strategy appears to be due to extrinsic limits imposed on the developmental system. *A. talpoideum* must reach a minimum threshold size for metamorphosis,
and extreme scarcity of food prevents most animals from reaching that threshold, dooming them to a developmental blind alley. Semlitsch and Wilbur (1989)

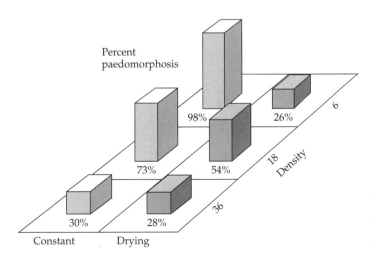

Figure 5.24 Effect of density
of conspecifics and drought
on the incidence of paedomorphosis in *Ambystoma*
salamanders. (From
Semlitsch, 1987.)

were later able to demonstrate that the propensity to metamorphose (when developmentally possible) is under genetic control and is variable within populations, since it is possible to obtain a response to artificial selection.

Examples of presumably adaptive phenotypic plasticity seem to be inextricably connected to constraints that act at the developmental level that either preclude the plastic response altogether or reduce its adaptive value. The quintessential example of the consequences of intertwining plasticity and development occurs when previous developmental "decisions" restrict flexibility at later points. For example, the ability of a plant to produce new fruits is highly dependent on current and previous fruit production and on the number of seeds produced in those previous fruits (Winsor et al., 1987; Stephenson et al., 1988). In the monoecious plant *Cucurbita pepo* (zucchini), both more fruits and more seeds per fruit have the effect of reducing the likelihood of a particular female flower maturing into a fruit (Figure 5.25A; Stephenson et al., 1988). In addition, the

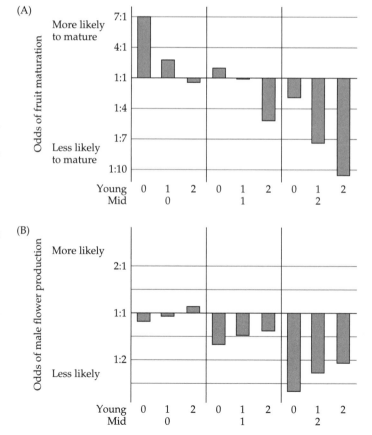

Figure 5.25   An example of ontogenetic contingency. The odds of maturation/production as a function of the number of fruits already present on a naturally pollinated zucchini plant (*Cucurbita pepo*). *Young* refers to fruits aged 0 to 5 days, and *mid* to fruits aged 6 to 15 days. (A) Odds of a pistillate (female) flower maturing into a fruit decrease as number of both young and mid pre-existing fruits increases. (B) Odds of a male flower being produced on any given day decrease as number of mid-fruits increases; but the presence of more young fruits *increases* the odds of male flower production within each mid-fruit category. (From Stephenson et al., 1988, and Schlichting, Devlin and Stephenson, unpublished.)

presence of fruits also alters the relative probabilities for the *production* of male and female flowers (Figure 5.25B; Schlichting et al., unpublished). This general phenomenon has been termed "ontogenetic contingency" by Diggle (1994, 1995; Watson et al., 1995).

Diggle studied sexual allocation in the plant *Solanum hirtum* (Diggle, 1993, 1994) by examining the relative production of staminate (stamens only) and hermaphroditic (stamens and pistils) flowers with and without resource limitation. The resource manipulated was the amount of maternal investment in the progeny. In this species, fruit-bearing plants produced a higher proportion of staminate flowers per inflorescence than plants of identical genotype that were not pollinated. The plant-level allocation to male or female function was determined by the total number of staminate and hermaphroditic flowers. The decisions to produce one or the other were local; each flower primordium was labile until it reached a size of about 10 millimeters, and its fate could be switched toward staminate or hermaphrodite, depending on resource availability. However, this plasticity was also contingent on the whole architecture of the plant. Only floral primordia in the distal portions of each inflorescence were developmentally plastic; basal ones always produced hermaphroditic flowers regardless of resource availability (a lack of plasticity). Differences among genotypes were explained by differences in the degree of architectural plasticity (i.e., what fraction of the inflorescence was responsive to environmental signals). Therefore, adaptive plasticity was limited by the specifics of the developmental system, even though these specifics can evolve when there is genetic variation for developmental flexibility in natural populations.

The animal analogue of ontogenetic contingency is the concept of "state-dependent life histories" developed by Houston and McNamara (1992; McNamara and Houston, 1996). They discussed the importance of the influence of the previous physiological status of the organism on the currently observed life history. They noted the significance of plastic responses in molding life histories, and concluded that natural selection acts on the reaction norm as a whole and not only on the focal life history phase. Prime examples of state dependence include work on guppies, *Poecilia reticulata* (Reznick and Yang, 1993); spadefoot toads, *Scaphiopus couchii* (Newman, 1994a); and dark-eyed juncos, *Junco hyemalis* (Rogers et al., 1994).

Ontogenetic contingency is a prime reason why plasticity should be studied throughout development and not just at the adult or sexually mature state. Pigliucci and Schlichting (1995) and Pigliucci et al. (1997) demonstrated this point in studies of the responses of two populations of *Lobelia siphilitica* to light and nutrient availability during stem elongation. Genetic differences in ontogenies were evident at different stages, but they also depended on the particular environment in which the plants were growing. Cheplick (1995) also showed the ontogenetic dependency of the amount of phenotypic plasticity. He subjected genotypes of the grass *Amphibromus scabrivalvis* to different amounts of fertilizer, and measured reaction norms for vegetative growth and resource storage. Examinations of the production of new ramets, for example, revealed that the amount of genetic variation did not vary much with age, but that the amount of plasticity leapt dramatically between 11 and 20 weeks, and increased only slightly between 20 and 26 weeks (Figure 5.26).

An alternative approach that highlights the developmentally imposed limits of adaptive plasticity was taken by Gedroc et al. (1996) in studies on the plasticity of root/shoot allocation in response to nutrient availability in the two annual species *Abutilon theophrasti* and *Chenopodium album*. They grew some plants under continuously high or low nutrient conditions, and switched others from low to high or high to low nutrient conditions midway through the experiment. When they compared the continuous high versus continuous low treatments, their results were in perfect agreement with optimal allocation theory—plants grown under low nutrients allocated relatively more biomass to roots than to shoots. The conclusion was very different, however, when the two switch treatments were considered. Both species were able to speed up growth when moved from low to high nutrients (but without altering the root/shoot ratio); but plants switched from high to low nutrients were unable to adjust to the new conditions. Again, both the developmental *and* environmental histories

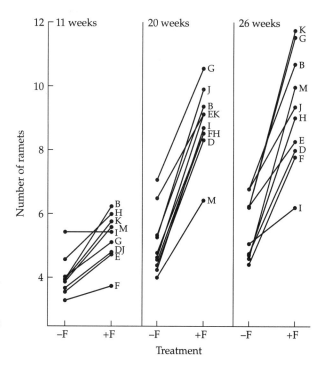

Figure 5.26 Developmental plasticity in genotypes (letters) of *Amphibromus scabrivalvis*. Plasticity and genetic variation for plasticity increase throughout development. +F, fertilized; –F, unfertilized. (From Cheplick, 1995.)

(i.e., the internal and external environments) affected the ability of the organisms to adapt to changing conditions (see also Chapter 9).

Wiltshire et al. (1994) compared the responses of wild type and mutant pea plants to changes in photoperiod (short versus long days). The lf^a mutant, for example, behaved paedomorphically when grown under long days. However, if raised under short days, the same genotype appeared peramorphic. A complete inversion in the type of heterochrony expressed by a gene was obtained by a simple change in environmental conditions

It is conspicuously clear from the preceding brief discussion that we can no longer consider development as a phenomenon independent from plasticity. The two interact closely, and sometimes provide us with explanations for the limits imposed by one upon the other; in other cases they yield useful hints about the mechanistic basis of adaptations that may be ecologically and evolutionarily significant. This is a field in rapid expansion, especially as the old "black box" of development is being (slowly) opened by modern molecular analyses (Chapter 8). However, we need to bear in mind the evolutionary and ecological contexts of development in order to achieve a truly comprehensive understanding of these phenomena, and thereby to avoid the trap of seeing the pixels while missing the picture.

CONCEPTUAL SUMMARY

1. The study of development, and of heterochronic phenomena in particular, has a long and venerable history that predates evolutionary concepts, and has continued largely independently of advances in evolutionary theory throughout the modern neo-Darwinian synthesis.

2. Ontogenetic trajectories and their differences are typically examined from one of two perspectives, either heterochrony or sequence models. Both have been identified empirically in many instances, and both are widely suspected to be important in ecological or evolutionary contexts.

3. Heterochrony can be described in terms of three fundamental parameters that bracket any particular segment of an ontogenetic curve: the onset, the offset, and the rate of growth. This results in six fundamental types that correspond to increases or decreases in the value of each of the above three parameters. Three types of heterochrony give rise to paedomorphic (juvenilized) phenotypes, and three to peramorphic ones. Studies of the genetic (isolation of mutants) and molecular bases of heterochrony have made astounding progress in recent years.

4. Sequence models examine ontogeny from the perspective of changes in developmental stages, for example, additions, deletions, and novel substitutions. This perspective is especially useful when new developmental stages arise.

5. Other mechanisms can be responsible for important alterations in developmental trajectories. Heterotopy, the differential spatial expression of genes, may be equally important to heterochrony, but it is more difficult to demonstrate and study in detail. Important classes of regulatory genes, such as the homeotic genes, can be interpreted as being either heterochronic or heterotopic, dependent on the specific situation. Changes in cleavage plane, in cell movements, and in receptors and transducers of signals can also have major effects on developmental change.

6. Homology is a fundamental and highly contentious concept. We prefer van Valen's definition that refers to the maintenance of information through evolutionary time. The continuity of this information, however, need not be genetic, due, for example, to phenomena such as genetic piracy.

7. Complex combinations of the different forms of developmental change have undoubtedly occurred to produce the diversity of living organisms.

8. Developmental trajectories can be affected by the environment, and the relationship between development and adaptive plasticity can be complex. Phenomena such as ontogenetic contingency or state-dependent life histories can explain the existence of limits to adaptive plasticity (beyond those imposed by reduced genetic variation).

CHAPTER 6

Constraints on Phenotypic Evolution: A Central Problem?

A BRIEF HISTORY OF THE CONCEPT

In our view, the evolution of phenotypes is ultimately a matter of balance between selective forces acting on the epigenetic system and the structure of the existing genetic machinery driving the epigenetic system itself. The limits and preferential routes that are superimposed on the action of selection have been collectively known as "constraints." We can already hear the groans of dismay of our colleagues who have become inured (justifiably) to repeated discussions on constraints; however, we must reconsider this concept in some detail to achieve a comprehensive picture of how developmental reaction norms (and hence phenotypes) change through evolutionary time.

155

Biologists have a bewildering selection of "constraints" to choose from: genetic, phylogenetic, mechanical, functional, developmental, selective, and ecological, to mention but a few. The literature and the spirited controversy about constraints is so vast that any attempt to synthesize the matter is doomed to be both incomplete and a very delicate operation (see, e.g., Antonovics and van Tienderen, 1991; Perrin and Travis, 1992; van Tienderen and Antonovics, 1994; Schwenk, 1995).

The genesis of the modern controversy appears to lie in Gould and Lewontin's 1979 paper "The spandrels of San Marco and the Panglossian paradigm: A critique of the adaptationist programme." In this paper the authors forcefully argue that not all features of organisms can or should be viewed as adaptations.[1] Organisms should be viewed as integrated wholes, whose *baupläne* are "constrained by phyletic heritage, pathways of development and general architecture." From this perspective, the constraints on adaptation acquire a cachet equivalent to that of the adaptations themselves.

According to Gould (1989), who reviewed the subject to some extent, there are basically two general definitions of constraint:

> (1) The description of any causal change (because any cause directs change on one path instead of any other one).

In this sense, as Gould pointed out, constraint acquires a very (clearly too) generalized meaning. We disagree that "we might as well speak simply of 'change'" and suggest that, at most, we might speak of "directed" changes as opposed to random ones.

Alternatively, constraints may be

> (2) The sources of those changes, or those restrictions upon change, that do not arise through the action of stated causes within a favored theory. One identifies the canonical causes within an accepted theory; the directors of other kinds of change and the preventers of change by the canonical cause are 'constraints.'

It is difficult to imagine crafting a more obscure definition of constraint. However, from what we can surmise, in this second sense a constraint is a relative concept. For example, in the context of natural selection as the favored agent of adaptive evolutionary change, a constraint is that which prevents selective pressures from changing the phenotype, or propels phenotypic change in a different direction. Along similar lines, Antonovics and van Tienderen (1991) correctly point out that, in order to effectively discuss constraints, it is necessary to refer to a null hypothesis. In this sense, they are using, in a more succinct and clear way, the same concept expressed by Gould's second definition.

A previous work by Gould contained a much better presentation of the concept of constraint in reference to the so-called "morphological space" (Gould,

---

[1]See Selzer (1993) for interesting and detailed analyses of the rhetoric employed by Gould and Lewontin, and by Maynard Smith (1979) in a rebuttal to their paper.

1980b). In that work, he points out that the array of forms that a particular kind of organism might achieve is a much larger set of those forms than they actually display (or have displayed, if a fossil record is available). Following Raup (1966, 1967), Gould refers to the "cube of life" (Figure 6.1) for coiled forms such as those presented by snails and clams. The shape of a shell can be defined mathematically by a relatively low number of parameters. The parameter space identified by all mathematically possible combinations of values is the potential morphological space for shells. In nature, however, organisms occupy only a very limited subset of this space. Why? This is a fundamental biological question, and one that has received very little theoretical or empirical attention.

The above considerations notwithstanding, things are in fact more complicated than the simple scenario proposed by Gould. First of all, we cannot simply

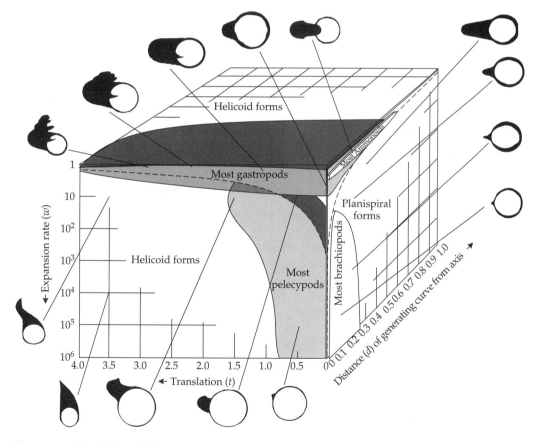

Figure 6.1   The "cube of life" representing all possible forms of a shell when its shape is determined by a simple equation in three parameters. Each axis represents the range of variation of the parameters. The actual shells arrayed around the outside are representative species. (From Raup, 1966.)

attribute the lack of forms in certain areas of the morphospace to "constraints." Some forms could have been there in the past, without us knowing it, because of an incomplete or missing fossil record. Second, the "gaps" in morphospace might be due to a combination of selection and constraints, and not just to constraints. In other words, it could be developmentally possible for an organism to generate a form in a currently empty portion of the phenotypic space, but this form would be selected against because of its misfit with the external environment. Third, what is now an empty zone might become occupied in the future, for either of two distinct reasons: (1) the appropriate genetic variation might appear in a population (the constraint would be released); or (2) a selective force that was keeping the population from entering that area might change or be nullified by an alteration of environmental conditions.

Other definitions of constraints are not difficult to find in the recent and not so recent evolutionary literature. Maynard Smith et al. (1985), for example, start their review by saying:

> A developmental constraint is a bias on the production of variant phenotypes or a limitation on phenotypic variability caused by the structure, character, composition, or dynamics of the developmental system.

If we substitute "genetic architecture" for "developmental system," we have an equally compelling definition of genetic constraints. The reader can play this game indefinitely, substituting various terms to produce a definition congruent with her/his "favored theory."

Following the efforts of Gould, Lewontin, and Maynard Smith et al., the quest has been open to find more quantitative ways to approach the problem, and to actually *measure* real or hypothesized constraints. In part anticipating both the quantitative genetic and the morphometric work that appeared later in the 1980s, Gould (1984) proposed using multivariate analyses such as principal components to quantify constraints. The underlying idea was that constraints are reflected in character correlations, that is, in preferential directions of variation in multivariate space. For example, he characterized the morphological variation in shell shape in the land snail *Cerion uva* from Aruba, Bonaire, eastern Curaçao, and western Curaçao. Gould identified sets of covariation among traits that clearly distinguished populations from the four geographic regions. He then interpreted his data as suggesting the presence of different types of constraints. The snails from western Curaçao and Aruba reached the same adult size, but the first group did it by producing few large whorls, while the second group was characterized by numerous small whorls. He concluded that the number *or* size of the whorls was free to vary, but not *both* simultaneously.[2]

---

[2]See however, Stone (1996), for the argument that use of a different parameterization results in a different view of the "constraints" in morphospace. Stone reanalyzed some *Cerion* data, and argued that one particular type of constraint identified by Gould (the "smokestack normal") is not due to geometric limitations, but is more likely related to ecological, life history, or biomechanical factors.

A recent approach to quantifying constraints from a morphological/developmental standpoint is the so-called "new morphometrics" (see Chapter 7). However, this approach still depends on the fundamental idea that constraints manifest themselves in a reduction of the variation associated with traits, or by allowing variation only in certain directions. This argument is also similar to the quantitative genetics approach to the problem. Cheverud (1984, 1988), among others, advocates that "it is through the genetic variance/covariance matrix that development constrains evolution." This is again an attempt to quantify constraints by measuring variation, this time genetic (as opposed to phenotypic) variation and covariation among a suite of characters. There is discussion in Chapter 7 on the limits of the use of the **G**-matrix and in Chapter 10 on the current studies of evolution of pleiotropy by assessment of the **M**-(mutation)-matrix that is germane to this debate on constraints (as well as to the debate on phenotypic integration).[3]

The problem with these quantitative approaches is that the correlations due to the effects of recent selection episodes cannot be distinguished from those traceable to actual constraints arising from the genetic and epigenetic machinery of the organisms studied. In other words, these methods highlight patterns, but tell us little about the underlying processes (Schwenk, 1995). When one considers the suggested morphological constraints in *Cerion*, the following question arises: Are these limitations on form due to a lack of genetic variation for the proper combination of traits, or are they the result of selection for a subset of the possible combinations (i.e., does it matter how you get a snail of a given size, as long as it is the "right" size)? Even if we can establish that the current constraint is due to lack of genetic variation, that still leaves open the possibility that this is simply the result of recent selection on the correlations among traits (which, according to the infamous "fundamental theorem of natural selection" is expected to reduce genetic variation), since correlation structures themselves are not a static property of a population, but can evolve (Cheverud, 1984, 1988; Wagner, 1996).

Other authors have noted that certain apparent cases of selection might be due to a more complex interplay between selection and constraints. Convergent evolution provides textbook examples in which a researcher might conclude that selection has been the main agent in the recurrence of the same morphology in distant branches of a phylogeny. However, Wake (1991) convincingly argued a complementary explanation. His position is that, starting from a given ancestral situation, only a limited subset of morphologies can evolve, due to genetic/developmental constraints; we perceive this as "convergent" evolution simply because of our ignorance of the underlying genetic/epigenetic events. A case in point might be the evolution of the number of digits in plethodontid

---

[3]Emphasizing the difficulty of writing a linear book about such a complex and non-linear topic, we might add . . .

salamanders (Figure 6.2). The ancestral condition was five digits, and at least three lineages independently reduced this number to four digits. Closer examination of the developmental details of digit formation in these vertebrates, however, indicates that selection for reduced body size might have had a correlated effect on the production of digits, since there is a relation between number of digits and size of the limb buds in these organisms.[4]

Other intriguing cases are offered by the evolution of mimicry in butterflies. Brower (1996) studied the evolution of several parallel races of mimetic butterflies in the species *Heliconius erato* and *H. melpomene,* and showed that the same patterns originated independently in the two species (Figure 6.3). This scenario suggests that strong selection induced repeated cases of convergent evolution. On the other hand, a genetic study in two closely related species of tiger swallowtail butterflies, *Papilio glaucus* and *P. canadensis,* demonstrated that the inheritance of the mimetic phenotype is controlled by a simple system of two loci (Scriber et al., 1996). The absence of mimicry in *P. canadensis* is attributable to a

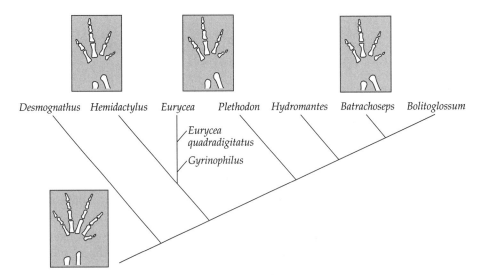

Figure 6.2 Convergent evolution of digit number in plethodontid salamanders. Each species with reduced digit number is also smaller than close relatives. This suggests that fewer digits may be a by-product of selection on body size due to reduced limb bud size. (From Wake, 1991.)

---

[4]A study of loss of function *Hox* alleles in mice complicates this scenario. Size and number of digits appears to be controlled in a dosage-dependent fashion: five toes in the wild type and no digits in mutants with all five loss of function alleles. However, the transition from five to zero toes passes through a six-toed stage (Zákány et al., 1998).

H. erato   *H. melpomene*

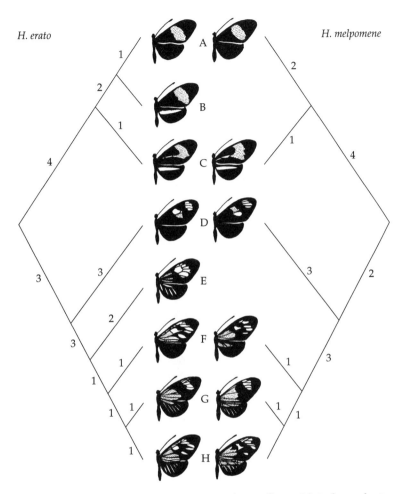

Figure 6.3   Parallel evolution in mimetic butterflies, with independent origin of the same pattern in *Heliconius erato* and *H. melpomene*. A–H represent different races of the two species; numbers refer to the inferred changes in wing pattern alleles along the branch. (From Brower, 1996.)

simple combination of genetic constraints. This species has a high frequency of the "pattern-suppressing" allele at the first locus, as well as a low frequency of the allele that determines the black pigment at the second locus. Once again, phylogenetic patterns might result from a combination of selection and constraints.

Scharloo (1987) advocated the use of selection experiments to test for the presence of constraints. This method would answer many questions concerning the availability of genetic variation and the strength of other epigenetic constraints.

Selection experiments may also suggest a role for selection, but would require the support of other tests. This approach is being increasingly employed to examine, for example, the genetic architecture of putative tradeoffs (see Chapter 1).

Some Thoughts on Process Structuralism, Baupläne, and Related Matters

A very different approach to the role of constraints in modern biology has been proposed by the so-called "process structuralists" who are calling for a rebirth of nineteenth-century rational morphology (originally based on ideas by Kant and rooted in Aristotle's philosophy). Recently, some mathematically oriented biologists have proposed similar ideas, based on the emerging fields of artificial life and complexity theory (Kauffman, 1993; Kauffman and Smith, 1986; Langton, 1986). We will briefly touch upon the similarities between these areas of research and the ideas of modern rational morphology. Resnik (1994) published a comprehensive review of this form of structuralism that we will use as the framework for our discussion.

Brian Goodwin's 1994 book, *How The Leopard Changed Its Spots*, brought the new structuralist perspective to a general audience.[5] He proposed that development, and morphogenesis in particular, are the keys to understanding the diversity of living beings. The fundamental argument is that innate properties of morphogenetic processes lead to a sharp constriction of the possible forms that organisms can take, due to the self-organizing properties of the developmental system. Extension of the concepts of innate properties and self-organization leads to a substantial diminution of the roles of genes and, ultimately, the role of natural selection. While the historical roots of "process structuralism" are certainly respectable (Buffon, Cuvier, and Owen were among its supporters, to name just a few), we are less than sanguine about its modern reincarnation. Below we examine Resnik's enumerated list of its four basic tenets.

(1) *There are general laws in biology.* Few biologists would deny that there are "general laws in biology." Aside from the trivial observation that the laws of physics apply to biological organisms (Scott, 1966; and including the second law of thermodynamics, despite recurrent claims to the contrary on the part of creationists), we can easily draw generalizations from the study of living beings. For one thing, we know that all organisms are based on carbon, that all need some kind of "information carrying" molecule, and that in all extant cases this molecule is a nucleic acid. We can also conclude that all individuals are subjected to natural selection, which is one of several forces catalyzing differentiation among populations, and so on. Population genetics and evolution textbooks are filled with these general principles. Whether such ideas can be identified with the "strong" meaning of the word "law" that is in use in the physical sciences

---

[5]See the comprehensive review by Price (1995), similar to our views in content, although superior in style.

remains to be seen. Because of the much higher complexity of organisms relative to atoms and quarks, that may not be the case, and biological laws may be open to a lot more exceptions than a physicist would feel comfortable with. To the best of our knowledge, process structuralists have not identified a single principle or law beyond those mentioned above that can serve as good evidence for supporting their contention, and, needless to say, the burden of proof is on them.

(2) *There are general forms of morphology achieved through development.* The point about the existence of general forms of development refers to the so-called baupläne or body plans (Hall, 1992a, 1996; Thomas and Reif, 1993). One definition of a bauplan is a "generalized, idealized, archetypal body plan of a particular group of animals" (Hall, 1992a; or plants, we might add; see Jones and Price, 1996). Even though the concept of bauplan can be applied to correspond to various levels of the taxonomic hierarchy, it is usually reserved for the highest ones. In general, the term has been applied to the levels of class, order, or phylum. For example, a giraffe certainly evokes a generalized, archetypal body plan—that defining the vertebrates. In plants, the general structure of angiosperms (the flowering plants) defines a bauplan distinct from the one identified with ferns or gymnosperms, for example. The apparent contradiction of defining baupläne at different levels of the taxonomic hierarchy has been circumvented by the idea of nested baupläne (Hall, 1992a). For example, the vertebrate bauplan is nested within the more general chordate bauplan. The concept loses clarity, though, when extended to even lower levels, for example, species level-baupläne.

If we accept baupläne as real features of the biological world and not as figments of our imagination, the real question becomes, What determines the differences between baupläne? Process structuralists would argue that baupläne represent a finite set of possible forms compatible with the epigenetic/developmental processes typical of living organisms, and that they are primarily a manifestation of *internal* forces. This is clearly an extreme position. Even though it is quite logical that there are internal constraints that limit the range of possible body plans, it should be equally clear that external agents (i.e., natural selection) also play an important role in the succession of macroevolutionary events. Attempts have been made to define *a priori* sets of organic design, such as the morphospace of the possible types of skeletal structures and organizations in vertebrates (Thomas and Reif, 1993). The idea is to derive from first principles and from considerations related to physical laws and available materials the whole range of possible outcomes for a given structure (not unlike the attempts of Raup and Gould to explore the possible morphospace of shelled animals referred to above). One can then compare the potential outcomes of evolution with the realized ones, in an effort to demonstrate the presence of universal constraints and laws of form.

Unfortunately, the "*a priori*" definition of morphospace is often informed by knowledge about the existing forms, which introduces an element of circularity to the whole endeavor. Kauffman and others have recently tried to circumvent this type of objection by abstracting the problem, removing all references to particular types of biological systems (Wolfram, 1984; Langton, 1986; Kauffman and Smith, 1986; Kauffman and Levin, 1987; Bak and Chen, 1991; Kauffman, 1993). These authors have argued that evolution on "rugged" adaptive landscapes can follow only those paths that connect proximate genotypes on the landscape. Even though these mathematical exercises are stimulating examples of insights that can be offered by nontraditional approaches to biological problems, their lack of resemblance to natural systems makes it very hard to even conceive of appropriate real-world testing. Although they are entirely self-consistent in mathematical terms, they do not attempt any "prediction" about biological entities, except in the most general and qualitative terms.

Two alternative, in some sense more "classic," approaches seem to us more likely to answer questions about the origins and evolution of baupläne. One such approach is offered by comparative morphology. As an example let us consider the macroevolution of the turtle body plan, which is a sub-bauplan within the reptiles (Lee, 1993, 1996). Lee argues that pareiasaurs (large anapsid reptiles of the Late Permian) shared a high number of derived characters with chelonians, and also displayed some traits previously thought typical only of chelonians (Figure 6.4). The most interesting conclusion of Lee's research, however, is that many of these shared character states evolved before the appearance of that most distinctive trait of the chelonians, the turtle shell. Lee suggests that the pareiasaurs' osteoderms (bony dermal plates) are in fact the precursors of the chelonian shell, a macroevolutionary transition that was accompanied by the migration of the shoulder girdle into the rib cage in a posterior direction.

A second, complementary, way to examine the problem of the evolution of baupläne is to look at the molecular genetic architecture responsible for current body plans. Some detailed examples from studies on the developmental and molecular events that led to the establishment of the body plan in *Arabidopsis* and *Drosophila* are reviewed in Chapter 8. The overall conclusion of these studies is that a few high-level regulatory genes seem to control the development of body plans, and that remarkably new body plans can evolve well below the class level (as in the case of chelonians). This is an exciting field of research that will certainly yield further insights into this ancient problem in evolutionary biology.

(3) *Organisms are unified wholes, and a holistic perspective is necessary to understand them.* The third point of process structuralism would have been controversial until a few years ago, and may still be among some molecular geneticists. We certainly agree that organisms are wholes (or highly integrated ensembles of complex parts), and that a whole-organism perspective is a necessary and integral part of any biological investigation. Had we thought otherwise we would not have written this book. However, too much emphasis on the whole-level analysis is dangerous when it precludes a detailed comprehension of mechanisms and proximal causes. Everything may depend on everything else, but the degree to which this is true varies enormously; and in some cases this

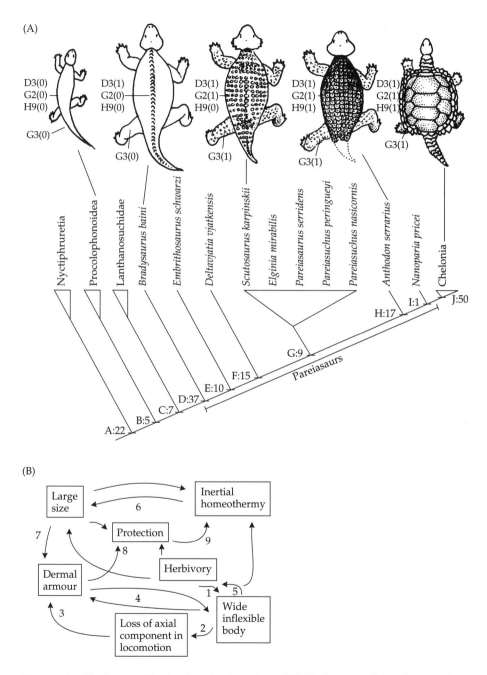

Figure 6.4 Evolution of body plans in chelonians. (A) Phylogenetic hypothesis with sketches of the body plans of representative groups. (B) A conceptual diagram of "correlated progression," in which particular evolutionary changes in an ancestral pareiasaur enable further changes in its descendants. (From Lee, 1996.)

interdependence is so small that some aspects of the whole can be provisionally ignored in order to gain a deeper understanding of the system.

(4) *Genes do not have any primacy in determining organismal forms and functions.* This is a fundamental debate that reaches back to the origins of natural philosophy. It has been unresolved and deemed unresolvable. We honestly think that the solution is very simple, and lies in the proverbial middle. Simply put, it is equally fatuous to advocate that genes represent the detailed "blueprint" of a living being, or to say that they are just another component of a complex ensemble and that focusing on them impairs progress instead of catalyzing it. On one hand, it would be foolish to deny that molecular biology has provided and will continue to provide some of the most fundamental insights into biological evolution since Darwin's work. Furthermore, dissecting a complex problem into manageable bits and pieces is a tenet of the scientific method, without which we might as well return to mystical approaches. On the other hand, most biologists recognize that organisms are characterized by a lot more than their genes; that one must consider their developmental systems (the epigenetic component) as well as their interactions with the surrounding environment. Epigenesis and plasticity were once considered "noise" to be minimized in order to focus on the genetic component. This attitude seems to be changing into one that is more flexible and comprehensive.

Admittedly, there still is a long road ahead, but the extreme arguments of rational morphologists do not seem to shed much light on that path. Our perspective is that there are indeed limits to evolution. These, however, arise from basic evolutionary principles concerning the availability of genetic variation and the role of the environment *sensu lato* in determining the success or failure of variants.

A PLETHORA OF CONSTRAINTS AND AN ATTEMPT TO DISTILL THEM

Constraints should be considered as limitations ("limitation imposed on motion or action," according to the Oxford Dictionary), and so as negative forces in evolution (*contra* Gould, 1989). They are what makes a Panglossian universe (Gould and Lewontin, 1979) impossible. Of course, if one is forced to take one route instead of another, new possibilities and unforeseen horizons will appear (as noted by Gould). But constraints are still a subset of the infinite range of possibilities in a constraint-less world. In the following section we discuss various current usages of the term *constraint*, then simplify the semantics and clarify the conceptual issues, and suggest that most categories should be subsumed within more general ones, or abandoned altogether. Notice that our attempt to "distill" the concept of constraints in order for it to retain biological interest generally follows the outline of Arnold (1992), but we make several fundamental modifications and merge some categories that he considered distinct.

GENETIC/EPIGENETIC CONSTRAINTS   These represent a legitimate source of constraint, which can be due to: (1) a lack (or limitation) of appropriate genetic variability (*genostasis*; Bradshaw, 1991); (2) linkage; (3) pleiotropy; (4) dominance; or (5) epistasis. Although many macromutational steps are clearly impossible, we know too little about the process or products of mutation to say whether there are inherent limitations to the exploration of phenotypic space by small sequential alterations. The last four categories represent manifestations of the genetic architecture that relates characters to each other. All can be demonstrated to place limits on the response to selection, and can be appropriately measured (e.g., Mitchell-Olds, 1996). However, we take issue with the common assumption that a standard genetic variance/covariance matrix can adequately describe genetic constraints (see Chapter 7).

SELECTIVE CONSTRAINTS   According to Gould (1989), selection cannot be considered a constraint, but should be viewed as an evolutionary force. We agree that selection is not a direct constraining force, and we will henceforth use the term selective pressure (or force) instead of constraint. However, given the possible regimes of selection—directional, disruptive, and stabilizing—it can play either a positive or a negative role in evolution (in the sense of either promoting or inhibiting change).

ECOLOGICAL CONSTRAINTS   An ecological constraint results from a particular kind of selective pressure that arises from the interactions of an organism with its abiotic or biotic environment (in contrast to those selective pressures arising from the necessity for a coherent developmental system, epigenetic interactions, etc.; see below).

DEVELOPMENTAL CONSTRAINTS   This is a mixed category that can include both selective pressures and genetic constraints. Developmental constraints, as generally envisioned, are restrictions on the potential evolutionary trajectory due to the fact that the developmental system must maintain an internal coherence. Each part has to work in the context of all other parts, which reduces the possibility of altering a particular aspect of the phenotype independently of other traits. However, this is clearly just a special form of selection acting on the internal environment through control of epigenetic processes (Cheverud, 1984). Since the objective is optimization of the whole organism, this is an evolutionary force and not a constraint. Of course, such selection can be at odds with other selective pressures (e.g., due to the external biotic or abiotic environment) and hence generate tradeoffs. On the other hand, some phenomena labeled developmental constraints might actually be genetic constraints, due either to the unavailability of the appropriate genetic variation or to limitations of the genetic architecture such that the coherent development of certain morphologies is impossible.

MECHANICAL AND FUNCTIONAL CONSTRAINTS   These arise within a given bauplan from the conflict between the actions of the developmental system and the realm of physical laws. We may not be able to rule out the evolutionary transformation of a 6 cm dragonfly into one the size of the Concorde, yet the alterations would have to be so comprehensive that it would no longer be considered an insect. These constraints result from the interaction between the biological and the physical worlds, and have been differentiated from developmental constraints that deal with the proper interactions of parts within the organism. Mechanical constraints, however, also arise due to selection and are thus a subcategory of selective pressures. If one chooses to consider either developmental or mechanical forces as constraints, then any other form of selection could also be considered a constraint, which leads again to the overgeneralization of the concept referred to by Gould.

PHYLOGENETIC CONSTRAINTS   The examples of putative phylogenetic constraints that we have looked at result from a number of different causes. They represent at best an observed pattern and seldom give us insights into processes. Phylogenetic constraints can be due to constant selective pressures or genetic constraints, or, most likely, to a combination of the two (McKitrick, 1993; Westoby et al., 1995).

This attempt to distill the concept of constraints is summarized in Figure 6.5. We recognize only two comprehensive categories of directional (i.e., indepen-

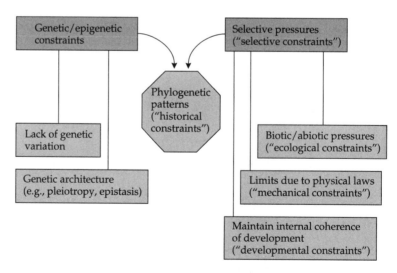

Figure 6.5   A conceptual summary of the relationships among various types of constraints. All known types of constraints can be reduced to two categories, genetic constraints or selective pressures. Most phenomena currently considered to be constraints are, in fact, instances of the operation of natural selection. (From Pigliucci et al., 1996.)

dent of drift) evolutionary forces—selective pressures (positive forces when selection is directional or disruptive, negative when stabilizing); and genetic constraints (negative forces). Evolution is then determined by the outcome of the interaction among all possible components of these two forces. Note that sometimes selection and genetic constraints can act in the same direction; for example, stabilizing selection tends to decrease genetic variation, and together they will keep the population in the current area of the phenotypic space.

We have suggested that constraint, like homology, is perhaps better envisioned hierarchically (Pigliucci et al., 1996). What are perceived as developmental "constraints" at one organismal level can be detected as genetic constraints or, at another level, as due to the action of selection. In Figure 6.6 we present a DRN view of constraint for the hypothetical example of the production of digits on a vertebrate limb (a favored location for developmental "constraints"). The diagram depicts the transition from the level of the gene through the epigenetic process to the phenotypic state. If there is only a single allele for a particular gene, then a genetic constraint exists. In our example there are multiple genes with multiple alleles that contribute to the expression of the continuous character cell number. We specify epigenetic rules that provide for a symmetrical bifurcation of cells when the cell number reaches a threshold of 300; this converts the continuous cell number distribution into a discrete pattern of 1, 2, 4, 8 . . . digits. In this example, we have also set limits on cell number through natural selection—combinations of alleles that result in <125 or >1000 cells are lethal due to disruption of the developmental system, restricting digit production to between one and four digits.

Although the fitness function specified suggests that two, three, or four digits would be equally fit, the epigenetic rule restricts phenotypic expression to two or four, but not three digits—this would typically be considered a "developmental" constraint. However, by examining the developmental reaction norm, we can see that this pattern is due to the invariant epigenetic rule of bifurcation of the cell mass that we specified. The expression of different sets of alleles at the lower level (i.e., cell number) is controlled in turn by the epistatic action of other genes at a higher level (the bifurcation rule). However, if we shift our focus to that higher level of the hierarchy, the observed pattern of one, two, or four digits results from a *genetic* constraint—there is no allelic variation at the gene loci determining the bifurcation rule. The production of three toes would require a trifurcation rule, or a rule that results in an unequal division of the cell mass.

SELECTION AND CONSTRAINT ON THE DEVELOPMENTAL REACTION NORM

One of the fundamental ideas of this book is that selection (as well as constraint) acts not just on the expression of phenotypes in particular environments or at particular ontogenetic times, but on portions of the developmental reaction

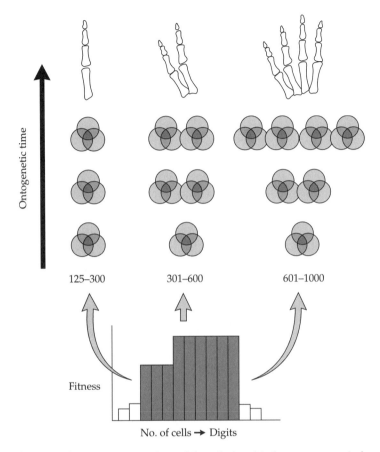

Figure 6.6   A developmental reaction norm view of the relationship between genetic/ epigenetic constraints, developmental "constraints," and natural selection. The lower diagram illustrates the relationship between initial cell number and fitness: cell numbers below 125 or above 1000 (open boxes) are lethal, those between these thresholds (shaded boxes) produce one, two, or four digits. Organisms with a single digit are less fit than those with two or four digits. The upper diagram depicts the operation of the epigenetic system during ontogeny. The fate of cell masses of different sizes is determined by an epigenetic rule that specifies an equal bifurcation of cell masses greater than 300 cells. Masses less than 301 cells do not bifurcate, and produce a single digit; masses between 301 and 600 cells bifurcate once and make two digits; and cell masses between 601 and 1000 cells bifurcate twice and make four digits. Within these rules there is no way to produce three digits, even though such an organism would presumably be fit. (From Pigliucci et al., 1996.)

norm (see also Scharloo, 1987). There are, however, few suitable examples of this in the literature, presumably because very few researchers have so far adopted this perspective, and because it is cumbersome to gather the appropriate data sets. In the following section we will present a few cases of constraints on reaction norms, and discuss how plasticity can play a role in population dynamics and modify evolutionary trajectories.

A case in which plasticity is both constrained by the genetic machinery of an organism and may act to divert future evolutionary pathways is found in a study by Young and Schmitt (1995) of characters that affect pollen dispersal and reception in *Plantago lanceolata*. These authors measured the efficiency of pollen dispersal and reception as a function of the height of the separate staminate (male) and pistillate (female) flowers. They found that pollen dispersal is more effective when staminate flowers are higher, while pollen reception is enhanced when pistillate flowers are lower. Furthermore, based on the genetic variation for inflorescence height in natural populations, they predicted that a regime of disruptive selection would increase the difference between the height of the male and female flowers. However, when Young and Schmitt investigated two natural populations, they uncovered a very strong positive genetic correlation between the heights of staminate and pistillate flowers. This would effectively constrain the simultaneous increase of male and female reproductive efficiency (although there can still be selection for one or the other, leading to the generation of a tradeoff between male and female function). Moreover, this constraint can interact with the phenotypic plasticity of these plants to affect population dynamics. Plants that grow in shady habitats tend to produce elongated inflorescences, a response that is probably mediated by the action of phytochrome. Because of the constraint, however, stem elongation translates into better pollen dispersal and worse pollen reception. This in turn implies an asymmetric gene flow between populations colonizing shady habitats and those living in more open environments (see also Stanton et al., 1997).

Constraints on reaction norms have been categorized into two major classes: physiological/developmental "costs" of plasticity (van Tienderen, 1990; Bergelson, 1994; Krebs and Loeschcke, 1994; Simms and Triplett, 1994); and inter-environment genetic correlations between the expressions of the same trait in more than one environment (Yamada, 1962; Via and Lande, 1985; Falconer, 1990; Andersson and Shaw, 1994; Pigliucci and Schlichting, 1996; Windig, 1997). Van Tienderen (1990; van Tienderen and Vandertoorn, 1991a,b) investigated the relationship between costs of plasticity and the evolution of genetically distinct ecotypes in *Plantago lanceolata* found in hayfields and pastures, habitats that presumably exert very different selection regimes (Figure 6.7). They conducted transplant experiments to study the plasticity of these two types when exposed to the alternate "alien" environment. Hayfield plants usually have more time to grow (though the surrounding vegetation is tall), while plants from pastures are continuously grazed and trampled. As a result, hayfield populations develop an erect growth habit and tend to invest more resources in shoot production; pasture populations show a prostrate habit, with an emphasis on early reproduction (Figure 6.7B). When genotypes from the two environments are planted in the reciprocal habitat, they tend to develop the "correct" phenotype; that is,

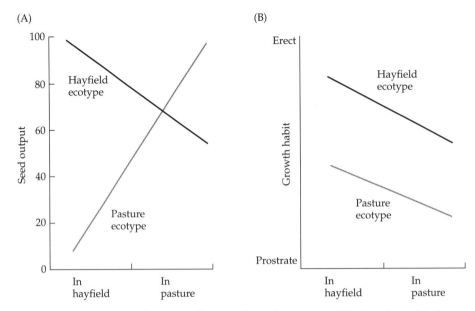

Figure 6.7 Responses of reciprocally transplanted ecotypes of *Plantago lanceolata* from a hayfield and a pasture. (A) Relative seed production of the transplants compared with the "native" population (scaled to 100), revealing ecotypic adaptation. (B) Parallel plastic shifts to a more prostrate growth form in the pasture environment, suggesting that plastic responses are adaptive (or both ecotypes are "constrained" in the same fashion). (After van Tienderen, 1990.)

pasture plants show a more erect habit, while hayfield plants display a more prostrate one. This adaptive plasticity, however, is not sufficient to overcome the gap completely, and the genotypes still perform best in their original environments (Figure 6.7A). This case demonstrates limits to plasticity and underscores the ability of plasticity to at least partially bridge the gap between two adaptive morphologies and to thereby allow the survival of plants colonizing a new or different habitat.

A very similar case at the interspecific level is represented by a study of sticklebacks by Day et al. (1994). They focused on two as yet unnamed *Gasterosteus* species that colonize the benthic and the limnetic habitats of Paxton Lake in British Columbia. Normally, the benthic species feeds mostly on worms, whereas the limnetic one feeds on plankton. When fed the alternate diet, the benthic form tended to converge toward the phenotype of the limnetic form for a wide range of traits, and vice versa (Figure 6.8). Transplanted populations showed tradeoffs in growth rates in the two habitats; each grew faster in the original habitat (Schluter, 1995). Neither of the plastic responses is strong enough to closely emulate the phenotype of the other species, but they are again in the

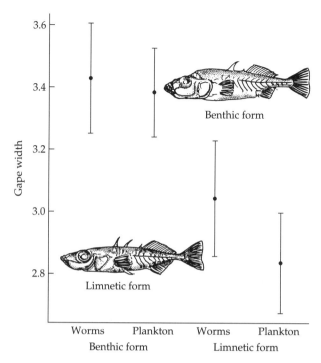

Figure 6.8 Plasticity of sticklebacks, showing that each of two species specialized for a particular diet converges toward the morphology of the other species when grown on the "wrong" diet. The benthic form typically feeds on worms, the limnetic form on plankton. Switching these to the other food type results in gape width phenotypes closer to the alternate form. (From Day et al., 1994.)

"correct" direction. The results here are similar to those for *Plantago*; both illustrate the limits to plasticity and the potential for plastic responses to "pre-adapt" organisms to novel environments without initial genetic change (*cf.* genetic assimilation, Chapter 10). For the sticklebacks, it is conceivable that the ancestor of both species was actually a generalist characterized by a broader plasticity that then evolved into two more specialized forms (Schluter and McPhail, 1992).

Interesting contrasts between the costs and benefits of plasticity have emerged from studies of the heat-shock response in *Drosophila melanogaster*. Conditioning is known to be necessary to activate the genes that control the heat shock response. Krebs and Loeschcke (1994) compared the survival and fecundity of four groups of flies: (1) unconditioned without heat shock, (2) unconditioned with heat shock, (3) gradually conditioned to increasing temperatures without heat shock, and (4) gradually conditioned to increasing temperatures with heat shock. These comparisons allowed a gauge for the cost of activating the heat shock resistance machinery. The results indicated that with the heat shock treatment, the conditioned flies with activated heat shock proteins had a much higher survival rate than unconditioned flies; but the fecundity rankings were inverted when there was no actual heat shock following the conditioning. Furthermore, these experiments were conducted under two regimes of food availability (presence or absence of yeast) to study the interaction between the two types of stress. In this case, the fecundity advantage due to the plasticity of the heat shock response increased under the no-yeast treatment, where the two stresses were compounded. To put it simply, plasticity was beneficial, especially in multiple stress situations; but the expression of the plastic phenotype in the "wrong" environment was costly and disadvantageous. In further studies, the amount of heat shock-induced protein hsp70 was found to be positively correlated with thermotolerance, but negatively correlated with survival at 25°C (Krebs and Feder, 1997). Flies engineered with more *hsp70* genes had higher eclosion success following heat shock (Feder et al., 1996).

Similar cases have been documented in plants that are susceptible and resistant to parasites or pathogens. Figure 6.9 shows that the fitness of parasite-resistant and susceptible lettuce genotypes is indistinguishable under high nutrient conditions, but that the resistant genotype grows relatively poorly under low nutrients (Bergelson, 1994). Another instance is the study by Simms and Triplett (1994), who measured resistance (the inverse of leaf damage) and tolerance (the slope of the line relating fitness to damage) of the plant *Ipomoea* to the fungal disease anthracnose. They found no general fitness costs, and that resistance to one of seven fungal isolates was a good predictor of resistance to the others. However, a detailed comparison of two of these isolates indicated a pairwise tradeoff in the ability to survive both isolates. There were also fitness tradeoffs between the disease and disease-free treatments, suggesting tolerance costs. These results imply that an appropriate plastic response would be favored in comparison to a genetically constitutive resistant phenotype, because of the latter's fitness disadvantage in any pathogen-free environments.

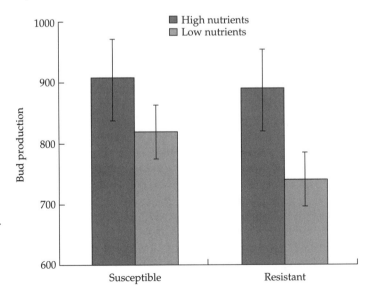

Figure 6.9 Fitness of susceptible and resistant genotypes of lettuce under high or low nutrient conditions. (From Bergelson, 1994.)

From a classical quantitative genetics standpoint, a measure of constraints on plasticity is offered by the across-environment genetic correlation (r_{ae}, see Chapter 3). Via and Lande (1985) adopted this idea in an evolutionary context to derive models for the evolution of phenotypic plasticity. More recently, van Tienderen and Koelewijn (1994) and de Jong (1995) have extended it to the case of many traits in many environments, and demonstrated the mathematical equivalency between this approach and the classical analysis of variance.[6] We will briefly discuss some empirical examples of genetic constraints on plasticity based on the application of r_{ae}.

Andersson and Shaw (1994) investigated reaction norms and inter-environmental genetic correlation in *Crepis tectorum*. They used F_3 and F_4 inbred families originated from two ecologically contrasting ecotypes. When they plotted the genotypic means measured under two light regimens against each other, they found a positive r_{ae} for almost all traits, which would indicate a propensity for correlated selection in the alternate environment. Since most of these traits are expected to be under divergent selection under field conditions, the prediction is that the populations would evolve toward ecotypic specialization, rather than give rise to plastic generalists. However, since the analysis was done using geographically isolated populations that had already differentiated as ecotypes, it is not clear if the results are better interpreted as a constraint on future evolution or as the outcome of past divergent selection that would have eliminated generalist genotypes.

Pigliucci and Schlichting (1996) examined plasticity to high versus low nutrients in the weed *Arabidopsis thaliana*, using 37 full-sib families belonging to three natural ecotypes (one early flowering and two late flowering) and a laboratory isogenic line (early flowering). In this study, even though some traits were strongly correlated across environments (e.g., growth rate), others were weakly or not significantly genetically related (e.g., number of branches; Figure 6.10). An investigation into the association of reaction norms and reproductive fitness uncovered directional selection for more branches combined with stabilizing selection for plasticity of branching, as well as stabilizing selection for both mean and plasticity of growth rate. The general conclusions that can be drawn from these and similar studies are that genetic constraints on reaction norms, as mea-

---

[6]Schlichting and Pigliucci (1995a) have pointed out that the across-environment genetic correlation (r_{ae}) is not biologically equivalent to the classical correlation between two traits within one environment (r_G). While r_G can be due to either pleiotropy or linkage, (r_{ae}) can also be affected by allelic sensitivity, that is, the same gene has different effects in different environments. This may arise, for example, because of interactions between the physical environment (e.g., temperature) and the characteristics of the coded protein or enzyme (e.g., reaction kinetics; see also Chapter 3). This is not, strictly speaking, pleiotropy; exactly the same genes could be acting in both environments and affecting the same traits, even though their presence would be associated with effects of different magnitude, or even sign, on those traits.

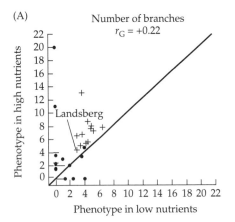
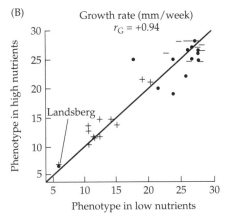

Figure 6.10 Genetic correlations between the expressions of the number of branches (left) and growth rate (right) in two nutrient environments for 37 families of *A. thaliana*. +, •, and − refer to high, medium, and low relative fitnesses. Number of branches is not constrained (i.e., r_G is not significant) compared to growth rate, which is highly constrained. (From Pigliucci and Schlichting, 1996.)

sured by across-environment genetic correlations, can exist for some traits but not others, and can depend on the particular organism, population, or environment under study. These conclusions can be confirmed via selection experiments.

A next step is to identify the mechanistic basis of these constraints. A promising avenue of research in this regard is offered by studies on the distribution of the phenotypic effects of mutations in known genetic backgrounds, such as that by Markwell and Osterman (1992) in *Arabidopsis*. They examined the effect of temperature (20°C and 26°C) on the production of chlorophyll a and b for the wild type Columbia line and for mutants obtained from Columbia by ethyl-methane-sulfonate (EMS) mutagenesis. The range of reaction norms that can be obtained by single mutations of a uniform genetic background is impressive (Figure 6.11). In particular, total chlorophyll production in the mutants could be either more or less plastic than the control, and several mutants were characterized by an increased production at 20°C, though none surpassed the control at 26°C.

Camara and Pigliucci (unpublished) explored the strength of several potential constraints in *Arabidopsis thaliana*, by comparing character correlations in three different ecotypes (two natural populations and an isogenic lab line) and in three populations of EMS mutants derived from the respective wild types. A well-documented "constraint" in *A. thaliana* is illustrated in Figure 6.12; there tends to be a positive relationship between the duration of the vegetative phase (measured as bolting time) and amount of vegetative growth (measured as the

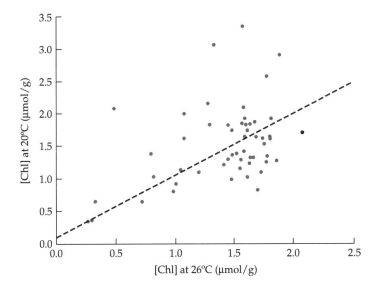

Figure 6.11 Plot of the chlorophyll concentration of *Arabidopsis* genotypes at 20°C and 26°C. Gray circles represent putative mutants of the standard line, Columbia (black circle). The dashed line indicates no plasticity. (From Markwell and Osterman, 1992.)

number of leaves in the basal rosette). Closer inspection of the figure suggests that there may be a second constraint at work, reflected by the lack of plants bolting earlier than six days, or with a rosette smaller than four leaves. The distribution of the mutants shows that it is possible to break both these constraints,

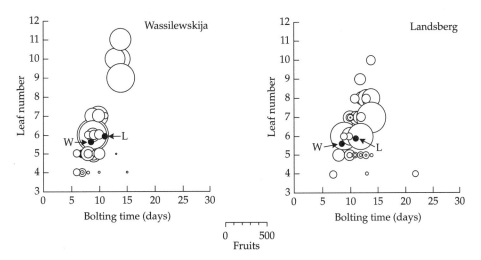

Figure 6.12 Phenotypic effects on two traits of two series of EMS mutants produced in two backgrounds of *Arabidopsis thaliana*. Notice the similar distribution of the effects, implying a fundamental genetic constraint for this species. L is Landsberg, W, Wassilewskija. The arrows indicate the wild types. Amplitude of the circles is proportional to reproductive fitness. (After Camara and Pigliucci, unpublished.)

as several mutants deviated from the linear relationship between the two characters. An *A. thaliana* with more leaves and unchanged bolting schedule, or with later bolting and unaltered rosette size, is possible starting from any of the three natural backgrounds (only two are shown in the figure). Mutants that showed earlier bolting than any wild type, or that produced up to two fewer leaves, were also obtained.

Camara and Pigliucci also examined the potential for an analogous constraint during the reproductive phase of growth by considering the relationship between the length of the reproductive period (time between bolting and senescence) and the allocation to inflorescence meristems other than the main shoot. The distributions of mutagenic effects are in this case much more restricted than in the previous example. Only two mutants of the Wassilewskija *Arabidopsis* had markedly different phenotypes. There was a high fitness mutant with several more inflorescences that delayed senescence 12 days longer than the wild type; and a very low fitness mutant that had the same number of inflorescences as the wild type but that senesced markedly later. Again, there was a strong constraint on the lower limit of the duration of the reproductive period, with few mutants characterized by a shorter time to senescence than the controls. On the other hand, *A. thaliana* may produce no basal inflorescences, which is the absolute lower boundary on meristem allocation to reproduction. Several mutations produced plants with much higher fitness than the wild types. This is probably due to the relaxation of selection in the wild types (all populations were previously selected under controlled conditions for very short life cycles). Nevertheless, it indicates that it is possible to dramatically and positively affect fitness with single-gene mutations, if the conditions are favorable, or with the appropriate environmental change. This underscores our contention that the DRN represents a context-dependent target for selection.

NULL MODELS AND *A PRIORI* EXPECTATIONS

In order to understand the roles of constraints and selective pressures in the evolution of the phenotype, we need to formulate a null hypothesis of how evolution would proceed in their absence (Antonovics and van Tienderen, 1991). As an aid in defining and quantifying alternative models, we employ the notion of transitional probability tables. Let us consider the evolution of a single character for which only three states (A, B, and C: see Figure 6.13) are possible. The simplest null hypothesis in this case is that all the cells in the transitional probability matrix are equiprobable, that the character is equally likely to move (via mutation) from any one state into any other, or to stay in the current state. This completely neutral model of phenotypic evolution can be extended to any number of characters and any number of states. One distinguishing aspect of this neutral model is that a population is *expected* to change, contrary to the predictions derived from the more simplistic assumptions of the classical Hardy-

NEUTRAL EVOLUTION					DIRECTIONAL SELECTION (Favoring A)			
1		From					From	2
	A	B	C			A	B	C
A	0.33	0.33	0.33		A	0.90	0.90	0.90
To B	0.33	0.33	0.33	To	B	0.05	0.05	0.05
C	0.33	0.33	0.33		C	0.05	0.05	0.05

DISRUPTIVE SELECTION (Favoring A and C)					STABILIZING SELECTION (Favoring B)			
3		From					From	4
	A	B	C			A	B	C
A	0.45	0.45	0.45		A	0.05	0.05	0.05
To B	0.10	0.10	0.10	To	B	0.90	0.90	0.90
C	0.45	0.45	0.45		C	0.05	0.05	0.05

Figure 6.13 A simple series of examples illustrating how common evolutionary phenomena such as neutral evolution, directional selection, disruptive selection, and stabilizing selection can be illustrated (and tested) with the aid of transition probability matrices. A, B, and C are traits. The "from/to" labels indicate the probability of transition from a given state to a different one. (1) Neutral evolution corresponds to a situation in which all state transitions have the same probability (notice that not only is there no selection, but there also are no genetic constraints). (2) Directional selection for state A implies that the probability of a transition from any state to A is higher than the probability of the complementary events (or of the probability of staying in the same state, with the exception of A). (3) Disruptive selection that favors both A and C implies a high probability of moving to either state A or C and a low probability of moving to state B. (4) Stabilizing selection for state B implies a high probability of either going to or staying in B, and a low probability of achieving or maintaining any other state.

Weinberg equilibrium. The null evolutionary trajectory of a real population is not a stable point but, due to mutation and drift, a random walk governed by non-directional forces. In this sense, the situation of a population is analogous to the situation of a body in Newtonian physics, in that it is assumed to be moving unless it is worked upon by some external force. In physics, the major force responsible for slowing down motion is friction; in biology, both selective pressures and genetic constraints can have the same effect.

The first step in testing the null model is to construct a table of probabilities derived from actual data. Deviation of the observed transition probabilities from those specified by the null model is equivalent to demonstrating that some other forces are acting on the population (measured, for example, by means of matrix comparison tests: Mantel, 1967; Smouse et al., 1986; Manly, 1986; Cheverud et al., 1989; Zhang and Boos, 1993; Steppan, 1997a,b). The simplest null hypothesis can be rejected for two reasons: either there is greater phenotypic change than expected, or there is less. A lack of change (or reduced change compared with the null expectation) results from genetic constraint or stabilizing selection. More change than expected must be attributed to positive selection (either directional or disruptive). More complex null hypotheses can be easily specified by using the same approach (Figure 6.13).

Selective forces, either positive or negative, can be estimated for both direction and intensity by multivariate regression analysis (or path analysis) of characters on fitness (Maddox and Antonovics, 1983; Crespi and Bookstein, 1989; Farris and Lechowicz, 1990; Kingsolver and Schemske, 1991; Jordan, 1992; Mitchell, 1992; Bergelson, 1994; Pigliucci and Schlichting, 1996). These measurements can constitute the basis for a new null model that assumes that, in a Panglossian world, wherever selection is pushing, the population will move. After the selection event, one can measure the response of the population; any movement (or restriction of movement) not accounted for by selective pressures must be due to genetic constraints.

A similar approach to the one advocated here has been used by Janson (1992) to examine phylogenetic trends in fruit dispersal mechanisms. Janson employed a Markov model to describe the evolutionary transitions between all possible combinations of eight types of fruit dispersal in flowering plants. The data were based on the putative phylogenetic relationships among 571 Neotropical genera. He set out to test four increasingly complex hypotheses describing different levels of constraints: (1) no constraints on any transition; (2) all syndromes are equally likely to change—their "evolutionary lability" was the same; (3) the probability of change is dependent on the descendant state, but not on the ancestral one; and (4) each pair of dispersal mechanisms has its own probability of transition, but this is reciprocal within the pair (e.g., the probability to evolve from wind to water dispersal is the same as that to change from water to wind). He then tested the four basic hypotheses sequentially, using the actual table of transition probabilities (Figure 6.14). He concluded that more complex scenarios matched the data better than simple ones, indicating the existence of "constraints" (*sensu lato*, encompassing both genetic constraints and specific selective forces) on the underlying macroevolutionary processes. From his data it is clear that some evolutionary transitions are much more likely than others. For example, it is "easier" to go from gravity or explosive dispersal to water dispersal than from water to wind dispersal. Note also that reciprocal transitions do not have the same probability—the switch from gravity to water dispersal is much more likely than the one from water to gravity dispersal. The limitation of this kind of study is that we still cannot distinguish the relative contributions of selection and constraint on the observed patterns, nor can we attribute the results to particular categories of constraints. Yet, we do have testable hypothe-

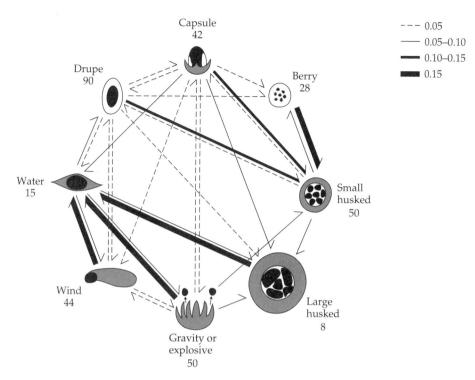

Figure 6.14 Transition probability diagram illustrating the evolution of types of seed dispersal in angiosperms. Increasing thickness of lines indicates higher transition probability between states. Notice that this is simply a visualization of an empirical variance/covariance matrix, and can therefore be directly compared with any equivalent theoretical matrix specifying the hypothesis to be tested. (From Janson, 1992.)

ses that can narrow down the possibilities and point to future research directions in quantitative genetics or functional ecology.

An example of how quantitative genetics and phylogenetic studies can further dissect the question of what "constrains" macroevolutionary patterns is the work of Futuyma et al. (1993, 1995) on host plant associations of phytophagous beetles in the genus *Ophraella*. Individual species in this genus feed on only one or a few congeneric species of host plant, but the plant genera hosting the different beetle species belong to four disparate tribes of the enormous plant family Asteraceae (Figure 6.15). Futuyma and colleagues estimated the amount of extant genetic variation for the survival of *O. communa* on its natural host as well as for its survival on seven other taxa (including several different plant families) that are host plants for related *Ophraella* species. The results that are most germane to our discussion are the following: (1) there was evidence of genetic variation in feeding responses to five of the alternative hosts; (2) in several cases

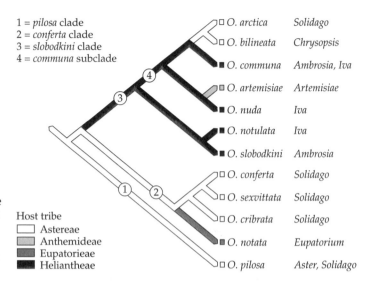

1 = *pilosa* clade
2 = *conferta* clade
3 = *slobodkini* clade
4 = *communa* subclade

O. arctica	*Solidago*
O. bilineata	*Chrysopsis*
O. communa	*Ambrosia, Iva*
O. artemisiae	*Artemisiae*
O. nuda	*Iva*
O. notulata	*Iva*
O. slobodkini	*Ambrosia*
O. conferta	*Solidago*
O. sexvittata	*Solidago*
O. cribrata	*Solidago*
O. notata	*Eupatorium*
O. pilosa	*Aster, Solidago*

Host tribe
☐ Astereae
▨ Anthemideae
▨ Eupatorieae
■ Heliantheae

Figure 6.15 Evolution of feeding on different plant hosts in beetles in the genus *Ophraella*. Each branch tip represents a species of beetle that feeds on the host plant listed above it (all within the family Asteraceae). Notice the general concordance between host plant tribe (indicated by shading) and herbivore phylogeny. (From Futuyma et al., 1995.)

genetic variation was found that allowed feeding on taxa that are host plants for close relatives of *O. communa*; and (3) there was no evidence of genetic constraints due to negative genetic correlations (antagonistic pleiotropy) between the performance on different hosts. These conclusions point to the role of one of the sources of constraint that we listed above (lack of genetic variation) and the apparent absence of another (pleiotropy), thereby leading the way to further dissection of the causal mechanisms associated with particular evolutionary pathways. Moreover, the observation of more genetic variation for performance on hosts of more closely related beetle species suggests that host switches (and possibly speciation) are not random. Instead, these are directed by (1) the availability of the appropriate quantitative genetic variation in concert with (2) the availability of similar host plant species (Futuyma et al., 1995).

As previously discussed, there are other examples in the literature of cases in which patterns usually interpreted as the result of natural selection are in fact reinterpreted to be due to some sort of constraint. Again, in these cases phylogenetic analyses are pivotal, as in the study of the association between fleshy fruit morphology in angiosperms and the corresponding seed dispersers. Jordano (1995) investigated the claim (or rather the assumption) that the characteristics of fleshy fruits (length, diameter, mass, energy content, protein content, etc.) evolve in response to the specific animals that disperse those fruits. He considered the putative phylogenetic relationships among 910 angiosperm species.[7] He found that among the 16 traits analyzed, 61% of the total variance

---

[7]Based on a hierarchical taxonomy derived from a comprehensive flora, not on direct phylogenetic analyses.

was explained by taxonomic relationships at the genus level. When he examined the correlations between fruit characteristics and type of disperser, most were found to be nonsignificant after accounting for phylogenetic effects. The overall conclusion is that historical contingencies (meaning, essentially, genetic constraints) are generally responsible for the observed macroevolutionary scenario, while selection played a comparatively minor role.

However, it is worth cautioning against over-interpretation of such analyses based on phylogenetic corrections, as explained by Westoby et al. (1995; see also Ackerly and Donoghue, 1995; Ricklefs, 1996). These authors observed that phylogenetic constraints are in fact a combination of ecological (or selection) and genetic (or "historical") forces. The word "combination" is usually taken in this context to mean "sum," so that current comparative methods (Felsenstein, 1985, 1988; Gittleman and Luh, 1992; Harvey and Pagel, 1991; Harvey and Purvis, 1991) are designed to "subtract" the historical component from the total variance. What is "left over" is assumed to indicate actual evolution in response to selective forces. However, the combination of selection and history might be not a simple sum of effects, but rather an interaction (Figure 6.16). If that is the case, there are actually three "components" to phylogeny-wide correlations among traits: effects due entirely to constraints; those ascribable only to selection; and effects that are due to their interaction (e.g., when phylogenetic and morphological divergence occur at different rates). The blind application of a phylogenetic correction removes not just the historical effects *sensu stricto*, but any joint effects as well.

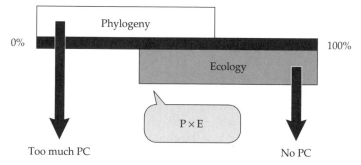

Figure 6.16   The danger of too much phylogenetic correctness in comparative method studies. The black bar indicates the total variation of a trait, part of which is phylogenetic and part dependent on the ecology. Simply subtracting the phylogenetic variance from the total, however, assumes that there is *no overlap* (i.e., *no interaction*) between ecology and phylogeny. This is equivalent to assuming that speciation is independent of the environment. If that is not the case, the phylogenetic "correction" might remove a relevant portion of the information, leaving the researcher with no actual "signal." (After Westoby et al., 1995.)

The problem can be envisaged as analogous to an analysis of variance of genotypic and environmental effects on a phenotype. In order to actually determine the magnitude of their interaction, one must have replicates of two sources of variation: a measure of environmental differences, as well as replicated genotypes within each environment. Before plasticity was considered a major player in evolution, the standard approach was to simply assume that any G by E interaction was negligible, dissect the observable variation into genotype and environment components, and then concentrate on the "important" component, that is, on the genetics. To avoid the same mistake in comparative studies, we need information about habitats (which is usually lacking) in addition to the phylogeny. Interactions would result whenever several species are found in similar habitats but belong to different clades (convergence), or when closely related species are found in dissimilar habitats (divergence). Until now, we have simply assumed, as Westoby et al. so clearly point out, that any interaction is negligible and that, between the two possible major effects, selection has some kind of ontological primacy over historical contingencies (which need to be "eliminated" or corrected for).[8] Unfortunately, comparative methods cannot at this point take into account this third source of variation, simply because phylogenies are much more easy to come by than the information about past and current selection and/or habitat constitution that is necessary to clarify the interaction effect. This is currently an entirely open field of comparative method studies, and one that is likely to revolutionize our perception of the role of constraints and selection.

Our view of the interaction between selection and constraints can be represented graphically in the case of the evolution of a population in a bi-dimensional phenotypic space (Figure 6.17). Let us consider a population in any given initial state, that is, at a particular set of coordinates in phenotypic space. The vector S represents the direction and intensity of natural selection along which the population would be pushed in a Panglossian world (Gould and Lewontin, 1979). Let us now suppose that there is some kind of constraint acting on the population. For example, if there is a genetic correlation between the two traits, we can represent this constraint as a different vector C, with its own direction and intensity. The resultant between S and C then gives us the realized evolutionary trajectory R. Obviously, both S and C in Figure 6.17 may combine different evolutionary phenomena (e.g., S can be due to internal coherence of

---

[8]Curiously, even though phylogenetic effects are considered to be ontologically less important, they are assigned a quite prominent role due to their removal *prior* to an analysis of selection. This seems odd, because, analogous to a type I analysis of variance, this attributes any interaction variance to phylogeny, and produces a quite conservative test of the possible effects of selection (i.e., it inflates the *type II error*). It could just as easily be argued that estimates of selection should be conducted first. One way to get a rough idea of the magnitude of the overlap could be derived from reciprocal type I analyses, alternately first "removing" historical or ecological effects. The difference between the estimates of variance in these two models would be akin to the interaction.

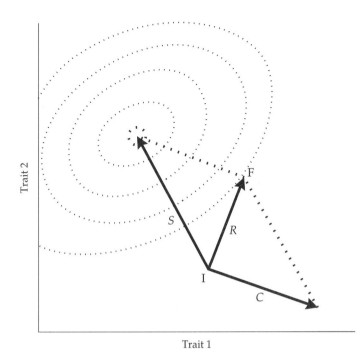

Figure 6.17 The fundamental forces acting in phenotypic space. The initial phenotype I of a population is "pushed" by two sets of forces (represented by vectors), genetic constraints *C* and selection pressures *S*. The resulting phenotypic vector *R* might lead to a suboptimal or maladaptive phenotype F. This represents a compromise resulting from a non-Panglossian world.

development, biotic pressures, etc.; and *C* can encompass a reduced amount of genetic variation, genetic correlation among traits, epistasis, etc.). In this connotation, either vector can be further dissected into more elementary components.

A simple way to think of the interaction between selection and constraints is to use the classical situation of selection pressure and selection response (Falconer, 1989; Lande and Arnold, 1983). The actual movement of the population in phenotypic space, the selection response, is due to a combination of selection pressure (the action of the selection vector in our diagram) and the friction caused by the limited genetic variation available (usually measured by the heritability, or the genetic variance/covariance matrix, and graphed as our genetic constraint vector). In the extreme case in which there are no constraints (the heritabilities, h^2, are 1 and the genetic covariances are 0), the response to selection is equal to the selection pressure, and the population's evolutionary trajectory is governed only by the selective forces. At the other extreme ($h^2 = 0$, and/or very high genetic covariances that produce complete genetic constraints), the population does not move, no matter how strong the selective vectors are. The real universe is usually somewhere in between, and includes nonadditive genetic factors and environmental effects.

We believe the important aspect to consider is that selection and constraints in particular environments can have ramifications not only on the phenotype produced in the focus environment, but for the entire reaction norm. This implies that in order to produce a realistic projection of an evolutionary trajectory, we need an understanding of (1) the range and frequency of actual environments in which evolution occurs, and (2) the existing environment-specific selective pressures and constraints (including the multivariate relationships among traits).

Unfortunately, as we have already pointed out, this kind of exhaustive data set is hard to find. For example, a study by Schluter and Nychka (1994) demonstrated an interesting form of selection that maintained a particular ratio (i.e., an allometric relationship) between two traits in sparrows (Figure 6.18). However, this study was carried out only in one environment, which leaves open the possibility that another form of selection might be operating under different conditions.

A more comprehensive study is offered by the work on the *Eurosta-Solidago* coevolutionary system by Weis and Gorman (1990). *Eurosta solidaginis* is a species of dipteran that induces galls in the plant *Solidago altissima* (Asteraceae). Weis and Gorman investigated selection on components of the reaction norm of gall size to the particular environment presented by the host (i.e., different plants of *S. altissima* represented the different "treatments"). They found strong genetic variation for gall size as well as genetic variation for plasticity of the trait. When they measured selection by regression analysis (using the standard Lande and Arnold approach; Lande and Arnold, 1983) they detected significant directional and stabilizing selection on the height of the reaction norm, as well as stabiliz-

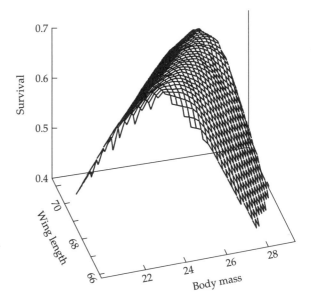

Figure 6.18 Three-dimensional representations of fitness surfaces used to infer the effect of natural selection on juvenile male sparrows. The two focus traits are body mass and wing length, and the fitness surface indicates selection along a ridge, favoring a given ratio of the two traits. (From Schluter and Nychka, 1994.)

ing selection on its slope (i.e., its plasticity). What makes this study particularly interesting is that the authors not only examined the reaction norms (Figure 6.19A), but also estimated the fitness function associated with the trait under study (Figure 6.19B), that is, they generated a curve describing the relationship between gall diameter and survival. Furthermore, they gathered information about the frequency distribution of a measure of the environment (in this case, lag time between oviposition and gall initiation; Figure 6.19C).

All these pieces of information can be brought together to yield a fairly complete picture of the biological system. Figure 6.20 shows how this is done in principle. By substituting the observed phenotypic reaction norms into the estimated fitness function, one obtains the so-called "fitness reaction norm" (upper panel). The fitness reaction norm is then multiplied by the observed frequency distribution of environments to gather the mean fitness of the different geno-

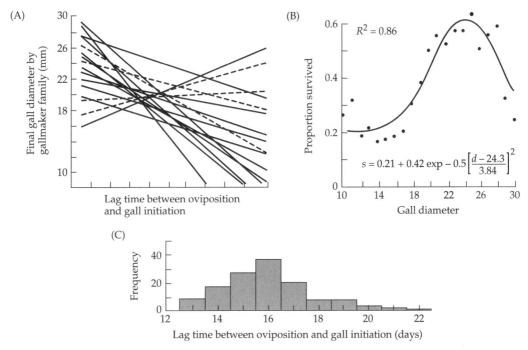

Figure 6.19 Reaction norms and fitness in the *Eurosta-Solidago* system between a gall wasp and its host plant. (A) Reaction norms of *Eurosta* fitness versus host environment (measured as the lag time between oviposition and gall initiation). (B) Fitness function relating the wasp-induced phenotype (gall diameter) to wasp survival. (C) Frequency distribution of the experimental environments (lag times). These are the elements needed for a complete analysis of the effect of plasticity on fitness. (After Weis and Gorman, 1990.)

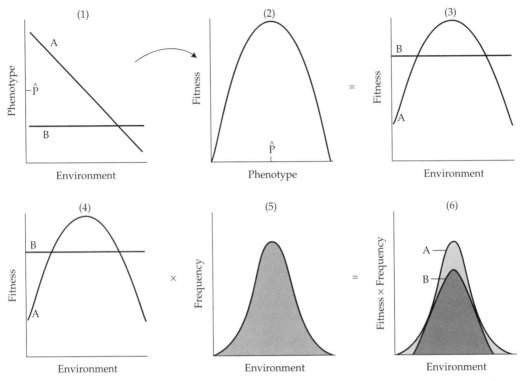

Figure 6.20 The theoretical procedure proposed by Weis and Gorman (1990) to study the relationships between plasticity and fitness. The reaction norms of different genotypes (A and B) for a given trait (1) are weighted by the fitnesses associated with those trait expressions (2) to yield the fitness reaction norms (3). These fitness reaction norms (4) are then multiplied by the distribution of environments (5) to obtain a diagram giving the distribution of actual selective pressures (6). The distribution of selective pressures depends on both the fitness of a given phenotype in a given environment and its frequency of occurrence. P̂ is the average phenotype. (After Weis and Gorman 1990.)

types, which can be incorporated into quantitative genetic models of phenotypic evolution. Many more data sets of this type need to be assembled before we can meaningfully generalize about the evolution of reaction norms. We have much more to gain by dealing with the complex realities of actual biological systems than by contemplating simplistic mathematical models.

CONCEPTUAL SUMMARY

1. Biologists recognize limits to the diversity of form, and believe that mutation alone will not allow unlimited access to other phenotypes. Such intuitions have become embodied in the term *constraint*. Unfortunately, so

many different forms of constraints have been invoked that the term has become almost meaning.

2. Process structuralists argue that innate properties of morphogenetic processes severely limit the potential forms that organisms can achieve, due to the self-organizing properties of the developmental system. Extension of the concepts of innate properties and self-organization leads to a substantial diminution of the roles of genes and, ultimately, to the role of natural selection in fostering phenotypic diversity. We believe these are extreme arguments.

3. We explicitly separate "constraints" arising due to genetic or epigenetic limitations from those produced by stabilizing selection. We retain the term *constraint* for the former, including the absence of appropriate genetic variation, pleiotropy, and epistatic effects. Other "constraints" (mechanical, functional, selective, ecological, and developmental) are in the majority of cases due to selection against deviations from a phenotype that works. We refer to these as selective forces, rejoining them with their already accepted compatriots, directional and disruptive selection. Phylogenetic and historical "constraints" are likely to be genetic in nature, but selection may be important here as well.

4. This categorization of constraint is for clarity and convenience. The processes responsible for phenotypic "stasis" are ongoing; for example, strong selective forces will eventually lead to a genetic constraint due to erosion of the genetic variability; and pleiotropy and epistasis may also be molded by selection.

5. The DRN perspective of constraint takes a hierarchical view. For example, patterns of invariance that appear, at higher levels, to be developmental constraints can be seen to be due to lack of appropriate genetic variation. This view also highlights the flux of processes that produce the snapshots of static patterns that we observe.

6. We discuss examples of constraint and selection on reaction norms. Costs and tradeoffs are commonly observed or inferred when organisms are raised in contrasting environments. Inter-environment genetic correlations may be useful here.

7. Null models are needed to serve as a basis for evaluation of the relative roles of constraint and selection. The null expectation should be one of change (a drift-based view) and not of stasis. The vectors of predicted and observed response to selection can be estimated in several ways, through experimentation and comparative analyses.

8. We add our concerns about too much phylogenetic correctness to those of previous workers; removing all phylogenetic signal is potentially far too conservative. We see the problem as akin to a two-way analysis of vari-

ance without replication: there is no way to estimate the importance of the interaction between the main effects. We desperately need a way to evaluate the extent of the interaction between ecological (selective) factors and the phylogenetic pattern.

9. In order to investigate selection and constraint on reaction norms, we need information on (1) the range and frequency of actual environments in which evolution occurs, and (2) estimates of the environment-specific selective pressures and constraints (including the multivariate relationships among traits). We highlight the work of Weis and Gorman in this respect.

CHAPTER 7

Phenotypic Integration

Much of the empirical work concerning phenotypic evolution has been done by focusing on one or a few traits, due to practical problems in handling large multivariate data sets. Evolutionary biologists do appreciate that phenotypes are complex ensembles of many parts, and that we need to understand how these parts are integrated and can change *inter se* during evolution. However, notwithstanding a long history of work on the relationships among traits, there is still disagreement on fundamental concepts. For example, the very definition of what constitutes a character has been open to much debate (see Gould and Lewontin, 1979). And, even if we agree on this issue, we are then faced with the problem of the homology of traits across related taxa (see

Chapter 5). And all of this before we can even begin to discuss ways to describe the relationships among characters and build a conceptual framework capable of yielding insights into their evolution. The key issue is, how do we study the development of separate traits and gain insight into the evolution of functional whole organisms?

In this chapter we consider how the phenotypes of organisms are integrated. Although "integration" generally connotes different phenomena to various authors, we attempt to provide a basis for unification of such terms as morphological, developmental, or plasticity integration by focusing on the commonalities underlying these terms. Foremost among these is the *necessity* that different parts of an organism must be coordinated (i.e., integrated) through development in order to yield a functional whole. Such integration has to be based on the properties of the genetic/epigenetic system specifying the developmental trajectory of each part and of the organism as a whole. In addition, the genetic and developmental components need to be attuned to the environmental variability the individual is likely to experience throughout its life. Our concept of phenotypic integration encompasses the genetic, structural, and physiological bases of the correlation and coordination of traits.

We briefly sketch some recent historical steps in the study of phenotypic integration, which take the form of distinctly different lines of research. We then discuss the advantages and limits of the prevailing conceptual framework of evolutionary quantitative genetics for the study of integration. The remaining sections of the chapter deal with the so far largely separate approaches to the study of phenotypic integration—classical and modern morphometrics; the developmental perspective, which is slowly regaining central stage in evolutionary theory; and the novel concept of plasticity integration.

THE HISTORY: FROM PHENOTYPIC GRIDS TO THE SIERRA NEVADA

What measurable characteristics are behind the intuitively obvious similarity between a human's and a chimpanzee's skulls? What justifies our perception that the crania of Eocene *Hyracotherium*, Oligocene *Mesohippus*, and Miocene *Protohippus* are very similar structures? How is it that we do not have any difficulty imagining intermediate structures linking the pelvis of *Archaeopteryx* with that of *Apatornis*, a toothed bird from the North American Cretaceous?

As pointed out by Coleman, Ghiselin, and Adams (in Mayr and Provine's *The Evolutionary Synthesis*, 1980), morphologists were either wary of or recalcitrant to embracing evolutionary thinking. Coleman (1980) argues that it wasn't until the 1960s that comparative anatomy clearly took an evolutionary perspective. Adams (1980) points out that the Russians Severtsov and Schmalhausen were among the exceptions. We will focus here primarily on those individuals that were interested in the issue of coordination or integration of the phenotype.

We recognize several approaches to understanding phenotypic integration. The first is epitomized by the non-evolutionary approach of D'Arcy Thompson. The second is the evolutionarily based tack represented by the separate development of concepts by Clausen, Keck, and Hiesey in their studies on environmental and genetic effects on character relationships in plants and by Miller and Olson in their work on vertebrate skeletal elements. A third approach, loosely allied to the second, is that of morphometrics, which we cover in a subsequent section.

The first modern biologist to tackle these and related problems concerning animal and plant form was D'Arcy Thompson, in *On Growth and Form* (Thompson, 1917). Thompson's beautiful diagrams of complex biological structures embedded in a reference grid that can be distorted to give rise to other similar forms still adorn modern biological textbooks (Figure 7.1). However, as pointed out by Peter Medawar (in J. T. Bonner's introduction to *On Growth and Form*, 1987), his influence on modern biological thought has been subtle and indirect (although at least one modern movement in morphometric analysis has been directly inspired by Thompson's work; see below).

The fundamental premise of *On Growth and Form* is that mathematics, physics, and chemistry can go a long way toward explaining a variety of biological phenomena. In this sense, Thompson's work is remarkably modern, in many ways anticipating by at least three decades the advent of contemporary reductionist biology. However, what Thompson considered an "explanation" is what a modern scientist would regard only as an interesting analogy. For example, his famous grid diagrams are based on the idea that mathematically defined transformations of coordinates "explain" how one biological structure can metamorphose into another. Yet, a modern developmental biologist would hardly be satisfied by an approach based entirely on description, with no role allowed for the exploration of how a developmental system would actually accomplish such a change in the coordinates of adult structures.

We see three problems with Thompson's work. First, even though he was well aware of Darwin's theory and did not deny some purifying role to natural selection, he did not think of it as the positive, creative force that most modern biologists recognize. Thompson thought that biological structures arise from the action of physical forces, quite independently of any selection imposed by environmental contingencies. In this sense, his reasoning comes close to the reemerging school of process structuralism (Resnik, 1994; see Chapter 6). Structuralists, or rational morphologists, maintain that there are general laws in biology governing the form of an organism, and that a primary goal of biology is to uncover these laws, in an effort to construct a non-historical biological science that mirrors the physical sciences.

Thompson's second limitation is that he did not mention the study of heredity or the role of mutations in his book. This omission reveals him to be clearly out of the mainstream of his day, which was dominated first by the rediscovery

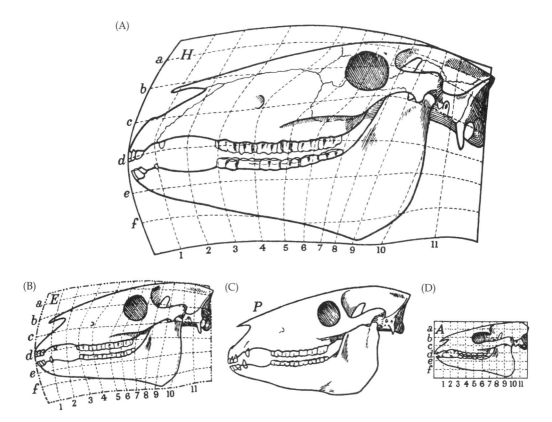

Figure 7.1   D'Arcy Thompson's coordinate transformation of the Eocene *Hyracotherium* (D) to the recent *Equus* (A), highlighting the relative changes in structures of the skull. (B) and (C) represent the transitional skull predicted by the transformations (B) and the actual skull of the Miocene *Protohippus* (C). The skulls are drawn to the same scale. (From Thompson, 1917.)

of Mendel's laws, and later by the attempt to unify genetics and evolutionary theory that gave rise to the neo-Darwinian synthesis. We view as a third conceptual oversight that concerning allometry, the differential (i.e., not directly proportional) growth of correlated structures. The theory of allometry was being developed during the time between the editions of Thompson's book (1917 and 1942), chiefly by J. S. Huxley (1924; Gould, 1966; Chapter 5). Since most of the changes described by his transformations of coordinates are, in fact, examples of allometry (albeit of a multivariate nature), his failure to appreciate the relevance of allometric theory is all the more surprising given the decidedly mathematical approach that he adopted in his work.

Despite these problems, Thompson's basic idea that complex morphologies can be connected in relatively simple ways has inspired generations of biologists seeking "missing links" in the evolution of baupläne. For example, when interpreted from an evolutionary vantage point, his diagrams of the skulls of different taxa representing fossil horses clearly show that transitions between some species were mostly a matter of change in size (very little grid distortion), while others implied marked changes in shape (Figure 7.1; compare D versus C and D versus A).

Opposed to Thompson's non-evolutionary approach was the body of work from the superlative ecological genetics research program of Jens Clausen, David Keck, and William Hiesey at the Carnegie Institution (Clausen et al., 1940; Clausen and Hiesey, 1960; see Chapter 2). Carrying out studies on the adaptation of plant species to local environmental conditions for more than two decades, they produced a body of work that is unsurpassed.

A central theme of Clausen and collaborators' research is that there are two opposing forces governing evolution: variation and coherence. Variation remains the focus of the overwhelming majority of studies in evolutionary biology. Then, as well as now, one of the questions that received the most theoretical and experimental attention was: How is genetic variability maintained in natural populations? Throughout the twentieth century, schools of thought have alternated or fought each other on the matters of the amount and meaning of genetic variation—from T. H. Morgan and the mutationists to Dobzhansky and the discovery of widespread variation in natural populations (Mayr, 1980); from the advent of modern molecular population genetics (Lewontin and Hubby, 1966) to the neutral theory of molecular evolution (Kimura, 1983). Clausen, Keck, and Hiesey pointed out that there is a second major force governing evolution, which they termed "coherence" of characters (Clausen and Hiesey, 1960; in our parlance, integration).

In the view of the Carnegie group, character coherence is due to the balanced action of genes (i.e., coadaptation) controlling the correlations among traits. Since genes affecting the same character may reside on different chromosomes, and genes controlling different traits may be linked on the same physical area of the genome, the degree of correlation among parts of the phenotype should be evolutionarily pliable. According to their pre-"post-modern" view, selection does not operate on single genes, but on gene complexes involved in the control of character correlations. Clausen, Keck, and Hiesey, therefore, considered much of the work done in classical genetics on the effects of major and rare mutations affecting visible traits to be of minor importance for the study of evolution. Their focus was on the tension between variation for character relationships and a stabilizing selective force that tends to maintain the coherence of the whole phenotype.

Experimental work carried out by these researchers centered on the change in character correlations during the evolution of members of several genera of plants, including *Achillea, Viola,* and *Potentilla* (see also Grant, 1979). CKH believed that the investigation of hybrids between closely related species, or even between local races, could provide invaluable information about the relative roles of variability and coherence in shaping the identity of a taxon. Figure 7.2 shows an example of their work in *Potentilla*. The diagram reveals the relationships between fourteen characters of the F$_2$ progeny of a cross between a foothill and a subalpine population. Many of these correlations were different when compared with those obtained from the progeny of a cross between a coastal and an alpine population. These changes were attributed to the genes that segregate in the different crosses. Therefore, gene flow between populations living in different habitats could create or break down different patterns of coherence, allowing trait correlations themselves to evolve.[1]

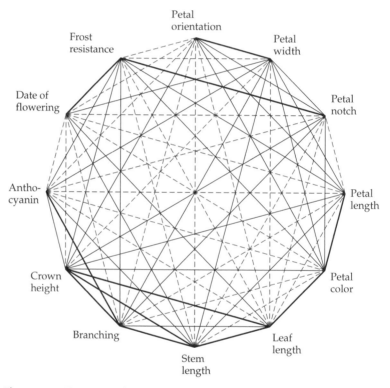

Figure 7.2 Diagram of the phenotypic correlations among characters for traits in the F$_2$ offspring of a cross between the Timberline and Oak Grove races of *Potentilla glandulosa*. Solid lines indicate significant correlations, dashed lines are nonsignificant. (From Clausen and Hiesey, 1958.)

---

[1] This focus on the evolutionarily relevant information provided by hybrids is all the more significant in light of the recent resurgence of interest in hybrid zones as much more than "evolutionary dead ends." Today, hybrids are studied because of insights that they can yield on gene interactions, ecological breadth, relevance to taxonomic theory and practice, and even in applied biology as models of the evolution of new weeds (Stace, 1987; McDade, 1992; Cruzan and Arnold, 1993; Bullini, 1994; Raybould and Gray, 1994; Arnold, 1995; Rieseberg et al., 1995).

Ultimately, according to Clausen and collaborators, the very existence of distinct taxa during the history of life on Earth is due to the necessity of maintaining coherence within complex phenotypes. This excursion into macroevolutionary biology is echoed by much of the more recent work by Gould on the existence of sets of covarying traits and their role in explaining the phenomenon of evolutionary stasis (although Gould's emphasis is on developmental constraints and non-adaptation; Gould, 1980b, 1984, 1989; Gould and Eldredge, 1993).

Another researcher who worked along lines similar to those of Clausen, Keck, and Hiesey was R. L. Berg. Berg's 1960 paper explored, for nineteen species of herbaceous plants, the relationships between correlation structure, life histories, and mating systems. In several cases she found evidence of two or more distinct groups of correlations—traits presumed to be functionally related to each other occurred in the same correlation "pleiade," while traits hypothesized to be functionally independent fell into different pleiades. The best example is that of *Chamaenerium* (= *Epilobium*) *angustifolium* (fireweed). Berg showed that all vegetative characters clustered in one correlation group, while all reproductive traits were in another. Although such separation is logical, it is by no means universal (as is demonstrated by Berg's own work on other plants). Overall, Berg concluded that twelve of the taxa showed distinct groups of correlations. In all cases, the reproductive structures (flower shape in particular) had become independent from the other characters and had evolved as a coherent unit. Each of these twelve species had rather specialized features, such as zygomorphic (i.e., not radially symmetrical) flowers, tubular parts, unusual positioning of the flower on the inflorescence, etc. Recent studies have picked up on these themes as well (Wagner, 1990; Pigliucci et al., 1991; Waitt and Levin, 1993; Conner and Sterling, 1995).

In the early 1950s Everett Olson and Robert Miller, motivated by interests in the evolution of form in fossil organisms, began to examine the intercorrelations among large sets of traits. Their goal was to consider as much information as possible in order to interpret patterns of evolution. Their conceptual views and a variety of illustrative analyses were presented in their book *Morphological Integration* (Olson and Miller, 1958). In this volume, they argued forcefully for a multivariate perspective, with the theoretical goal of characterizing the whole organism (what they referred to as "the total animal"). The necessity for this stems from the fact that organisms cannot be considered as sets of disconnected traits. This implies that an understanding of phenotypic evolution requires insight into how traits are integrated with one another, and how different trait groupings are integrated into a whole.

They proposed that correlations (ρ) among parts of organisms are determined by the function of the trait and its developmental history

$$\rho = \alpha \,(\text{function}) + \beta \,(\text{development}) + \gamma \,(\text{residual}) \quad (\text{Equation 7.1})$$

where α, β, and γ represent the proportional contributions of the components to the correlation. In their analyses, they formulated *a priori* hypotheses about groups of functionally or developmentally related traits (F-groups), and compared those with the groups of traits that are statistically correlated (ρ-groups). Intersections between these two groups yield ρF-groups; these are of the most evolutionary interest. Olson and Miller introduced an index of morphological integration that took into account both the ratio of significant correlations to the total possible and the number of different groups of correlated traits. Although their statistical analyses were crude by modern standards, extending only to partial correlations, their insights were keen; and along with the methods promulgated by the early numerical taxonomists, they paved the way for future multivariate thinking and analysis.

This brief survey shows that several types of phenotypic integration have been investigated since the beginning of the twentieth century.[2] Yet, it has so far played only a secondary role in evolutionary biology, either as an indirect source of inspiration or as a recurring nightmare of complexity, depending on the viewpoint of the researcher. During the last two decades, however, more and more students of evolution have focused their attention on this rather thorny subject, directly or indirectly building on the ideas summarized above.

Still, current studies on phenotypic integration are the result of somewhat independently evolving research agendas that reflect the specific interests of the persons involved. The conceptual glue tenuously binding these disparate approaches is that supplied by evolutionary quantitative genetics. In most of the rest of this chapter we will discuss this current situation and some examples from the primary literature. Later, we will argue that it is time for a conceptual as well as an experimental unification of these approaches in order to make progress on some of the very same issues that plagued the minds of D'Arcy Thompson, Olson and Miller, Clausen, Keck, and Hiesey, and Berg, as well as dozens of other biologists throughout the twentieth century.

A CONCEPTUAL FRAMEWORK: EVOLUTIONARY QUANTITATIVE GENETICS

Most current approaches to the study of phenotypic integration are based on extensions of classical quantitative genetic theory (Falconer, 1989), known as evolutionary quantitative genetics (Hartl and Clark, 1989; Roff, 1997). It is an inherently statistical treatment of complex data sets, whose limitations are the same that apply to standard quantitative genetics (see Chapter 1; Turelli, 1988; Barton and Turelli, 1989; Pigliucci and Schlichting, 1997). The basic goals of quantitative genetics as applied to the study of multivariate phenotypes are: (1) to describe the variation in character correlations in terms of genetic and environmental components; (2) to use this description to either make predictions

---

[2]See Sultan (1992) for a discussion of the related debate between holists and reductionists.

about future evolutionary trajectories of the system, or to infer past selective pressures. The fundamental tool of this approach is the genetic variance/covariance matrix (**G**), used for its ability to describe genetic tradeoffs and constraints.[3]

A correlation between any two variables X and Y is the ratio between their covariance and the product of the two standard deviations. The genetic correlation, r_G, measures this relationship for covariances and variances of the additive genetic effects

$$r_G = \frac{\text{cov}_{G(XY)}}{\sqrt{\sigma_{G(X)}\sigma_{G(Y)}}} \qquad \text{(Equation 7.2)}$$

If we adopt the standard quantitative genetic distinction between environmental and genetic components of phenotypic variation, then the observed phenotypic correlation between two characters becomes

$$r_P = h_X h_Y r_G + e_X e_Y r_e \qquad \text{(Equation 7.3)}$$

where h is the square root of the heritability, e represents the environmental effects ($e^2 = 1 - h^2$), and r_e is the correlation of environmental effects (Falconer, 1989). Comparing these equations, we can see that the genetic correlation between two traits depends on the heritability of the two traits and on the magnitude of environmental influences, quantities that eventually determine the actual variances and covariances of interest. The **G** matrix, then, is simply a collection of the variances and covariances that statistically describes the pairwise relationships among traits summarizing the phenotype, behavior, or life history of an organism.

But what exactly does a genetic correlation mean in biological terms? The typical view is that there are two genetic phenomena that can underlie a correlation between two characters: linkage disequilibrium and pleiotropy (Falconer, 1989). In the case of linkage, different genes affect the two traits, but they are physically near each other on a chromosome. As a consequence, alleles at these loci tend to segregate together because of reduced recombination, which generates the impression that the two characters share a similar genetic basis. Pleiotropy, on the other hand, occurs when a single gene affects more than one trait, and the magnitude of the genetic correlation should reflect the degree to which the traits are controlled by the same genes.

There is a third fundamental variable influencing genetic correlations, one usually implied but rarely mentioned in empirical or theoretical studies. Genetic

---

[3] In the following, we use "constraint" in the sense meant by quantitative geneticists, i.e., as a restriction on possible evolutionary trajectories. See Chapter 6 for a full discussion of the various meanings of constraint and for the limitations of the concept itself.

variances and covariances (and therefore correlations)—as population parameters—also reflect the underlying allele frequencies. If the allele frequencies change (because of migration, mutation, selection, or drift) then the genetic correlation is bound to change (Mitchell-Olds and Shaw, 1987; Shaw et al., 1995a). It is easy to see this in the simplest case. If the alleles for the set of loci affecting a pair of traits go to fixation in a population because of genetic drift, then the heritabilities of those traits will go to zero and the genetic correlation would be undefined (even though the two traits are obviously still under genetic control and are in fact perfectly coupled).[4] As we will discuss in the following sections, there are multiple ways to achieve most values for a genetic correlation, in fact so many that we have pointed out that the concept itself might have limited applications in evolutionary theory (Schlichting and Pigliucci, 1995a; Pigliucci, 1996b; Pigliucci and Schlichting, 1997).

The inherent theoretical as well as empirical interest in **G** as a descriptor of phenotypic integration lies in the assumption that a significant genetic correlation between two traits is an indication of a constraint on the future response to selection (Falconer, 1952; Cheverud, 1984, 1988; Via and Lande, 1985). Simply put, if two traits have a positive genetic correlation, then selection for an increased value of the first character should automatically select for an increase in the value of the correlated one. Alternatively, if the two traits are negatively correlated, then this represents a tradeoff, in which instance selection that increases the value of one trait should lead to a simultaneous decrease in the value of the second one. Of course, tradeoffs and constraints (in the sense of a genetic correlation) can actually be *favored* by selection; for example, if the optimal phenotype can be achieved by an increase in both trait values, or by the increase of one and the decrease of the other. In other words, constraints are *context-dependent*. However, the presence of a significant correlation is a constraint in the sense that, in response to selection, the population will move only (or more readily) in certain directions, as opposed to a situation in which the two traits are uncorrelated, allowing any combination of responses by the two characters.

An elegant example of the use of genetic correlations to predict response to selection is the work of Cheverud et al. (1983) on the heritability of tail length and body weight in rats. They measured quantitative genetic parameters (genetic variances and covariances) that describe the ontogeny of these two traits in a laboratory population. They considered the same trait expressed at different times as two genetically correlated traits, a device first proposed by Falconer (1952) for a trait measured in two environmental conditions. Cheverud et al. found evidence for the existence of three classes of gene effects: (1) genes that cause an increase or decrease in character value throughout the ontogeny; (2)

---

[4]This once again underscores the fact that genetic correlations are attributes of *populations*, not individuals.

genes that are associated with an increase early in ontogeny and a decrease later (or *vice versa*); (3) genes that have an influence only during very short periods of time during development. Cheverud and coworkers then used the estimated genetic variances and covariances to simulate response to selection for increased weight or tail length (Figure 7.3). They found that selection at different times would alter either the height or the shape (or both) of the ontogenetic trajectory, consistent with the activity of different classes of genes.

Despite the intuitive appeal of genetic correlations for the quantification of constraints highlighted by Cheverud's work, there are theoretical as well as empirical issues that impinge on their applicability. Two basic questions arise in the use of genetic correlations in evolutionary studies. First, are tradeoffs and negative correlations actually observed? Second, what do genetic correlations actually measure, and therefore, to what extent can we rely on their estimation in order to explain and predict multivariate organismal evolution? As we shall see, the two questions are actually tightly connected, and the answers depend on both the careful evaluation of empirical data and the particular framework that we adopt in order to interpret those data.

Because real organisms have access to only a limited supply of resources, they must "decide" how to invest those resources in order to maximize lifetime fitness, choosing alternative strategies such as early versus late fecundity, vegeta-

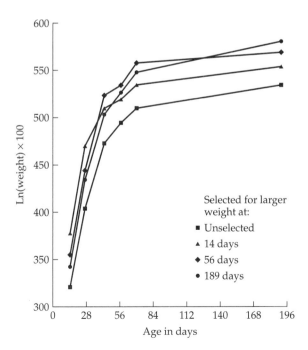

Selected for larger weight at:

- ■ Unselected
- ▲ 14 days
- ◆ 56 days
- ● 189 days

Figure 7.3 Ancestral growth curves and predicted curves resulting from the response to selection at three different ages for increased weight in rats. Notice that selection at age 189 results in an alteration of both the height and shape of the curve, while selection at ages 14 and 56 only shifts the height of the curve. (From Cheverud, 1983.)

tive versus sexual reproduction, survival versus reproduction, and so on. A single individual cannot maximize all of these components of fitness, precisely because its resources are limited; and the decision on how to optimally allocate these resources depends on the life history and habitat typical of the organism in question. For example, because an annual plant does not have a choice of vegetative versus sexual reproduction, we would predict that it should invest most of its resources to maximize flower and fruit production, and therefore seed set. On the other hand, a perennial plant might have a wider spectrum of opportunities and, depending on environmental conditions, may "choose" to delay sexual reproduction to the next season in order to enhance the probability of survival or vegetative propagation. Despite these considerations, life history theory predictions about the ubiquity of tradeoffs are rarely confirmed in empirical studies (see Roff, 1992, and Stearns, 1993, for reviews).

Van Noordwijk and de Jong (1986) proposed a simple explanation for the apparent incongruence between the general expectation of tradeoffs and the widely observed positive correlations among fitness components in real organisms. According to these authors, there are two phases of the process: resource acquisition and resource allocation. Much theoretical debate up to that point focused on allocation, with the consequent expectation of visible tradeoffs. But, van Noordwijk and de Jong argued, the observable correlation between two traits actually depends also on the efficiency of acquisition. If there are plenty of resources because of very efficient acquisition (or benign environmental conditions), the tradeoff can be masked because the value of both traits (say, early and late fecundity) will increase. This model was framed in terms of optimization theory, with no reference to the genetic component underlying the tradeoff (included later: de Jong and van Noordwijk, 1992).

Along similar lines, but from a strictly quantitative genetic standpoint, Houle also recognized the importance of *functional architecture*, and reached generally similar conclusions (Houle, 1991). Let us assume that the phenotypic traits z_1 and z_2 represent different life history strategies, and that allocation of resources to these strategies is determined by a group of loci, P. The pool of resources itself is provided through the action of a second group of loci, R (Figure 7.4). The P loci will tend to build negative covariances between z_1 and z_2, because they will partition whatever resources are available. However, the R loci will tend to build positive covariances, simply because the more resources are made available, the more *both* traits will benefit. The observed correlation will therefore represent a balance between the contributions of both types of loci, possibly masking the presence of a tradeoff (for example, if there are a lot more R than P loci). Houle's investigation has therefore placed the crucial question dead center in the debate: what *do* genetic correlations actually measure? (See Chapter 3.) This bears directly on how informative they are, and therefore on how much of a role they can play in evolutionary predictions (Pigliucci, 1996b). A series of simulations by Gromko (1995) expanded on Houle's work

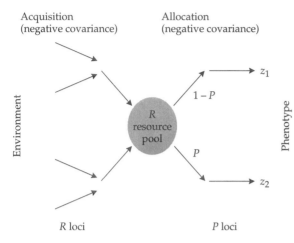

Acquisition
(negative covariance)

Allocation
(negative covariance)

Environment

R resource pool

$1 - P$

P

z_1

z_2

Phenotype

R loci

P loci

Figure 7.4 Dependence of the genetic correlation between two traits, z_1 and z_2, on the underlying genetic architecture. The R loci determine acquisition of resources that can be channeled to both traits, thereby contributing positively to the genetic correlation. The P loci partition resources between the two traits, generating a negative correlation. The observed correlation depends on the number of as well as action of the P and R loci. (From Houle, 1991.)

by showing that several combinations of pleiotropic effects yield the same value of the genetic correlation, and that for a given value of the correlation, these differences in pleiotropy can markedly alter the realized correlated response to selection. Gromko also showed that the significance of a genetic correlation does not necessarily correspond to the existence of a constraint, and that substantial constraints on future evolution can be present even in cases of no genetic correlation (i.e., $r_G = 0$).

The work of Biere (1995) represents an excellent example of questions about what underlies functional architecture. He looked at the tradeoff between survival and fecundity in the perennial hay-meadow species *Lychnis flos-cuculi*, growing cloned individuals at four sites along a nutrient gradient. He found significant variation in allocation among clones, but this reflected differences in acquisition of resources, not partitioning. Genetically based tradeoffs were observed only at the less productive sites, thereby implying a dependence of the functional architecture on the particular environmental conditions (Figure 7.5). Results of artificial selection experiments in *Plantago lanceolata* by van Tienderen and van Hinsberg (1996) echo this environment dependence: the correlated responses to selection on leaf length depend on whether selection has been carried out under high or low red/far red light conditions.

This and similar empirical results lead to two more fundamental questions related to genetic correlations and phenotypic integration: What are the mechanistic bases of a correlation between two traits, and how does such a correlation evolve? Some light has been shed on both questions by the elegant work of Clark on the physiological basis of the genetic correlations for energy storage traits in *Drosophila*. He reckoned that the correlations between phenotypic traits such as live weight, triacylglycerol (lipid) content, and glycogen content must

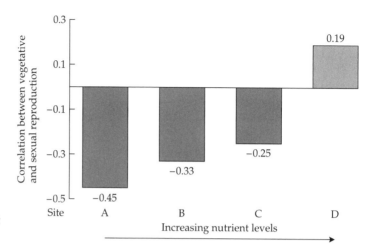

Figure 7.5 Environment-dependent tradeoff between vegetative and sexual reproduction in *Lychnis flos-cuculi* in four sites (A–D) characterized by differences in nutrient levels (Biere, 1995).

depend on the correlations between the activities of enzymes involved in the lipid and glycogen accumulation pathways. He therefore calculated the genetic correlations between the activities of several enzymes for flies grown with standard and sucrose media (Clark, 1990; Figure 7.6). The patterns of correlations between enzymes were dramatically altered by the exposure to the different

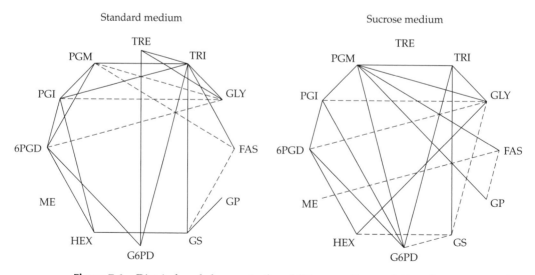

Figure 7.6 Diet-induced changes in the additive genetic correlations for enzyme activities underlying energy storage in *Drosophila melanogaster*. Solid lines indicate positive correlations, dashed lines indicate negative correlations. (From Clark, 1990.)

environments. Some changed from positive to nonsignificant (for example, the one between TRE and G6PD), others from negative to nonsignificant (FAS↔GS), and some even switched sign (HEX↔GS).

Clark and Wang (1994) looked at the evolution of these correlations at the species level within the genus *Drosophila*, using a phylogenetic framework. They obtained a phylogenetic tree by pooling sources of molecular data separate from the enzymes studied in the correlation analyses (Figure 7.7A), and examined the correlated changes in activity of enzyme pairs. For example, changes in activity

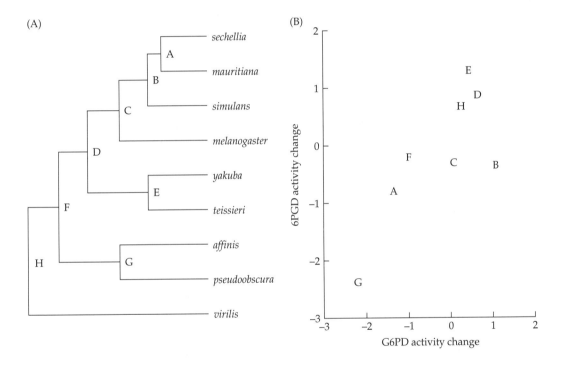

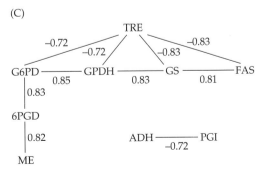

Figure 7.7 Comparative analysis of enzyme activity evolution in *Drosophila*. (A) UPGMA molecular phylogeny of nine species. (B) Example of the comparative analysis of pairwise enzyme evolution for G6PD and 6PGD. Data points are the nodes from A, not the species means themselves. (C) Result of all pairwise comparisons, showing those pairs of enzymes whose activities are evolutionarily correlated. (From Clark and Wang, 1994.)

of 6PGD from one species to another were significantly correlated with changes in the activity of G6PD (Figure 7.7B), and this correlation remained significant even after adjusting for the phylogenetic relatedness of the taxa (Figure 7.7C) using Felsenstein's method of independent comparisons (Felsenstein, 1985, 1988). This suggests that enzyme activities are genetically correlated, independent of their phylogenetic history. Clark and Wang also showed that enzymes that are closely connected metabolically tend to change activity in synchrony during evolutionary time, while enzymes that are more distantly related metabolically vary more freely (i.e., their activities are less "constrained" with respect to each other).

What is emerging is an awareness that genetic correlations cannot be considered as static representations of the relationships between traits, or at face value as confirmation of constraints. This is because: (1) the magnitude of a genetic correlation can be altered by various factors such as changes in allele frequencies, degree of linkage or pleiotropy, developmental stage, or environmental conditions; (2) the observable sign and magnitude of a genetic correlation do not necessarily reflect any underlying tradeoffs because they depend on the precise, and usually unknown, genetic architecture of the traits in question (Leroi et al., 1994a); (3) genetic correlations can vary even among closely related species (e.g., Merila et al., 1994; Paulsen, 1994). This lability should not be taken to indicate that the correlations are uninteresting. As shown by the work of Clark and colleagues, it is possible to derive substantial information on the mechanistic basis of genetic correlations, as well as to use comparative analyses to gain insights into both the physiology and molecular biology of character correlations and into the patterns of macroevolutionary change in the correlations themselves. In fact, as we will argue later, the integration of traits and the lability of these patterns of integration should be important targets of selection, and therefore of primary interest to evolutionary biologists.

MORPHOMETRICS: MEASURING PHENOTYPIC INTEGRATION "CLASSICALLY AND REVOLUTIONARILY"

There are currently two main approaches to the study of multivariate phenotypes, the so-called "classical" and "new" morphometrics. We will briefly examine the theory and practice of both. Classical morphometrics is based on an array of tools loosely related to multivariate statistical techniques, such as principal components or factor analysis, but also including hierarchical techniques, such as path analysis, and nonparametric ones, such as multidimensional scaling or correspondence analysis (Sneath and Sokal, 1973; Dillon and Goldstein, 1984). What these approaches have in common is the attempt to reduce the multidimensionality and complexity of a data set to a smaller number of independent axes that can be visualized and comprehended by the human brain. That is, the *main* reason to adopt any of these techniques is simply because we cannot deal with information if the number of dimensions exceeds three (James and McCulloch, 1990).

The basic logic underlying multivariate analysis is as follows: for the simplest, bivariate case, we are measuring two variables and are interested in simplifying our understanding of the system, or in finding out if the two variables actually belong to a single "set of covariation." The ellipsis in Figure 7.8 represents the distribution of the actual data. A simple 45° rotation of the original axes brings about a situation in which one axis (the first principal component) explains most of the variation in the data, while a second axis perpendicular to the first one (the second principal component) accounts for only a small fraction of the variation in the sample, and can therefore be provisionally ignored. The dimensionality of the system has therefore been effectively cut in half. This same technique can be extended to any number of variables, with the intention of summarizing the most information with the fewest principal components. The *eigenvalues* associated with each principal component are a measure of the percentage of variance explained by that component, and are used by the investigator to determine how many components need to be retained to avoid the loss of too much information. The *eigenvectors* associated with each component provide the means for interpreting the biological significance of each component by indicating the strength of the correlation (or weight) of each original variable with that component. For example, if the first principal component is associated with number of flowers, fruits, and reproductive branches, but not with number and size of leaves, we can "interpret" that component as summarizing information pertinent to sexual reproduction. The second component might then describe the variation due to traits associated with vegetative growth.

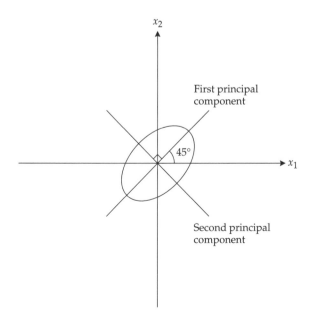

Figure 7.8   The principle of principal components analysis. x_1 and x_2 are two traits. Individuals in the population are distributed within the ellipse. A simple rotation of the axis allows the identification of a major principal component that summarizes most of the information originally yielded by two variables.

As an example of a biological application of multivariate analysis, let us consider the study of geographic variation in the land snail *Cerion*. Gould (1984) measured 19 characters in 135 samples of 20 snails each from Aruba, Bonaire, and Curaçao. He found that the second and fifth axes were particularly effective at separating populations from Bonaire from those in Aruba (Figure 7.9). The corresponding eigenvectors indicated that factor 2 was strongly associated with the number of ribs on the second and sixth whorls, and rib density on the first whorl. Factor 5 was correlated with shell-suture distance, aperture tilt, shell width, and umbilicus width. Furthermore, shell-suture distance and aperture tilt were *positively* correlated with factor 5 (a higher score on a factor corresponded to higher values of the trait), while the other two characters were *negatively* correlated with this factor (and therefore with the first two traits). The separation along axis 5 allowed Gould to identify this set of four variables describing geographic differentiation of shell features.

Gould interpreted his findings as evidence of nonadaptive variation in these organisms. He argued that even if selection were to favor an increase or decrease in the value of one character, the allometric intercorrelations would "drag" several other characters along. Even though Gould's particular conclusions can be debated (and probably should be; for example, there are no data on the possible selective pressures acting on these traits, at either the univariate or the multivariate level), the method is clearly powerful, and it has been widely used for all sorts of problems in biology, from evolutionary to taxonomic ones (Sneath and Sokal, 1973; James and McCulloch, 1990).

A major problem with principal components and related techniques, however, is that the derived variables (i.e., the "components" or "factors") are mathe-

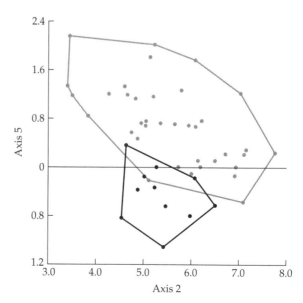

Figure 7.9 Plot of population samples of *Cerion* against the 2nd and 5th axes from a factor analysis of shell characteristics. The polygons represent a group of samples from Aruba (black) and Bonaire (gray), separated almost completely by this combination of variables. (From Gould, 1984.)

matical artifacts, which the investigator hopes can be more or less directly related to the original variables through careful inspection of the eigenvectors (James and McCulloch, 1990). A single analysis including several groups may be measuring an artificial covariance structure, while separate analyses of the taxa can result in dramatic differences in the composition of the eigenvectors, and hence the "meaning" of the components. This lack of correspondence among data sets makes it particularly difficult to treat the components as biologically homologous entities, so that it becomes arduous to do ecological or comparative studies using them. This problem, coupled with the fascination with D'Arcy Thompson's transformation grids, eventually led to the establishment of the new morphometrics in the late 1980s and early 1990s.

The foundations for the new morphometrics were firmly established with the publication of a series of books during the 1990s (e.g., Rohlf and Bookstein, 1990; Bookstein, 1991; Marcus et al., 1996). The basic idea is simple, even though the mathematical tools necessary to realize it are anything but elementary. As in D'Arcy Thompson's transformation grids, a given structure of interest (say, the skull of a mammal; see Figure 7.10) is scanned for suitable landmarks or positions on the structure that can easily be identified in different specimens, possibly even in distinct species. The goal of the analysis is then to trace how the relative positions of these landmarks change through ontogeny (for a single individual) or phylogeny (when comparing different populations or species); that is, to track how the shape of the structure changes through developmental or evolutionary time.

The new method is based on three major features (Rohlf and Marcus, 1993). (1) The raw data are the coordinates of the landmark points in two or, more rarely, three dimensions (as opposed to "characters" defined by the researcher). (2) The geometric relationship among the coordinates is summarized by fitting an appropriate nonparametric function to the data (typically thin-plate splines; see Brodie III et al., 1995); the parameters describing the function can then be analyzed with standard multivariate statistics. A main advantage of the thin-plate spline transformation is that changes in the relative positions of the landmarks can be described as *affine* and *non-affine*. Affine changes influence all features of the structure at hand in the same way (i.e., they are size changes), while non-affine alterations represent local modifications or deformations (i.e., they are shape changes). This brilliantly solves one of the oldest problems in morphometrics, the often blurred distinction between changes in size and shape (Chapter 4). Furthermore, non-affine transformations can occur at different scales, for example, the shape of the whole skull or of a particular part of it; and each transformation can be summarized by *principal warps* (mathematical entities equivalent to the sequential powers of a polynomial curve fitted in a Fourier analysis). Warps are orthogonal to each other, and therefore describe independent types of changes of shape. (3) In traditional multivariate analyses, a researcher was forced to measure as many variables as possible with no guidance about which would then turn out to be of biological importance, and would be left with derived mathematical constructs of uncertain biological inter-

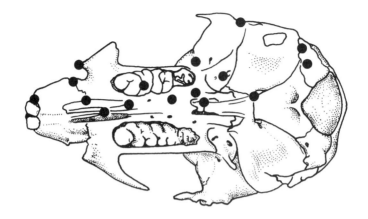

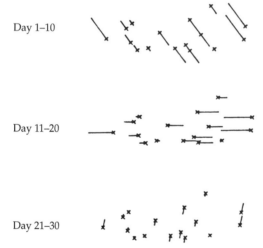

Day 1–10

Day 11–20

Day 21–30

Figure 7.10 Diagram of the skull of *Sigmodon fulviventer*. Dots represent landmarks used in warp analysis. The lower diagrams show changes in the first principal warp for these landmarks during the first three 10-day periods of development. For example, the general direction of movement of the landmarks (orientation of the vectors) changes abruptly between the first and the second period, while the intensity of the change (length of the vectors) decreases dramatically between the second and third periods. (After Zelditch et al., 1992.)

pretation. The statistics employed by the new morphometricians identify those combinations of landmarks that vary globally or locally during ontogeny or phylogeny, thereby identifying integrated biological units and their "natural" range of variation. The scientist concentrates on those combinations of landmarks that appear most interesting for the problem at hand, and will (theoretically) have less trouble identifying homologous reference points across a wide variety of specimens.

A typical example of the use of these methods is the work of Zelditch and collaborators (1992) on the ontogeny of phenotypic integration in the skull of the cotton rat, *Sigmodon fulviventer*. They identified 16 landmarks and described the affine and non-affine changes from day 1 to day 30 during development.

The first principal warp (i.e., that corresponding to the largest scale of geometrical changes) is diagrammed for three intervals (day 1–10, 11–20, and 21–30) in Figure 7.10. The vectors associated with each landmark represent the direction and intensity of distortion of each position during each 10-day period. One can see that some parts of the skull change in similar directions (identifying biologically integrated units, or "characters"), while others—even those physically adjacent—hardly change at all. Also, it is clear that the direction as well as the intensity of the changes are themselves altered during ontogeny, thereby identifying discrete phases of the development of the skull. Notably, the integrated units of development uncovered by this method do not correspond to the "characters" used in classical developmental studies of the cotton rat.

The tools offered by the new morphometrics are certainly very powerful, and we will return to the example of the cotton rat when discussing developmental integration. However, the limitation of these methods is that they deal only with actual physical structures, such as skulls or leaves. If one wants to investigate the covariation of life history traits, or of a collection of substructures from different parts of the body, it is impossible to define any coherent set of landmarks, and one is forced to rely on more traditional analyses. On the other hand, the variables used in new morphometric studies could be used in calculations of heritabilities, variance/covariance matrices, and selection coefficients on the changes of the landmarks. This could lead to a merger of D'Arcy Thompson-like methods with modern quantitative genetics. So far, however, there have been no attempts in this direction.

THE ONTOGENETIC COMPONENT: DEVELOPMENTAL INTEGRATION

Notwithstanding the impressive progress of mechanistically oriented approaches to the study of development, most evolutionarily oriented developmental genetics is still framed in the statistical terms of quantitative genetics. In part this is due to the fact that substituting a mechanistic for a statistical representation requires molecular studies encompassing many more taxa than those likely to be carried out in the near future. Quantitative geneticists have recently attempted to explicitly include development in models of phenotypic integration (Atchley and Hall, 1991; Cowley and Atchley, 1992). Atchley and Hall suggested that a phenotypic correlation between two traits can arise not just because of the usual genetic effects, but from environmental and epigenetic sources as well. They provided a conceptually interesting explanation of the distinction between pleiotropic and epigenetic effects (Figure 7.11). Suppose that the expression of a gene (A) affects several traits in a pleiotropic manner, while the simultaneous expression of another gene (B) affects another set of characters. Further suppose that the two sets of traits represent the differentiation of two cell types, and one cell type can start its genetically controlled differentiation process only after tissue/tissue-mediated induction from the other cell type.

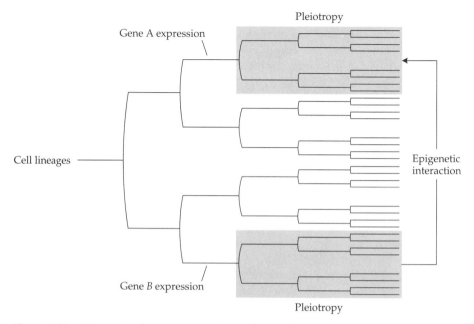

Figure 7.11   Distinction between genetic correlation due to pleiotropy or to epigenetic effects. Genes *A* and *B* have direct effects on eight cell lineages (each shaded region). The indirect interaction of genes *A* and *B* occurs through tissue-tissue communication. Both contribute to genetic correlation. (After Atchley and Hall, 1991.)

Then the characters associated with the induced cell type would actually be influenced by *both* sets of genes, one directly through pleiotropy and the other indirectly through epigenetic interactions. Both effects will show up in genetic correlations between the two character groups, providing yet another example of the potential problems in interpreting these correlations.

   As much as an increased awareness of development is vital to evolutionary theory and practice, there are two major limitations to incorporating it into a statistical approach. First, environmental effects are treated as non-genetic, while they clearly can affect genetic correlations in a way similar to epigenetic effects; the literature on the genetic basis of plasticity is now well developed (see Chapters 3 and 9). Second, the approach proposed by Atchley and Hall expands the classical quantitative genetic partition of phenotypic variation into genetic, environmental, and interaction effects,

$$V_P = V_G + V_E + 2\text{cov}_{G*E} \qquad \text{(Equation 7.4)}$$

to something like

$$V_P = V_{G(t)} + V_{\varepsilon(t)} + V_{M(t)} + V_{E(t)} + 2\text{cov}_{G\varepsilon} + 2\text{cov}_{GM} + 2\text{cov}_{\varepsilon M} \qquad \text{(Equation 7.5)}$$

where ε represents epigenetic effects, M stands for maternal effects, and the ts identify different times during ontogeny (this equation should also include more covariance terms for the genetically based environmental components). The complexity and size of the experiment necessary to adequately estimate all of these terms would be gargantuan. Furthermore, the value of such estimates is unclear, given the difficulty with which any mechanistic interpretation could be assigned to the above terms (under the best scenario, we would be able to say that, yes, there *are* maternal or epigenetic effects).

Theoretical limitations aside, there is good empirical evidence that phenotypic integration changes through development. Atchley (1984), for example, reviewed a series of studies on the genetic correlation between body weight and tail length during the ontogeny of mice and rats. In rats, there is a marked difference in the trajectory of the correlation between males and females. In the case of males, the correlation starts out and remains high and positive, justifying the conclusion of a strong constraint relating the two characters, independent of ontogeny. This is not the case for females: the correlation again starts high, but it drops rapidly within the first 40 days, and ends up slightly negative by the time the adult stage is reached. In this second case, the existence of a "constraint" is age-dependent.

An even more dramatic example of age-dependent genetic correlations is provided by Roach's (1986) study of the plant *Geranium carolinianum*. She measured the genetic correlation between fecundity and total weight (a measure of plant size) in three populations at three stages of the life cycle: early juvenile, late juvenile, and adult (i.e., reproductive). The results were consistent across populations, yielding a complex picture in which the correlation between the two components of fitness was negative early, nonsignificant before the adult stage, and finally became positive at reproduction (Figure 7.12). A study conducted only at the adult stage would have concluded that there are no tradeoffs, despite the strong negative correlation detectable at the very early juvenile stage.

What are the implications of these findings? If the genetic correlation can inform us about the ability of a population to respond to selection, and this correlation is labile during development, then it becomes crucial to know not only the time course of the correlation, but also when selection is operating. Although most studies and theoretical models assume that selection is on adult, reproductively capable individuals, that is clearly not the case. Selection can act at different phases during the life cycle (Clegg and Allard, 1973; Christiansen and Frydenberg, 1973; Searcy, 1992; Bell, 1997), and the response of the population can vary markedly according to the pattern and degree of expression of genetic constraints. It thus becomes an important empirical matter to judge how often reversals of genetic correlations occur, and what their likely effect on the evolutionary trajectory of a given species or population will be (Turelli, 1988). As we shall see, exactly the same reasoning applies to the environmental lability of character correlations.

Not all studies of developmental integration are conceived in quantitative genetic terms. A small but increasing number of researchers are using the tools

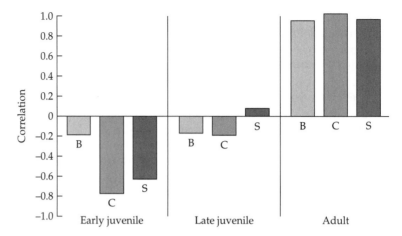

Figure 7.12 Changes in genetic correlations (of family means) between fecundity and size for three populations (B, C, S) of *Geranium carolinianum* at three ontogenetic stages. (From Roach, 1986.)

of the new morphometrics discussed above to approach similar problems. Zelditch et al. (1993), for example, framed the question of changes in means versus variances through ontogeny in terms of non-affine (i.e., shape) deformations at different spatial scales during the development of the skull of the cotton rat. Using 16 landmarks (see Figure 7.10), they found evidence for changes in the shape of the skull from day 0 to day 30 at 14 different spatial scales, and uncovered a series of distinct patterns in the relative alteration of means and variances. They were able to classify these patterns into three major types of constraints: canalization (constancy of form from age to age), a reduction of the variation orthogonal to the mean trajectory, and a reduction of age-specific variance *along* the mean trajectory over ages. What this all means in terms of underlying genetics is anybody's guess. The existence of three distinct ways in which variances of character complexes can be altered (as well as combinations of the three), however, suggests that a simple bi-dimensional rendition of these processes, such as that embedded in the **G** matrix, is likely to be a substantial oversimplification of the biological reality.

THE ENVIRONMENTAL COMPONENT: PLASTICITY INTEGRATION

Character correlations and, more generally, phenotypic integration are labile not only through evolutionary time or development, but also in response to some changes in environmental conditions. There are two fundamental types of relationships between environmental effects and correlations. (1) Trait correlations can be plastic (i.e., the correlation between two characters can be altered by the environment); we introduced this topic in Chapter 4, but here we review studies looking at the integration of many traits as environmental conditions change. (2) Plasticities themselves can be correlated (i.e., the reaction norms of different traits can be correlated to yield coherent multivariate phenotypic plasticity).

However, we wish to first make clear that there is no necessary relationship between the plasticity of correlations and the correlation of plasticities, although it might seem intuitive that there should be one. The plasticity of correlations refers to environmentally induced changes in the *within-environment* relationships among traits, whereas the correlation of plasticities refers to the pattern of covariation of two traits *across environments* (i.e., the correlation of their reaction norms). Figure 7.13 depicts these two facets of integration. The within-environment correlation of traits x and y changes from positive to negative as we move from environment A to C. The across-environment correlation of the responses of traits x and y is negative; as the average value of x increases from A to C, that of y decreases.

Kawano and Hara (1995) planted rice at densities ranging from 11 to 4445 plants/m^2. They found that the strength of phenotypic correlations between characters changed radically from high to low density (see Figure 4.16). The number of significant correlations (an indication of integration) declined precipitously from complete integration at the highest density (4445 plants) to only 8 of 21 correlations significant at 400 plants (Figure 7.14). The rate of decline of integration was shallower at the lower densities.

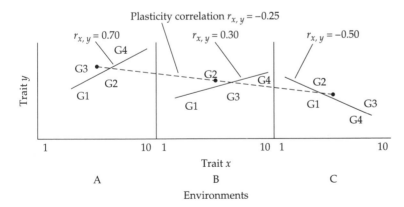

Figure 7.13 Two measures of integration associated with environmental change. In this example, there are three separate within-environment correlations between traits x and y that can be measured, and these may differ. The across-environment plasticity correlation can also be calculated from the mean values (•) of the two traits in each environment.

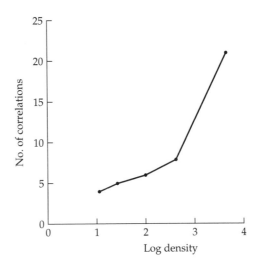

Figure 7.14 Relationship between planting density and phenotypic integration for rice. (From data in Kawano and Hara, 1995.)

Schlichting (1989a) exposed populations of three species of the annual plant *Phlox* to five biologically relevant stresses. He found marked differences in the patterns of integration among species, but these differences were also dependent on the environmental context (Figure 7.15). Similar results were obtained from a study of variation of correlation patterns within *P. drummondii*, when seven populations were exposed to five treatments (Schlichting and Pigliucci, 1995b); the phenetic relationships among populations were dramatically altered from one environment to the other (e.g., see the position of Jayton in the phenograms under herbivory versus low nutrients; Figure 7.16).

In a fascinating example of adaptive allometric changes, Johnson and Koehl (1994) discovered that the giant kelp, *Nereocystis*, modifies traits depending on the hydrodynamic habitats it occupies (differing in waves, tides, and currents). By changing different traits in different habitats, it maintains a constant ratio of force to cross-sectional area, so that maximum stresses are similar for kelp of different sizes and in different environments. For example, blade shape and "ruffliness" can be adjusted to change the drag coefficient, stipe diameter to alter stress, and stipe material properties to alter strength. Figure 7.17 depicts a contrast of the principal components for size and shape at three sites. Size and shape are not correlated in the protected site, negatively correlated in the current-swept site, and positively correlated in the wavy site.

Correlation of Plastic Responses

A complementary view to that of the plasticity of character correlations (Chapter 4) is a consideration of the possibility that the plasticities of different traits, or even

(A) *Phlox drummondii*

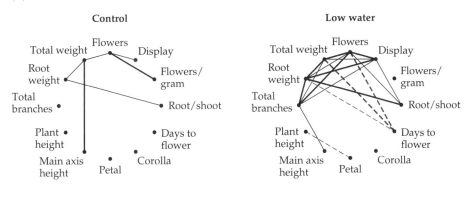

(B) *Phlox cuspidata*

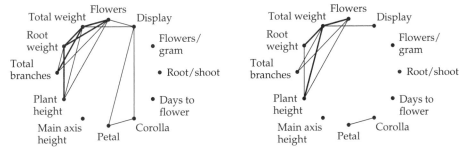

Figure 7.15 Environment-dependent phenotypic correlations among traits in two species of *Phlox* exposed to two treatments. Solid lines, positive; dashed lines, negative correlations. (From Schlichting, 1989a, © Springer–Verlag.)

suites of traits, may themselves be correlated. The concept was first proposed by Schlichting (1986), and it has received some attention (Schlichting, 1989b; Scheiner et al., 1991; Newman, 1994b), despite the experimental and statistical difficulties in measuring it. From a biological standpoint, it makes sense that the reaction norms of some traits should be correlated. For example, let us consider a classical scenario of environmentally determined tradeoffs in resource allocation between male and female reproduction in plants (e.g., Solomon, 1985; Delph and Lloyd, 1991). In such cases, it can be argued that an emphasis on female fitness is best under good environmental conditions (because the female function requires a larger investment of resources); while production of more male components is the appropriate response to a deterioration in habitat quality (males are cheaper . . .). Accordingly, one would expect the reaction norm for male function to follow a monotonically decreasing trend with environmental quality, opposite to the reac-

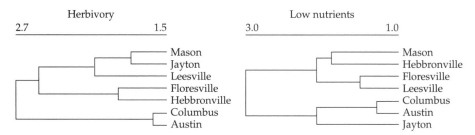

Figure 7.16 UPGMA phenograms depicting the multivariate morphological differences among seven populations of *Phlox drummondii* grown in two environments. The axes show Mahalanobis distances among populations. Notice the environment-specific relative positions of taxa; for example, Jayton groups with Mason in the herbivory treatment, but is in the most distant cluster in low nutrients. (From Schlichting and Pigliucci, 1995b.)

tion norm for female function, and leading to an expectation that the two reaction norms should be negatively correlated.[5]

Schlichting (1986) first depicted plasticity integration in three species of *Phlox* exposed to six treatments by diagramatically representing the relationships between the *plastic responses* of different traits as correlations on a single diagram. Although purely observational in nature, several interesting facets emerged from these diagrams (Figure 7.18). First, the three species differed substantially in their patterns of plasticity integration. Second, the plasticities of some traits appeared strongly correlated regardless of the taxon (e.g., plasticities of flower production and total weight, or plasticities of floral display and root weight). In a later study, Schlichting (unpublished) documented the plasticity integration of seven populations of *Phlox drummondii*.[6] There was as much variation within a single species as there was in the comparison of three species! Additionally, when the differences in plasticity integration were plotted against differences in mean flower production, a surprisingly strong positive correlation emerged (Figure 7.19), even though the highest level of integration was only 51% (Schlichting, 1989b.) Schlichting hypothesized that this pattern is a reflection of a balance between the obvious requirement for *some* integration of responses across environmental conditions, versus the need for an ability for independent response when conditions change substantially.

---

[5]Note that it is possible to calculate a correlation between *two* reaction norms because each one is described by many points (one point per environmental treatment). Each point, not the reaction norm as a whole, is used as an entry in the calculation of the correlation (Schlichting and Levin, 1984).

[6]Populations from Austin, Columbus, Floresville, Hebbronville, Jayton, Leesville, and Mason, Texas. Plants of each population were grown in five treatments: control, low water, low nutrients, herbivory, and small pots.

(A)   (B)

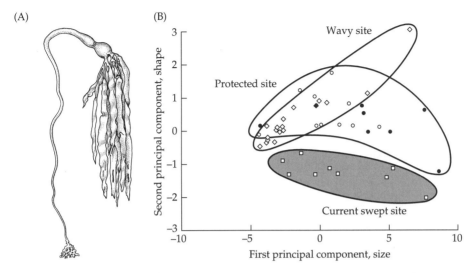

Figure 7.17   Integrated responses to environmental differences in the giant kelp, *Nereocystis luetkeana*. (A) The "leafy" structure is the blade, the "stem" is the stipe, and the pneumatocyst is a gas-filled bulb at the blade/stipe junction used for buoyancy. (B) Plot of principal component scores from three sites. PC1 (size) has positive loadings of blade width, length, and area, stipe diameter and length, and pneumatocyst diameter; PC2 (shape) has positive blade width, stipe diameter, and pneumatocyst diameter, and negative loadings for blade area and length, and stipe length. Stipe diameter and length are independent at all sites. It is clear from the diagram (compare orientation of the ellipses) that the correlation between size and shape varies as a function of the environment. (From Johnson and Koehl, 1994, © The Company of Biologists.)

The correlation of plastic responses can also be extended to comparisons of the form of plastic responses of two taxa (Schlichting and Levin, 1984, 1990). For example, Wimberger (1991) evaluated the plastic responses of two species of cichlid fish, *Geophagus brasiliensis* and *G. steindachneri*, to two food types, chironomids and brine shrimp. He reported that there was little divergence in the plastic responses for 20 traits (examined individually). However, the plasticity correlation for all 20 traits, calculated from his data, is $r = -0.05$, indicating a general *dissimilarity* in the pattern of plastic responses of these two species.

Previous studies of the genetic basis of plasticity integration (antecedent to the actual formalization of the concept by Schlichting) were framed in terms of research on the adaptation to heterogeneous environments. One example is the work of Khan et al. (1976) in *Linum usitatissimum*. They plotted the plasticity of capsule number versus the plasticity of height in two varieties and in 60 F_3 lines (Figure 7.20). The pattern obtained is consistent with the action of several genes

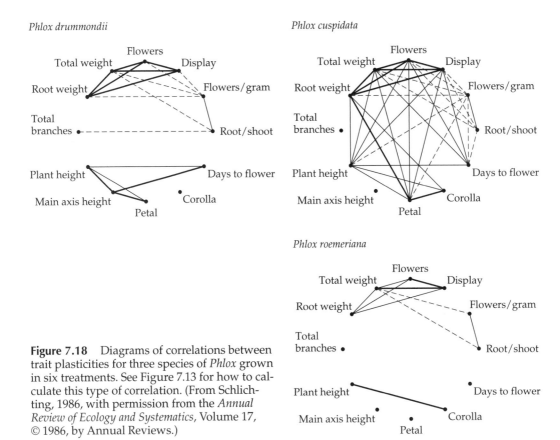

Figure 7.18 Diagrams of correlations between trait plasticities for three species of *Phlox* grown in six treatments. See Figure 7.13 for how to calculate this type of correlation. (From Schlichting, 1986, with permission from the *Annual Review of Ecology and Systematics*, Volume 17, © 1986, by Annual Reviews.)

segregating in the two parental types, and indicates the extent of potential genetic variation for the combined responses of the two traits and the constraints acting on them. For example, it is clear from the plot that one can obtain an F_3 line with reduced plasticity for height, but not for capsule number. Analogously, it would appear feasible to select for a line with increased plasticity of capsule number, but not of height.

The work of Chippindale et al. (1993) on *Drosophila melanogaster* represents an excellent example of the relationship between plasticity integration and tradeoffs. Families of flies were fed different levels of nutrients, while starvation time (a component of viability) and number of eggs deposited per day (a component of fecundity) were recorded. They found that in low nutrient treatments the insects allocated more resources to fighting starvation, thereby generating a tradeoff with fecundity. At high levels of resource availability, the strategy was reversed; the tradeoff was still present, but now increased fecundity

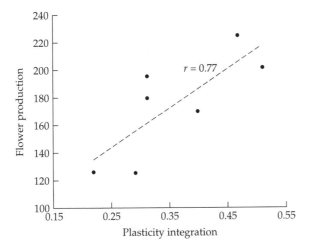

Figure 7.19 A link between plasticity integration and fitness? The graph depicts the correlation between the level of integration (measured as the fraction of the total possible number of significant correlations) and the average flower number for seven populations of *Phlox drummondii* grown in five treatments. (From Schlichting, unpublished.)

was favored over starvation resistance. Most interesting, however, were the results for medium-low nutrient levels, where allocation to the two strategies converged (Figure 7.21). This work was followed up with ten generations of selection, during which the tradeoff disappeared due to adaptation of the stocks to subtle differences in culture regimes, and by evolution of increased acquisition ability in the late fecundity populations (Leroi et al., 1994b). They also found that the selection responses were not predictable from the sign of the genetic correlations (Leroi et al., 1994a).

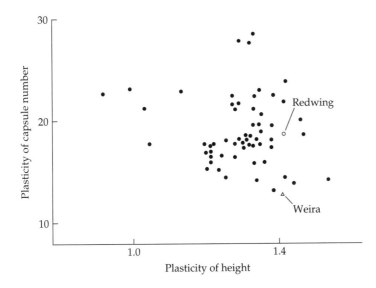

Figure 7.20 Plot of the relative plasticities of two traits in *Linum usitatissimum* in response to density. The circle and triangle represent the parentals, and the points are their F₃ hybrid offspring. (After Khan et al., 1976.)

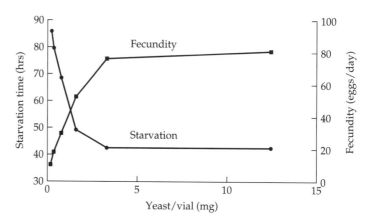

Figure 7.21 The tradeoff between fecundity and starvation resistance in *Drosophila* is a function of diet: increased allocation to fecundity is favored at higher food levels. This O population had been selected for later fecundity. (From Leroi et al., 1994b.)

Scheiner et al. (1991) studied phenotypic plasticity in *Drosophila melanogaster* exposed to different temperatures, and were able to demonstrate the significance of both broad and narrow sense genetic correlations of plastic correlations, thereby implying some sort of genetic control for plasticity integration. A similar result was seen by Newman (1994b) while studying reaction norms to food supply in spadefoot toads (*Scaphiopus couchii*); he found that plastic responses of the duration of the larval period were negatively correlated with plastic responses of size.

Candidates for the genetic control of correlations of plasticities can again be sought among environment-dependent regulatory loci; for example, the genes controlling what are referred to as generalized stress responses. Although response to "stress" has been considered by some authors as a special category of organismal reaction to environmental heterogeneity (with a voluminous literature reviewed in Hoffman and Parsons, 1991), it is just a subset of the more general phenomenon of phenotypic plasticity. Parsons (1987) proposed the existence of generalized reactions to stress in living organisms, and pointed to the widespread occurrence of heat shock response as an example (Maresca et al., 1988; DeRocher et al., 1991; Heikkila, 1993a,b). According to Parsons, both phenotypic and genetic variation are enhanced under stressful conditions, although the jury is still out on this particular generalization (Lewontin, 1974a; Blows and Sokolowski, 1995; Pigliucci et al., 1995; Bennington and McGraw, 1996). Furthermore, and perhaps more interestingly, recombination, mutation, and transposition have been found to increase in frequency when habitat quality deteriorates, implying the existence of concerted genetic mechanisms to deal with environmental degradation (McDonald, 1983; Sniegowski et al., 1997).

Parsons also suggested that most of the variation in quantitative traits that respond to stressful situations is ascribable to a few loci with major effects.

Another suggestion of the existence of generalized responses to stress comes from Chapin (1991) in relation to the physiological integration of plant phenotypes. Chapin considers the four basic abiotic factors influencing plant growth and performance: light, carbon dioxide, nutrients, and water. Light and CO_2 influence photosynthesis, nutrient supply affects nutrient uptake, and water supply affects water uptake in the roots. The observation that general growth depends on a balance among all of these components leads to the formulation of the hypothesis that a stress—defined as a sudden reduction in the availability of any fundamental factor—should trigger a generalized response due to the physiological integration of the plant phenotype. This response would adjust different components of the phenotype and tend to maintain a positive growth rate.

What is the evidence for the overall importance of generalized responses to stress, versus reactions to specific challenges posed by a complex and constantly changing environment? A review of the literature indicates that both phenomena exist, although it seems that the number of specifically targeted responses that can be discovered may be limited by the experimental efforts devoted to this area. For example, Lois and Buchanan (1994) isolated mutants of *Arabidopsis thaliana* sensitive to UV light. These mutants, however, did not respond differently from the wild type to heat, hypoxia, salt, or cold stress. Thus the mutation represented a deficiency in a genetic element expressly targeted for coping with UV light (Figure 7.22).

Results compatible with the existence of both specialized and generalized responses to stress were found by Takimoto et al. (1994), who studied the induction of flower formation in the aquatic plant *Lemna paucicostata*. Many aquatic plants reproduce vegetatively most of the time, resorting to sexual reproduction only under extreme environmental conditions. Stressed individuals of *L. paucicostata* released a chemical into the water in response to drought, osmotic stress, and heat stress, inducing flower formation in nearby plants, which suggests a generalized reaction to unusual environmental conditions (see also Braam and Davis, 1990; Nordin et al., 1991).

That the genetic architecture of an organism can be dramatically altered when exposed to a novel, and perhaps stressful, environment has been suggested by Service and Rose (1985) and by Holloway et al. (1990), and attributed to the immediate effects of selection in the new environment. Genes creating a positive correlation in the old environment were favorably selected, and therefore fixed; and, at equilibrium, the residual genetic variance for correlations among fitness components will be attributable to alleles causing a negative covariance to be expressed. The new environmental regime jump-starts the selection process, causing a sudden shift toward positive genetic correlations. As we have proposed earlier, though, changes in genetic correlations can come about also

UV light	Heat	Hypoxia	Salt	Cold

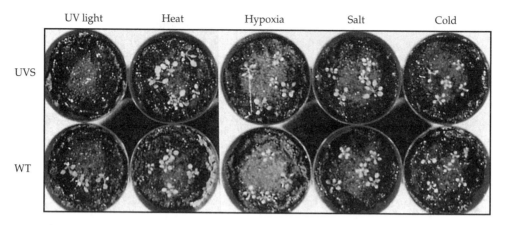

Figure 7.22 Responses of wild-type and UV-sensitive *Arabidopsis* mutants to UV exposure and four other stresses. Notice that the mutant is negatively affected by UV light; it does not respond differently from the wild type under other stresses. (From Lois and Buchanan, 1994, © Springer–Verlag.)

because of changes in gene expression (in the absence of selection) induced by the "novel" environment (or *any* different environment; Schlichting and Pigliucci, 1995a).

Even though plasticity integration is a potentially important phenomenon in the production of coordinated phenotypes in response to environmental heterogeneity, there are very few studies actually addressing the problem. All we know at this point is that correlations among plasticities are demonstrable in different systems, and that they are caused by some sort of genetic machinery. We do not know much about the ecological significance of plasticity integration, very little about what kind of genes control it, and nothing at all concerning its evolution. Part of this dearth of information is due to the fact that plasticity integration is inherently difficult to study. One needs to raise clones or families of many genotypes belonging to different taxa under a wide range of environmental conditions in order to obtain the raw data. The statistical analyses of this kind of data are far from being simple, and most often run into limitations of power due to reduced sample sizes. We believe, however, that part of the problem is simply a lack of appreciation for the potential implications of the phenomenon. After all, it took evolutionary biologists the better part of the twentieth century to start considering simple phenotypic plasticity as an interesting object of study, as opposed to a nuisance while estimating heritability . . .

The study of phenotypic integration, either from a developmental or environmental perspective, involves a complex interaction of all major disciplines within evolutionary biology—from molecular and developmental genetics to ecology, from physiology to evolutionary systematics and the comparative method. It should be apparent from the preceding discussion and examples, however, that it is a subject matter certainly within the grasp of modern biological inquiry, and that the increasing number of researchers working in the field in the last two decades is possibly just the beginning of a true integration between mechanistic and organismal biology. We have been waiting for such a synthesis since the rediscovery of Mendel's laws, and perhaps the current generation of biologists will finally see it realized.

CONCEPTUAL SUMMARY

1. The concept of character coordination has a long history, originating in a non-Darwinian context and largely unincorporated in the neo-Darwinian synthesis. We use the term *phenotypic integration* to refer to those sets of traits that are genetically or functionally interrelated.

2. Quantitative genetics is the most widely used conceptual tool in the study of phenotypic integration. The major limitation of this approach is that it does not deal with the genetic machinery actually underlying phenotypic complexity.

3. Recent studies that originate within the framework of quantitative genetics have achieved new insights either by incorporating physiological or molecular information or by approaching the problem of phenotypic integration from a comparative standpoint.

4. The description of patterns of morphological integration is done with classical multivariate statistics (e.g., principal components or factor analysis), or by means of the so-called "new" morphometrics, which relies on the analysis of changes in the positions of trait landmarks. Both are useful. While classical analyses are usually difficult to interpret in terms of biological homology, the new tools apply strictly to shape analysis, and cannot include variables representing heterogeneous structures or life history traits.

5. A fundamental component of the study of phenotypic integration is the representation of the developmental aspect. Modern molecular genetics, coupled with comparative methods, are making this one of the fastest growing fields of inquiry in evolutionary biology. More detailed quantitative genetic studies are being carried out and point to complex changes in genetic variances and covariances during ontogeny. These changes might markedly alter future evolutionary trajectories, depending on when (at what life stage) selection actually operates.

6. A second essential aspect of phenotypic integration is represented by its environmental lability. There is conclusive evidence that genetic and pheno-

typic correlations can change in magnitude and sign from one environment to another, thereby markedly altering predicted responses to selection.

7. The reaction norms of traits can themselves be correlated, suggesting the existence of genetic mechanisms coordinating phenotypic plasticity at the multivariate level. Measurement of these relationships (plasticity correlations) may provide insight into mechanisms of physiological or genetic control of the responses of suites of traits.

CHAPTER 8

Epigenetics

"There's something happening here, what it is ain't exactly clear . . ." (Stills, 1967)

In 1995, an extraordinary report was published, documenting the ability of researchers to induce eyes in a variety of decidedly non-eyelike locations on the body of a *Drosophila* (Barinaga, 1995; Halder et al., 1995). The *eyeless* (*ey*) mutation is characterized by either deformed eyes or a lack of compound eyes. It has pronounced homologies with both mouse (*Pax-6*) and human (*Aniridia*) genes; mutations of each of these genes also result in phenotypically defective eyes. By inserting a transcriptional activator into the *Drosophila* genome, Halder et al. were able to induce the expression of *eyeless* in imaginal disks and ultimately to form ectopic eyes in wings, legs, and antennae. Furthermore, they transplanted the mouse gene *Pax-6* into *Drosophila* and again induced eye formation (repeated subsequently with squid *Pax-6* as well: Tomarev et al., 1997). This coup led to the proclamation that *eyeless* represents the "master control gene for eye morphogenesis" as the initial regulator of the subsequent cascade of gene action leading to eye formation in metazoans (Halder et al., 1995). Wider searches for *eyeless/Pax-6* homologues have been carried out, revealing two *Pax* genes even in species of Cnidaria, the most basal phylum of metazoans with eyelike structures (Sun et al., 1997).

This chapter is devoted to epigenesis and its role in phenotypic evolution. We already possess an enormous amount of *factual* information about organismal development both in animals and in plants. Yet, despite various endeavors at synthesis (de Beer, 1930; Hall, 1992a; Rollo, 1994; Raff, 1996; Gerhart and Kirschner, 1997), most biologists would agree that, in general, our *conceptual* understanding of this area of inquiry is still limited (Wray, 1994).

Although developmental biology has a very long history, it has in the twentieth century nearly always operated at the margins of evolutionary biology, as seen especially during the formulation of the neo-Darwinian synthesis of the 1940s (Mayr and Provine, 1980; Chapter 2). Gould's 1977 *Ontogeny and Phylogeny* was intended as a wake-up call for evolutionary and developmental biologists alike, and the more recent flourishing of modern molecular developmental genetics (e.g., Dickinson et al., 1993; Akam et al., 1994b; Ambros and Moss, 1994; Smith and Hake, 1994; Brakefield et al., 1996; Almeida et al., 1997; Kanki and Ho, 1997) has opened possibilities that will likely change the century-old relationship between the two disciplines.

Why should an evolutionary biologist care about development, or a developmental biologist adopt an evolutionary perspective (Callahan et al., 1997)? This might seem a rhetorical question, but although most biologists would agree that a more direct relationship between the two fields is desirable, there are still few serious attempts at rapprochement. Developmental biology is burgeoning thanks to new molecular tools. The time is ripe for evolutionary theory to actively integrate development into its conceptual and experimental arsenal, lest we squander a second opportunity at a true synthesis.

Evolutionary biologists (at least on the phenotypic side) need to care a great deal about development because our ultimate goal is to understand how different forms have evolved. Obviously, adult forms result from a complex series of events that occur during development. Although the "rule" that early development is relatively impervious to evolutionary tinkering (de Beer, 1930) appears to hold true generally, there are striking instances of the breakdown of this maxim. An excellent example comes from the work of Wray and collaborators on sea urchin evolution. Wray and McClay (1989) compared the spatial and temporal patterns of expression of three proteins of several species of sea urchins belonging to five different genera. They found instances of marked heterochrony and heterotopy during early development that must have accompanied the post-Paleozoic radiation of echinoids (see Chapter 5, Figure 5.18). Wray

(1995, 1996) has also pointed out the existence of patterns of stasis at the adult morphological level, accompanied by dynamic and rapid changes in larval behavior (feeding to non-feeding) and morphology (Figure 8.1). In this case, examining only adult stages would actually hide a rich and fascinating evolutionary history reflecting fundamental aspects of the ecology of sea urchins.

If the interest of an evolutionist in developmental biology should be clear, what is the benefit to a developmental biologist who embraces an evolutionary perspective? Molecular developmental biologists strive to understand how organisms are progressively built through the differential action of regulatory and structural genes. However, the currently observed process is undoubtedly the result of a long, and probably tortuous, evolutionary history. Many features of current developmental systems may not make any sense if analyzed in the context of an ahistorical, function-oriented framework. Questions such as how families of regulatory or structural genes come into existence are inherently evolutionary. Even applied problems, such as how broadly (from a phylogenetic standpoint) a gene engineered in one taxon can still function when transferred to other taxa, are answerable only by knowing about how gene function may have diverged during evolutionary time. An example of the kinds of insights

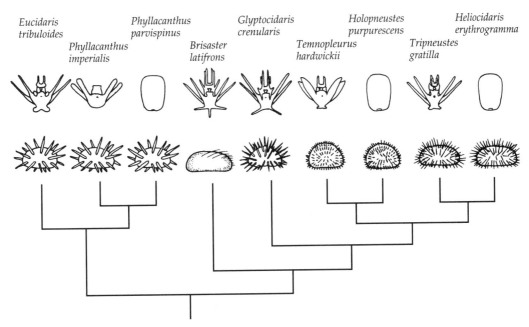

Figure 8.1   Larval and adult morphologies of sea urchins mapped onto a phylogeny. There is clear evidence of multiple origins of non-feeding larvae and associated radical morphological alterations, even though adult forms have changed little. (From Wray, 1996.)

that can be gained in this respect is provided by Doyle's (1994) study of the evolution of the MADS-box homeotic gene family in plants. Many members of this family are involved in the regulation of steps in flower development. By using phylogenetic as well as developmental genetic information, he reconstructed the likely evolutionary history of the patterns of gene expression of this family in *Arabidopsis thaliana* (Figure 8.2). He distinguished floral versus vegetative expression, early versus late floral expression, and expression in petals/stamens versus stamens/carpels, and concluded that the ancestral gene probably was a late floral gene expressed in stamens and carpels. At least three shifts in the timing of expression and nine duplications need to be hypothesized to account for the modern expression patterns.

Medawar and Medawar (1983) wrote that "genetics proposes, epigenetics disposes," to emphasize that developmental processes play a much more fundamental role in determining the phenotype of an organism than allowed by the prevalent gene-based view of life. But what *is* epigenetics? Waddington defined epigenetics as the "causal interactions between genes and their products which bring the phenotype into being" (Waddington, 1975, p. 218).[1]

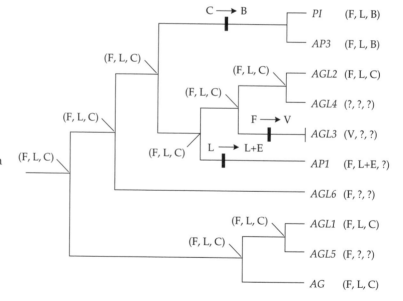

Figure 8.2 Evolution of the MADS-box gene family in *Arabidopsis thaliana*. Gene expression has changed in terms of general location (floral, F versus vegetative, V), timing (early, E versus late, L), and specific floral location (petals and stamens, B versus stamens and carpels, C). (Unknown character states, ?). (From Doyle, 1994.)

---

[1]Introduction of the term is attributed to Waddington, who was inspired by Aristotle (Hall, 1992b; Waddington, 1966). Aristotle distinguished two theories of animal development: *preformation* posited that the fertilized egg contains structures corresponding to each organ in the adult organism; the opposing view was *epigenesis*, in which adult structures form by the continuous interaction of a much smaller number of simple units throughout development.

Although this and other statements might be construed as blurring the distinction between the direct effects of gene actions (genetics *sensu stricto*) and the indirect (epigenetic) effects due to interactions of genetic and non-genetic factors, it seems clear that Waddington maintained the distinction between the two; for example, in his concept of the epigenetic landscape (Figure 8.3). The landscape represents "a description of the general properties of a complicated developing system in which the course of events is controlled by many different processes that interact in such a way that they tend to balance each other" (Waddington, 1966). The depth of valleys and steepness of slopes and ridges represent the extent to which a particular developmental progression can be perturbed by internal or external influences (i.e., the level of canalization).

Waddington (1975, p. 219) proposed that there are two key aspects to epigenetics: "changes in cellular composition (cellular differentiation, or histogenesis) and changes in geometrical form (morphogenesis)." Rachootin and Thomson (1981) focused on the consistency of developmental outcomes and proposed that epigenetic systems have properties that ensure such results—self-assembly, feedback loops, and alternative pathways to compensate for irregularities. Many other authors have seized upon epigenesis as the key to producing morphological novelty.

As Nijhout (1990) eloquently pointed out, the concept of epigenetics has led us away from views that "genes 'control' development and genomes embody 'programs' for development." He proposed as an alternative view, that "genes act as suppliers of material needs of development and as context-dependent cat-

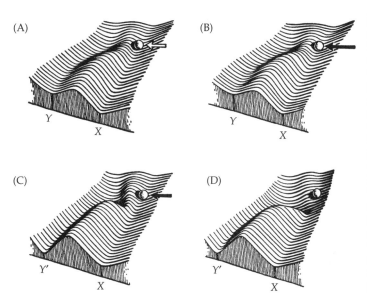

Figure 8.3 Waddington's epigenetic landscape. The ball represents the organism, rolling down an ontogenetic valley. Ridges separate alternative "canalized" trajectories, some (e.g., deep valleys) more stable than others (e.g., slopes). (A) Development can be altered by an environmental stimulus (white arrow), which may divert the default X pathway to the alternative Y pathway. (B) Y can become the default by a single mutation (black arrow), or by mutation and selection of "modifier" genes progressively altering the ridge dividing the two paths (C, D). (From Waddington, 1957.)

alysts of cellular changes." That genetic and structural complexity need not be closely related is a direct result of the properties of epigenetic systems and the role of genes.

The temptation to close this introduction without giving our personal definition of epigenetics is great, but we will resist it nonetheless.

> Epigenesis is an ensemble of processes that propagate phenotypic characteristics throughout development. These processes derive from either indirect effects of gene action (emergent properties) or from non-genetic phenomena (e.g., cell-cell or hormone-target communications).

THE HUNT FOR DEVELOPMENT'S MASTER SWITCHES: THE HOMEOBOX GENES

Over the past decade, tremendous interest has developed in the roles of genes or groups of genes such as *eyeless*[2] in coordinating basic developmental plans. The major phenotypic effects of homeosis (replacement of one segment or organ type by another) had long been recognized in animals and plants, but now have been demonstrated to be due to particular genes (Lewis, 1978). It was later discovered that many homeotic mutants contained a conserved sequence of approximately 60 amino acids, the homeodomain (or homeobox), found in a variety of proteins influencing transcription. The ability to affect transcription derives from the physical structure of the homeodomain: it lies along the major groove of the DNA double helix and binds to the phosphate backbone and certain nucleotide bases (but there is *not* a lot of specificity for the different genes; Biggin and McGinnis, 1997). A remarkable feature of these homeodomain-containing genes is that they have been found in all phyla examined to date.

Homeobox-containing genes have been separated into two groups, the *Hox* genes, which are represented in clusters in the genome, and a variety of classes of dispersed non-*Hox* genes. Homologies of *Hox* genes have been demonstrated across cnidarians, flatworms, nematodes, arthropods, and chordates, indicating an ancient evolutionary origin pre-dating the divergence of protostome and deuterostome lineages (Valentine et al., 1996). There are classes of non-*Hox* genes shared even between sponges and chordates (Kappen, 1996).

Similar "development genes" and associated homeotic mutants are also known in plants. One conserved gene sequence has been labeled the MADS-box (Doyle, 1994). MADS-box genes have been identified as having major effects on the identity of floral organs (Purugganan et al., 1995; see below). Other home-

---

[2]Although the initial reports suggested a role for *eyeless* as the master controller of eye production, work on another eye-formation mutant in *Drosophila* suggests that there may be other genes nearly as prominent in the hierarchy (Roush, 1997). Flies with *dac* (for dachshund, due to their stubby legs) also produce ectopic eyes, but those with mutant *dac* are unable to.

odomain-containing proteins have been found in yeasts. Although Kappen (1996) proposed that the homeodomain classes of yeasts, plants, and animals do not appear to have a common evolutionary origin, Bharathan et al. (1997) suggest that there is a common ancestral protein. Their results also indicate that at least one gene duplication event occurred prior to the divergence of animals, plants, and fungi, because the two major homeodomain subgroups each include representatives from all three kingdoms.

In a comprehensive review, Manak and Scott (1994) detail the structure and operation of the *Hox* and non-*Hox* homeobox genes in animals. In *Drosophila*, for example, the homeobox genes of the Antennapedia and Bithorax complexes are involved in segment identity, that is, the fate of cells in particular segments. There is notable evidence of their conservative nature, in that regulatory sequences can operate reciprocally between *Drosophila* and mouse (references in Manak and Scott, 1994), and *Drosophila* homeobox genes can substitute effectively in the nematode *Caenorhabditis* (Hunter and Kenyon, 1995).

Although the idea that individual genes could be identified that determine the formation of particular organs or structures was certainly enticing, current thinking has retreated from this overly simplistic idea. At present it is generally agreed that the origins of structural diversity more likely derive from changes in the regulation of homeobox genes, or from modifications of the interactions of homeobox proteins with subsequently expressed genes and their gene products (Carroll, 1995; Beckers et al., 1996; Biggin and McGinnis, 1997).

The number of regulatory sites and target genes can be substantial. Using a transcriptionally activated reporter, fragments of 53 genes regulated by *Ultrabithorax* were tagged in *Drosophila melanogaster*, and the authors estimated a total of 85–170 targets throughout the genome (Mastick et al., 1995). Graba et al. (1997) review data on how *Hox* genes regulate morphogenesis and identify three classes of targets: structural genes, transcription factors, and cell signaling molecules. For example, detailed work by Castelli-Gair and Akam (1995) revealed that a single *Hox* gene, *Ultrabithorax*, could specify both a thoracic and an abdominal segment due to variation in its spatial and temporal expression.

Comparative Studies of *Hox* Gene Expression

Evolutionary perspectives on the distribution and expression of *Hox* genes in a variety of organisms appeared suddenly in 1994 with several papers (e.g., Akam et al., 1994a; Carroll, 1994; Patel, 1994; Ruddle et al., 1994) in the edited volume *The Evolution of Developmental Mechanisms* (Akam et al., 1994b). More recent broad surveys of *Hox* homologues among the metazoans reveal the importance of large-scale gene duplication events (Valentine et al., 1996); for example, the quartet of vertebrate *Hox* clusters seems to have formed by a minimum of two duplications (Zhang and Nei, 1996). Such duplication events have been speculated to provide the raw material for radical alterations in body plan such as that postulated for the Cambrian explosion and the evolution of vertebrates

(Holland and Garcia-Fernandez, 1996). Following this reasoning, one might predict that there should be fewer genes in the *Hox* clusters of the more basal taxa; this is apparently true in the chordates. Grenier et al. (1997) examined this prediction for arthropods and found, however, that both myriapods and onychophorans have the complete set of genes found in arthropods. Thus, an increase in the number of *Hox* genes is not necessary to explain the rapid radiation of the arthropods (or the insects; Akam et al., 1994a).

The comparison of patterns of expression of *Hox* genes relative to phylogenetic hypotheses is a promising approach to understanding the evolution of morphological diversity. In crustaceans, Averof and Patel (1997) show that differences in the expression of the *Hox* genes *Ubx* and *AbdA* are correlated with the various modifications of anterior thoracic limbs into maxillipeds (feeding appendages). The authors suggest that such alterations of regulatory control of homeotic genes may be part of the normal process of adaptive evolutionary change. Averof and Cohen (1997) isolated crustacean homologues of two wing-specific genes in insects (*POU domain protein* and *apterous*) from the brine shrimp *Artemia*, which has multi-branched appendages. These genes were expressed only distally in the second epipod, which serves respiratory and osmoregulatory functions in crustaceans, suggesting that the insect wing and these epipods are homologous structures. Are such extrapolations prudent? Jockusch and Nagy (1997) pointed out the conflicting nature of patterns of expression of other genes, and the general problem of extrapolating from gene homology to structural morphology (Bolker and Raff, 1996).

Panganiban and coworkers (1995) found that similar genetic pathways underlie the formation of various appendages and body wall outgrowths in the Arthropoda. Later, they reported similar patterns of expression of the gene *Distal-less* (or its homologues) in the appendages of other protostomes and in deuterostomes, which led them to suggest that there are commonalities among cellular processes leading to appendage formation in general (Panganiban et al., 1997). They also make note of the distinction between levels of homology: "... while some of the appendages/outgrowths described here certainly are not homologous in the classic sense (i.e., directly derived from a common structure), our findings suggest that they are homologous *at a more fundamental level*."[3]

Rogers et al. (1997) examined the correspondence between the expression of the gene *Sex combs reduced* (*Scr*) and the origin of key features in insect morphological evolution. *Scr* is required for suppression of wings on the prothorax of *Drosophila* and is also found expressed in a site consistent with this function (dT1, Figure 8.4) in representative Orthoptera and Hemiptera. However, it is also expressed at dT1 in the thysanuran *Thermobius*, a primitively wingless

---

[3]Our italics: perhaps we are oversensitive about molecular chauvinism, but we see the phenotype as just as fundamental as the genotype.

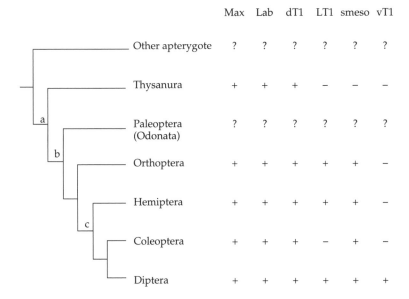

	Max	Lab	dT1	LT1	smeso	vT1
Other apterygote	?	?	?	?	?	?
Thysanura	+	+	+	–	–	–
Paleoptera (Odonata)	?	?	?	?	?	?
Orthoptera	+	+	+	+	+	–
Hemiptera	+	+	+	+	+	–
Coleoptera	+	+	+	–	+	–
Diptera	+	+	+	+	+	+

Figure 8.4 Hypothesized scheme for the evolution of arthropod appendages based on the zones of expression of the homeobox gene *Sex combs reduced* (*Scr*) in arthropods. **a** marks the position where *Scr* gained some control over prothoracic development, coincident with the appearance of *Scr* expression in dT1. **b** represents the evolution of wings, coincident with the development of the leg spot. **c** indicates the origin of leg combs. Max, posterior maxillary expression; Lab, labial expression; dT1, expression in dorsal anterior prothorax; LT1, expression in the lateral prothorax including the leg spot; smeso, expression in the somatic mesoderm of thoracic legs; vT1, expression in the ventral prothorax. (From Rogers et al., 1997, © The Company of Biologists.)

insect. Similarly, *Scr* also promotes comb formation in *Drosophila*; it is found at LT1 in Hemiptera, which have a leg comb, but also in Orthoptera, which lack leg combs.

Weighing in with findings that raise questions about the pan-conservative nature of homeobox genes and their functions, Lowe and Wray (1997) examined the functions of *Distal-less* (*Dll*), *engrailed* (*en*), and *orthodenticle* (*otd*) in species representing four of the five extant classes of the radially symmetrical echinoderms, sister clade to the chordates. Their results indicate that each one of these genes has evolved the gain or loss of expression domains during the diversification of this phylum. They argue that several cases represent co-option of a gene to a new developmental role. For example, *otd* and *Dll* are expressed in elements of the water vascular system (unique in this phylum), but not in all classes. The *engrailed* gene is expressed in the calcitic endoskeleton, another derived feature of echinoderms. These results indicate that there is a profusion of additional information locked up in these genes, and doubtless in many other genes and gene families.

The vast majority of work on evolutionary developmental genetics has been focused on a few model organisms. We will provide some vignettes of representative findings in three of these (the fly *Drosophila melanogaster*; the nematode *Caenorhabditis elegans*; and the flowering plant *Arabidopsis thaliana*) to render some of the flavor of the approaches taken and the discoveries that are being made.

THE FLY: *DROSOPHILA*

There have been substantial advances in the understanding of the mechanisms of enzyme gene expression in *Drosophila melanogaster* and relatives.[4] These have been achieved by studies of mutants or by means of interspecific comparisons of both the genes and their regulation. For example, in the *virilis* group, only *D. virilis* itself is ethanol-tolerant, and this has been related to both regulatory and structural differences of the *Adh* gene (Ranganayakulu and Reddy, 1994). In the *melanogaster* subgroup, *Gld* activity varies little among species in non-reproductive tissues, but quite a lot in reproductive tissues, and at least some of this can be ascribed to variation in the promoter region (Ross et al., 1994). Tamarina et al. (1997) discovered that the regulatory regions of homologous esterase genes in *D. melanogaster* and *D. pseudoobscura* were vastly different, despite the apparent conservation of expression patterns. Notwithstanding the increase in our understanding of the functioning of what were until recently classified as "structural" genes, there are still cases in which the patterns revealed are surprising. Andruss et al. (1997) discovered strikingly complex patterns of calmodulin gene expression, with marked instances of tissue-specific up- and down-regulation (Figure 8.5).

Remarkable progress has been made in dissecting the roles played by the diversity of genes involved in the determination of the basic body design of *Drosophila*. This research initiative has been so successful that the basics of the roles of these pattern-determining genes are now readily available in general texts (see, e.g., Lewin, 1997). These genes have been classified into three major groups: maternal, segmentation, and homeotic (Figure 8.6). Maternal germline genes code for receptors and transcription factors that initiate the earliest developmental processes in the egg. Following fertilization, segmentation genes (many are transcription factors) control the division into segments and para-segments, as well as further compartmentalization; gap genes regulate pair-rule

---

[4]We will not even pretend to be in command of the *Drosophila* literature. The outpouring of research results is astounding; e.g., 107 papers with *decapentaplegic* or *dpp* in their titles listed in MedLine from 1995 to 1997. A superb, and frequently updated, source for information on the genes involved in *Drosophila* development is 'The Interactive Fly'—http://sdb.bio.purdue.edu/fly/aimain/1aahome.htm (Brody, 1997). For access to current literature go to MedLine—http://www.ncbi.nlm.nih.gov/PubMed/.

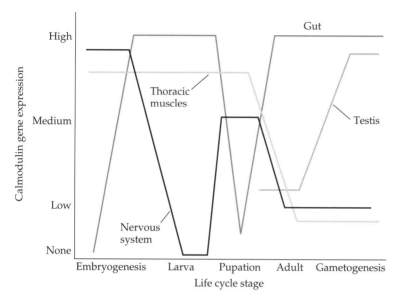

Figure 8.5 Complicated patterns of regulation of calmodulin during the life cycle of *Drosophila melanogaster*. (From data in Andruss et al., 1997.)

genes, which in turn regulate segment polarity genes. Homeotic genes are involved with determining the identity of segments and are activated or repressed by gap and pair-rule gene products.

We will provide a few examples of the types of details being uncovered, but if you recall the diagram of the regulatory network from Chapter 1 (Figure 1.1), the interactions among gene products are multifarious. Early studies examined fairly direct effects, but subsequent research, building on these foundations and employing increasingly sophisticated techniques, have begun fine-scale analyses of the feedbacks among these genes (e.g., Rivera-Pomar and Jackle, 1996; Micchelli et al., 1997; Lecuit and Cohen, 1997).

The maternal gene *dorsal* (*dl*) initiates dorsal-ventral patterning in the pre-cellular *Drosophila* embryo by determining different regions of gene expression. Regulation of *snail* (*sn*), *decapentaplegic* (*dpp*), and *short gastrulation* (*sog*) by *dl* is hypothesized to be responsible for delimiting tissue and cell types in mesoderm and ectoderm (Rusch and Levine, 1996). The *dpp* gene is a member of the transforming growth factor-β (TGF-β) gene family and is expressed in both embryonic and adult tissues. Both the dorsal-ventral gene *dpp* and the segment polarity gene *wingless* (*wg*) have been implicated as cell-signaling genes in many facets of dorsal-ventral patterning, and their protein products have been proposed as prime candidates to be true morphogens (i.e., substances with direct,

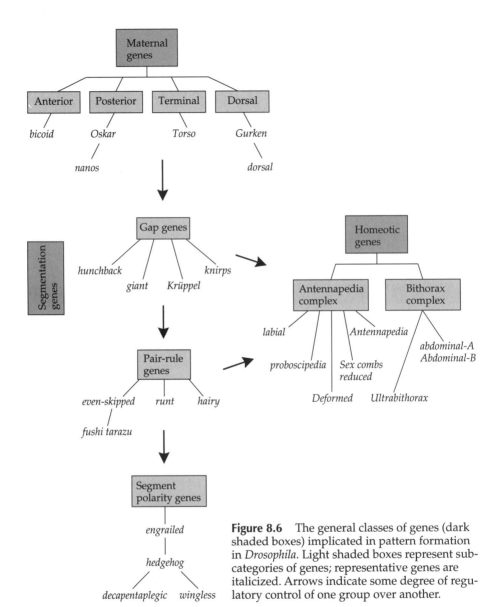

Figure 8.6 The general classes of genes (dark shaded boxes) implicated in pattern formation in *Drosophila*. Light shaded boxes represent subcategories of genes; representative genes are italicized. Arrows indicate some degree of regulatory control of one group over another.

concentration-dependent effects on target cells; Kosman and Small, 1997; Neumann and Cohen, 1997). The *dpp* gene induces expression of its own antagonist, *Daughters against dpp* (*Dad*), whose overexpression blocks *dpp* activity (Tsuneizumi et al., 1997). Interestingly, information derived from the concentration signal of *dpp* appears to be retained by descendant cells after they are no longer near the signal source (Lecuit et al., 1996). Maves and Schubiger (1998) reported that *wg* and *dpp* signaling pathways cooperate to determine the wing fate of imaginal discs by inducing the expression of *Vestigial* (*Vg*). Dorsal leg disc cells express high levels of *dpp*; ventral leg disc cells can be "dorsalized" by targeting of *dpp*, allowing *wg* to induce *Vg* expression and produce a wing cell fate.

Rivera-Pomar et al. (1995) examined the control of posterior gap gene expression and found that the maternal gene *bicoid* determines an anterior-posterior gradient that activates the gap gene *hunchback*, which in turn activates the gap gene *Krüppel*. Hulskamp et al. (1994) found that particular mutants of *hunchback* (those lacking the second zinc-finger domain) had adverse effects on *Krüppel* expression but not on that of the gap gene *knirps*, while another class of alleles resulted in gain of function phenotypes. Both *Krüppel* and *knirps* are thought to act as segmentation repressors that function over distances as small as 50–100 bp by inhibiting closely linked upstream activators (Barolo and Levine, 1997). *Krüppel* has been shown to be required for determination of muscle identity: having its expression switched on or off is enough to direct individual cells between alternative muscle cell fates (Ruiz-Gomez et al., 1997).

It is abundantly clear from a variety of results that distinct developmental functions can arise not only by alterations in the coding region of a gene, but also by modifying the *cis*-acting regulatory sequences up- or downstream from the gene itself. Li and Noll (1994) demonstrated the acquisition of distinct functions by three *Drosophila* homeobox genes. One of these, the pair-rule gene *paired*, is involved in the activation of segment polarity genes; the second, *gooseberry*, controls genes specifying denticle pattern; and the third, *gooseberry neuro*, is involved in neural development. These genes show substantial coding sequence divergence as well as differences in timing of expression. However, each produces a pair-rule phenotype if induced by heat shock between 3 and 6 hours of embryogenesis The proteins of these three genes have conserved the same fundamental regulatory function, with the essential differences among them traceable to their *cis*-regulatory regions.

THE WORM: *CAENORHABDITIS*

The nematode *C. elegans* has become one of the most widely used model organisms for understanding epigenetic events because of its relatively small genome and size (Riddle et al., 1997). The deterministic developmental program of the wild type leads to a newly hatched larva with 558 cells, and eventually through four larval stages (L1–L4) to an adult with fewer than 1000 cells. The lineage of each of these individual cells has been documented (Figure 8.7), and this detailed knowledge of cell fates has permitted the analysis of the causes of the direct phenotypic effects of various mutations.

A system of particular interest in the context of epigenetics and the developmental reaction norm is that of dauer larva formation. The non-feeding dauer larva (an arrested L2 stage) is produced under ecological conditions of nutrient deprivation, cued by limited food supply or crowding (related to the concen-

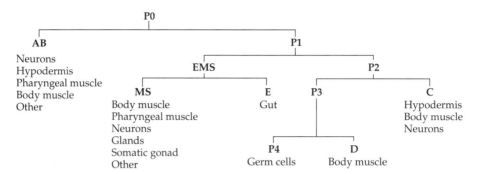

Figure 8.7   Cell fate map for wild type *Caenorhabditis elegans*. *(From Strome, 1995.)*

tration of dauer-inducing pheromone and a decrease in internal insulin levels; Kimura et al., 1997). This form is specialized for dispersal. Given appropriate cues, development resumes at the L4 stage and the larval to adult transition is made, characterized by a switch of the hypodermal cells from a proliferating to a differentiated state.

Ren et al. (1996) showed that the gene *daf-7*, encoding a TGF-β protein, is inhibited by dauer-inducing pheromone, but reactivated by food. The *daf-7* gene is expressed in chemosensory neurons whose processes are exposed to the external environment; killing these induces a dauer phenocopy (Bargmann and Horvitz, 1991). A complementary study (Zwaal et al., 1997) found that two G protein alpha subunits (GPA-2 and GPA-3), with promoter activity in subsets of chemosensory neurons, act in the process leading to dauer formation. Mutations activating these G proteins result in constitutive, pheromone-independent dauer formation.

Ruvkun and Giusto (1989) found that the gene *lin-14* creates a temporal gradient during development: it is expressed early but is not produced after the early larval stage (Figure 8.8A). This determines the switch from the early cell fate (L1) to the later cell fates (L2 . . . L4). A gain of function mutant that maintains expression of *lin-14* throughout development also causes an indefinite reiteration of the L1 fate. Interestingly, a loss of function mutation that causes the *lin-14* gene never to be expressed translates to an absence of cell fate L1 even from the early larval stage. Further studies have uncovered an ensemble of *lin* genes, organized in a hierarchy of heterochronic switches. Some of these control a wide array of developmental events; others specifically affect the switch between the larval and the adult stage, and yet others specify hypodermal cell fates.

It is also worth noting that the effects of these mutations are context-dependent. A null allele of the *lin-28* gene causes the nematode to skip the second stage of larval development, thereby producing a shortened ontogeny. However, this stage deletion occurs only if there is continuous development. If a dauer

(A)

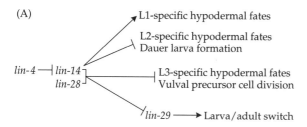

L1-specific hypodermal fates

L2-specific hypodermal fates
Dauer larva formation

lin-4 ⊣ *lin-14* ⊤
         lin-28 ⌐

L3-specific hypodermal fates
Vulval precursor cell division

lin-29 ⟶ Larva/adult switch

(B)

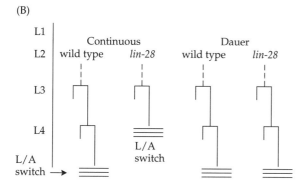

Figure 8.8 Expression of *lin* genes of *Caenorhabditis elegans*. (A) Proposed positive (→) and negative (⊣) regulation cascade. (From Ambros and Moss, 1994.) (B) Expression of *lin-28*, showing the larval to adult switch under normal and nutrient-deprived (dauer) conditions. L/A, larva/adult switch. (From Liu and Ambros, 1991.)

stage is induced, the *lin-28* mutation has no effect, demonstrating that the heterochronic gene is necessary to achieve stage L2 only when the organism follows one developmental trajectory, but not the other (Figure 8.8B; Liu and Ambros, 1991; Ambros and Moss, 1994). Although the presence of a dauer stage is correlated with longer life span, the genetic pathway controlling dauer development is *not* identical to that determining overall longevity (Larsen et al., 1995).

Two other well-studied systems in *Caenorhabditis* are the specification of the anchor cell from gonadal precursors, the Zn cells, and the formation of the vulva from a series of precursors in the ventral epidermis, the Pn.p cells (n refers to the cell position, numbered anterior to posterior; p or a denotes posterior or anterior). Sternberg and Félix (1997) reviewed the data on anchor cell (AC) formation, noting that in *C. elegans* the AC can be formed by either of two cells, Z1.ppp or Z4.aaa, and the decision appears to be stochastic. The alternate cell forms the ventral uterus (VU) cell. In other species, however, Z4.aaa is either biased toward or fixed to become the AC. Species also vary in the competence of Z1.ppp or Z4.aaa to produce AC or VU cells when one of the two cells has been ablated.

In *Caenorhabditis*, six of the twelve ventral epidermal precursor cells (VPCs), P(3–8).p, are competent to form the vulva (Figure 8.9A). Typically, three of these, P(5–7).p, are induced by a signal from the AC (due to the gene *lin-3*) to form the vulva; the others remain undifferentiated. Lateral signaling (via the product of the gene *lin-39*) among the VPCs is also important for cell fate specification

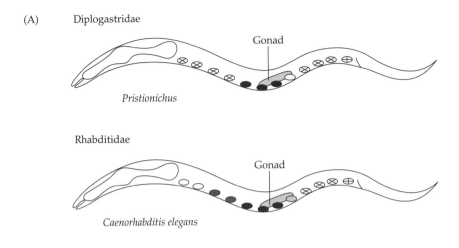

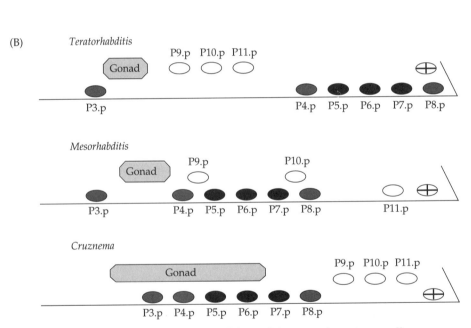

Figure 8.9 Depiction of cell positions and fates of the ventral precursor cells (P1.p–P12.p, left to right) in species of nematodes. (A) Comparison of *C. elegans* and *Pristionchus*. (B). The fates of VPCs P(3–12).p in three other species in the Rhabditidae. Shading: black, vulval progenitors; gray, competent; white, undifferentiated; +, hyp12; ×, cell death. (After Sommer, 1997b, and Sommer and Sternberg, 1994.)

(review in Sommer, 1997a). Other species have also been analyzed, and important variations upon this theme have been found: posterior displacement of the vulva due to cell migration or differences in the source of the inducing signal;

changes in competence of the VPCs; and programmed cell death of seven of the VPCs in some species (Sommer, 1997a,b).

In the family Diplogastridae, P(1–4).p and P(9–11).p undergo programmed cell death and have no descendants,[5] and in *Pristionchus*, P8.p has no competence for vulval formation (Figure 8.9A). Comparison of other members of this family reveals similar cell fates, but distinctive patterns of competence in cell ablation experiments (Sommer, 1997b). Species in three other genera in *C. elegans'* family, the Rhabditidae, all have posteriorly displaced vulvas, although the VPCs responsible are the same (Figure 8.9B; Sommer et al., 1994; Sommer and Sternberg, 1994). In *Cruznema*, the gonad grows posteriorly and the AC contacts the VPCs, inducing vulval development. In *Mesorhabditis*, there are only two cells competent to take on the primary cell fate, and the AC never contacts the VPCs, nor is it necessary for fate specification.

Of course, drosophilids and nematodes are not the only animal systems being genetically dissected; humans, mice, and sea urchins are also giving up their secrets to focused research programs. One of the resulting fascinating discoveries is the diversity of upstream and downstream sites that interact with the genes (Arnone and Davidson, 1997). Davidson's laboratory has done detailed work on the regulatory control of several genes in sea urchins. The *CyIIIa* cytoskeletal actin gene has three *cis*-regulatory modules with at least 20 binding sites that recognize nine different transcription factors (Kirchhammer and Davidson, 1996). Yuh and Davidson (1996) examined the regulation of the *Endo16* gene (coding for a cell surface glycoprotein), which is produced at different sites in the blastula, gastrula, and pluteus larva. They mapped a 2300 bp *cis*-regulatory region with over 30 high-specificity binding sites and at least 13 transcription factors acting as on-off switches or "volume controls" (Figure 8.10). These sites are organized into six modules, three with positive and three with negative control of expression. This diversity of regulatory sites clearly validates the potential for intricate control of gene expression.

THE PLANT: *ARABIDOPSIS THALIANA*

Thanks to a very intense recent effort by a large number of laboratories, we now know more about the developmental biology of the weedy mustard plant, *Arabidopsis thaliana*, than we did for any other plant in the past two centuries. Here we summarize the framework of our knowledge of different aspects of the bauplan of this organism. For example, we know that the few genes so far uncovered that "control" the development of the flower act as very high level regulatory elements in an intricate regulatory cascade—however, they do not actually *make* the flower.

---

[5] In *lin-39* mutants of *Pristionchus*, cells P(5–8).p also die. In *lin-39* mutants of *C. elegans*, P(5–8).p share the undifferentiated fate with the other VPCs (Eizinger and Sommer, 1997).

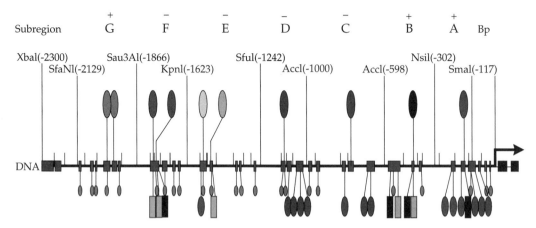

Figure 8.10 Map of the multiple binding sites for transcription factors on the *cis*-regulatory region for the expression of *Endo16* in *Strongylocentrotus purpurea*. Ten binding sites (ovals above the DNA) are subregion-specific. Four types of binding sites (large and small ovals, black and gray rectangles below the DNA) operate in multiple subregions. +/– indicates whether the subregion is a promoter or repressor of expression. (From Yuh and Davidson, 1996; see also Yuh et al., 1998.)

Establishing the Body Plan

The wild type development of *A. thaliana* follows a series of stages common to dicotyledonous flowering plants (Figure 8.11). After zygote formation, the two-celled stage 3 of development establishes polarity within the embryo, with the lower cell much longer. At stage 6, the fundamental domains of the body plan are recognizable: the upper portion forms the cotyledons (the embryonic leaves) and the shoot meristem; the central group of cells produces the hypocotyl (the embryonic stem) as well as the root system; the lower portion develops into the basal part of the embryo. At stage 7, the epidermis is a distinct layer, and at stage 13 the basic structure of the seedling is recognizable, consisting of the apical meristem, the two cotyledons, the hypocotyl, and the radicle or embryonic root.

It is quite easy to disrupt this developmental pattern by means of artificially induced mutations. EMS-mutagenesis has been widely used to systematically study the genetic basis of embryo development (Meinke and Sussex, 1979; Patton et al., 1991; Berleth and Jurgens, 1993). Some mutations affect only the radial patterning in the embryo (Laux and Jurgens, 1994). This pattern consists of the following tissues (from outside inward): the epidermis (protecting the plant from outside injuries), the "ground" tissue (which functions in mechanical support), and the vascular tissue (through which nutrients and water are distributed to all parts of the plant). These tissues are present in all plants, but in some they differentiate further into more specialized structures such as trichomes (from the epidermis) or tracheids (from the vascular tissue). The mutant

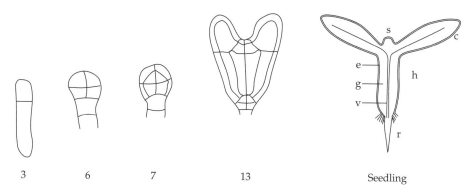

Figure 8.11   Basic wild type development of the *Arabidopsis* embryo. Stage 3 is made of two cells, after division of the zygote. Stage 6 is known as the octant stage, stage 7 the dermaton, and stage 13 the mid-heart stage. The last panel shows the fully formed seedling (not to same scale). s, shoot meristem; c, cotyledons; h, hypocotyl; r, root primordium; v, vascular strands; g, ground tissue; e, outer epidermis. (After Laux and Jurgens, 1994; Mayer et al., 1991.)

short root deletes an entire cell layer (the root endodermis) in the embryo. The *keule* mutant shows greatly enlarged cells in the external embryonic layer, related to the epidermis, and the result is an embryo with a distorted radial pattern. The *knolle* mutation causes an altered radial pattern related to cell enlargement that affects all layers, not just the epidermis.

Considerably more information has been gained from the study of mutants that alter the apical-basal pattern (Mayer et al., 1991; Laux and Jurgens, 1994). An extensive series of single mutations deleting specific sections of the seedling has been described (Figure 8.12). Nine alleles at the *gurke* locus effectively erase the apical region; the resulting seedling has a fully formed root and shoot system, but lacks cotyledons and an apical meristem. Five alleles at the *fackel* locus delete the central portion of the embryo, producing a seedling that completely lacks the shoot. Mutant alleles at the *monopteros* locus result in the lack of the basal region, with seedlings showing only the apical meristem and cotyledons (this locus is therefore complementary to *gurke* and partially overlapping *fackel*). Finally, fifteen alleles at the *gnom* locus have the ability to delete *both* the apical and the basal region of the embryo, resulting in seedlings with only a hypocotyl (complementary to *fackel*).

This ensemble of mutations still yields only a very rudimentary characterization of the components of embryo development in flowering plants. However, they demonstrate that the fundamental pieces are, at least to some extent, separable from one another, enabling further empirical work to be pursued. There has not been much analysis of double or multiple mutants (to investigate epistatic interactions among these genes), or comparative analyses using *Arabidopsis*

Region affected

Mutant phenotype

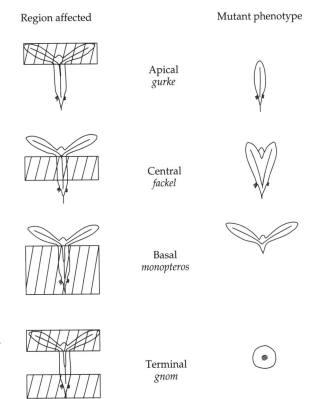

Apical
gurke

Central
fackel

Basal
monopteros

Figure 8.12 Four types of mutations affecting the apical-basal pattern formation in *Arabidopsis thaliana*. Each class of mutations effectively deletes a specific portion of the seedling. (From Mayer et al., 1991.)

Terminal
gnom

genes as probes to uncover similar sequences in distantly related species. Nevertheless, this body of work does stand out as an invaluable complement to the classical descriptive work on plant embryogenesis carried out over the last two centuries.

Roots

Plant root systems are a much understudied component of a viable phenotype, in relation to the importance of their roles in seeking and taking up water and micronutrients from the soil. The study of root development has proceeded along the same lines outlined above, with the discovery and description of a series of mutants affecting different aspects of root formation. In fact, some of the same mutations directly affect the root system, for example, *gnom* and *monopteros*. Other mutations result in less dramatic outcomes: *hobbit*, for example, interferes only with the formation of the root meristem, but not the rest of the embryonic root (Benfey and Schiefelbein, 1994a). Another group of mutations alters the regular pattern of differentiation of the root system.

Root hairs seem to be particularly easy to study by mutagenesis analysis. The genes *rdh1* and *reb1* both change the size of the initial primordium from which root hairs develop. *rdh2* root hairs appear anatomically normal but do not elongate, while changes in the shape of the hair can be caused by mutations at the *tip1*, *rdh3*, and *rdh4* loci (Benfey and Schiefelbein, 1994a). Some of these mutations have unexpected and potentially very interesting pleiotropic effects. For example, *tip1* is also characterized by disruption of pollen tube growth (is there some homology of pollen tubes and root hairs?). Researchers have also uncovered anatomical mutants that affect the morphogenesis of the root system. Some affect the degree of lateral expansion of a root, but act on different segments: *cobra* expands the epidermis, *pom-pom* enlarges the epidermis and cortex, *sabre* expands only the cortex, and *lion's tail* augments the stele (i.e., the central vascular network of roots). *short-root* has no endodermis: the roots initially develop normally (and continue to do so in agar), but soon stop because of the absence of an elongation zone at the tip (Benfey and Schiefelbein, 1994a).

A fascinating series of mutants have been discovered that affect the response of the root system to external stimuli, in particular, to gravity and light. The wild type root grows with a gravity gradient—positive gravitropism—and against a light gradient—negative phototropism (Benfey and Schiefelbein, 1994b). The two responses are therefore normally parallel in a natural setting. However, it is relatively easy to screen for mutations decoupling these responses by growing wild type seedlings and potential mutants in agar on a transparent plate (Figure 8.13). Rotation of the plate by 90° will alter wild type growth accordingly; mutants will follow the original direction. The position of a light source can also be varied so as to create a 90° angle between light and gravity vectors, allowing the experimenter to discern gravitropically versus phototropically defective mutants.

Vegetative and Leaf Development

During the vegetative phase *A. thaliana* grows a basal rosette of leaves prior to the onset of reproductive growth. Surprisingly, the entire vegetative phase of growth can be skipped by altering a single gene. The *embryonic flower* mutation yields a plant that essentially flowers as a seedling, with no recognizable basal leaves[6] (Sung et al., 1992)! Once the transition from seedling to vegetative growth has been achieved, however, there are other processes that can disrupt normal leaf development (Hake and Sinha, 1991; Reiter et al., 1994; Jenks et al., 1996). Tsukaya (1995) has classified the known mutations affecting leaf morphology into four classes. First, there are genes that affect only cotyledons. A second class of mutations is complementary in that it affects only mature leaves

---

[6]This plant does produce cauline leaves, small leaves on the reproductive shoot normally produced only during the reproductive phase in *A. thaliana*, and that presumably contribute little to photosynthesis.

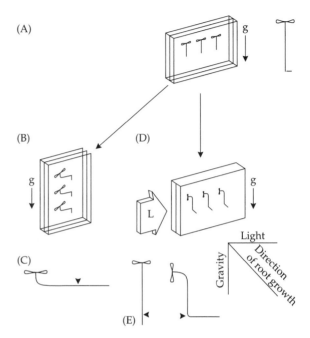

Figure 8.13 Mutational screening to uncouple the effects of gravitropism from those of phototropism. (A) Wild type seedlings grown on a vertical plate experience gravity and light in the normal way and serve as controls; (B) the plate is rotated 90°, and wild types (on plate) respond to the change; (C) mutants (enlarged) do not and continue growth in the same direction; (D) light comes from one side, creating a 90° angle with the gravity vector; wild type (on plate) reorient their cotyledons and root. (E) One mutant does not respond to light (enlarged, left), the other does not respond to gravity (enlarged, right). (After Benfey and Schiefelbein, 1994.)

but not the cotyledons. This suggests that the two types of leaves are determined in part by genetically independent systems. A third class of mutations affects both cotyledons and adult leaves. The fourth class comprises a miscellaneous group of mutants that do not specifically affect leaf morphology, but rather disrupt the overall pattern or rate of growth, altering leaf morphology pleiotropically. Further mutations affect only the leaf blade or only the petiole, with others altering both (Tsukaya, 1995). Some mutations mechanistically link leaf and flower development (Hofer et al., 1997), perhaps speaking to the theory that flower parts are simply modified leaves. For example, the *acaulis2* mutant, although originally described as shortening inflorescence length, also reduces the leaf petiole, but not the leaf blade.

Leaf expansion (width and length) is controlled by three genes (Tsukaya, 1995; Figure 8.14). The *angustifolia* mutation reduces width, but not length; the *rotundifolia* mutation, on the other hand, alters only length.[7] However, a third type of mutation (at *ACAULIS1*), interferes with both. Again, there is evidence of a hierarchical architecture of the phenotype, reflected to some extent in the dif-

---

[7] Leaf length and width have been operationally treated as independent characters, although correlations between them are often quite strong. This evidence suggests that there are mutations that can affect them independently.

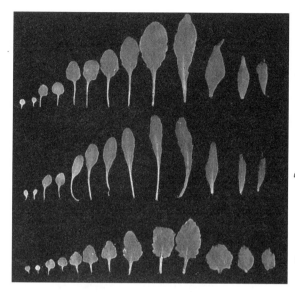

Wild type

angustifolia mutant

rotundifolia mutant

Figure 8.14 Mutations affecting leaf expansion in *Arabidopsis thaliana*. From the left are the two cotyledons, followed by eight rosette leaves, and then by three cauline leaves. Notice the opposite effects of the mutants *rotundifolia* and *angustifolia* on the overall shape of the leaf. (From Tsukaya, 1995.)

ferent levels or domains of action of regulatory genes. Other fascinating leaf-related mutants have been described in the primary literature and summarized in recent reviews (Garrido et al., 1991; Hake and Sinha, 1991; Meinke et al., 1994; Reiter et al., 1994; Smith and Hake, 1994; Tsukaya, 1995; Jenks et al., 1996).

Inflorescences and Flowers

The transition from the vegetative to the reproductive phase in *A. thaliana* has been intensively studied, and a combination of mutagenesis and molecular analysis has now given us a relatively detailed picture of the steps involved in the initiation of reproduction, production of inflorescences, and formation of flowers (Ma, 1998; Figure 8.15).

Flowering time in this species is markedly affected by almost any environmental factor (Pigliucci et al., 1995). The reactions to photoperiod and vernalization are particularly well described (Martinez-Zapater and Somerville, 1990; Koornneef et al., 1991; Bagnall, 1992, 1993; Karlsson et al., 1993; Chandler and Dean, 1994). Of 22 loci known to directly affect flowering time, 12 are sensitive to photoperiod only (flowering is delayed under short days), 7 are sensitive to vernalization only (exposure to cold accelerates flowering), and none are sensitive to both (Coupland, 1995). These mutations can be divided into those caus-

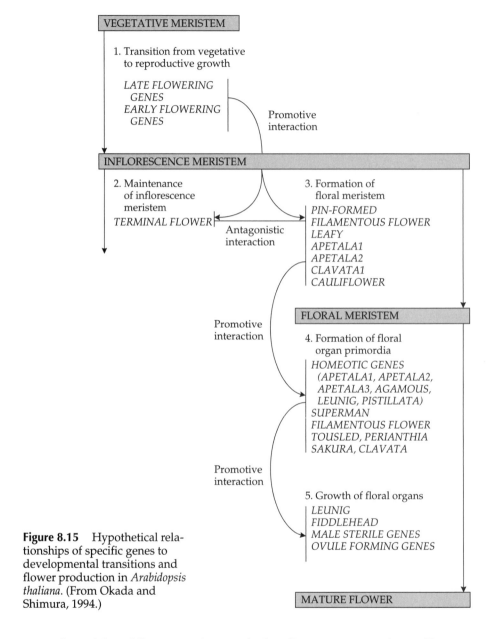

Figure 8.15 Hypothetical relationships of specific genes to developmental transitions and flower production in *Arabidopsis thaliana*. (From Okada and Shimura, 1994.)

ing early or delayed flowering. Among the late flowering genes, those of known function are transcription factors, light receptors, or enzymes involved in the biosynthesis of the hormone gibberellin. Much less is known of the proteins coded by early flowering genes, and the only two that have been studied in any detail are a light receptor and a transcriptional repressor (Coupland, 1995).

Despite this elaborate genetic machinery, and all its environmental feedback, *A. thaliana* will eventually flower under normal photoperiod (although not all eco-types do so under very short days).

Commitment to the floral fate appears to be progressively acquired, rather than a single switch (Ma, 1998), and is followed by three further phases of development, each characterized by a specific pattern of gene expression (Schultz and Haughn, 1993). The so-called FLIP (FLoral Initiation Process) genes are activated by unidentified factors named COPS (Controllers Of Phase Switching, which integrate environmental signals), and the genes *LEAFY* (*LFY*), *APETALA1* (*AP-1*), and *APETALA2* (*AP-2*) (Simon et al., 1996). During the second phase, *LFY* blocks the production of leaves on the inflorescences. Eventually, all three (*LFY, AP-1* and *AP-2*) act in concert to catalyze inflorescence formation, the first step toward the activation of the floral program. Simultaneously, these same genes inhibit the action of the homeotic genes responsible for the formation of the sexual organs within the flower because the flower has yet to form the perianth organs, sepals and petals. During the third phase, the expression of all FLIP genes becomes weaker, and the Class C genes (see below) are expressed, resulting in formation of anthers and stigmas (male and female sexual structures) (Schultz and Haughn, 1993).

Several other genes have been identified that affect inflorescence formation, but do not affect flower development, suggesting that the two sequential (and then partially simultaneous) processes are controlled semi-independently. The most striking mutations affecting these processes are *terminal flower* and *cauliflower*. The *terminal flower* gene inhibits the formation of the inflorescence altogether, forming a plant with a single flower emerging from the rosette (Shannon and Meeks-Wagner, 1991). This is the normal phenotype of a close relative of *A. thaliana, Cardaminopsis petraea*, when grown under a long photoperiod (M. Pigliucci, pers. obs.). The *cauliflower* gene, in concert with *AP-1*, causes the inflorescence production program to loop repeatedly, without ever initiating flowers. The resulting phenotype resembles a miniature version of the commercial cauliflower (*Brassica oleracea*, a distant relative of *A. thaliana*). When the *A. thaliana cauliflower* gene was used as a probe, it was found that the cauliflower cultivar does indeed contain a nonfunctional homologue of the *A. thaliana* gene, thereby providing a molecular explanation for one of the oldest flower abnormalities known (Kempin et al., 1995).

The basic account of flower formation, the ABC model, is becoming well known, and we will present only a simplified version. Several more genes and layers of regulation have now been discovered, and we are still far from a complete molecular picture of the events catalyzing the development of this crucial structure in the plant kingdom (Coen and Meyerowitz, 1991; Ma, 1994b; Meyerowitz, 1994b; Okada and Shimura, 1994). A normal flower of *A. thaliana* has four whorls: the external whorl of four sepals, a second one of four petals, a third of six stamens, and the innermost "whorl" of one carpel. Mutation analysis, coupled with techniques capable of tracing temporal and spatial gene expression, have identified three regions of influence. These regions correspond to the so-called ABC classes of genes. Class A genes, such as *AP-2*, are necessary for the formation of sepals in the first (outermost) whorl. When they act in concert with Class B genes such as *AP-3*, petals are formed, normally in the second whorl. The third whorl is characterized by the activity of Class B and Class C genes (e.g., *AGAMOUS—AG*), but not Class A, which are shut off. The result is the formation of stamens. Finally, only Class C genes are normally active in the fourth whorl, producing the carpel.

This scenario has been confirmed several times by creating a series of single and double mutants (Irish and Sussex, 1990; Bowman et al., 1991, 1993; Jack et al., 1992; Crone and Lord, 1994). For example, suppressing *AP-2* causes the expansion of *AG* expression into the first and second whorls (implying interactions of spatial gradients within the flower). The resulting plant has a sequence of organs that reads: carpel, stamen, stamen, carpel. Similarly, elimination of Class C function causes the following sequences of structures in the four whorls: sepals, petals, petals, and sepals. Apparently, a single mutation replaces stamens with petals and carpels with sepals. It is thought that Class A genes are repressed by the onset of Class C genes. Without C genes, A genes simply keep working and produce exactly the phenotypes that they normally specify (i.e., sepals when alone, petals when associated with B genes).[8] The details of the control of these processes are still being unraveled (e.g., Sakai et al., 1995; Goodrich et al., 1997).

The evolutionary importance of all these mutations lies in the fact that, in the cases in which researchers have looked, they are remarkably conserved throughout (and sometimes beyond) the plant kingdom. For example, genes conferring early flowering in *A. thaliana* have a similar effect in the very distantly related aspen (which is a tree, not a small herbaceous weed; Weigel and Nilsson, 1995). The ABC genes have demonstrated homologues in several other flowering plants (Endress, 1992). In fact, application of a molecular clock to MADS-box genes suggests divergence of the various lineages of floral homeotic genes before 400 MYA (Purugganan, 1997), that is, before the appearance of flowering plants. The evolution of phytochrome genes (involved in light perception) has also been traced back to a pre-angiosperm lineage (Furuya, 1993). Soon we hope to be able to draw a map of comparative developmental biology based not only on the similarity and homology of the morphological structures, but also of the genes regulating their appearance.

EPIGENETICS AND PHENOTYPIC PLASTICITY

An increasing number of studies have begun to address a fundamental topic previously unexplored in experimental systems due to its inherent complexity,

---

[8] It could be argued that the A genes are now expressed in the wrong place (the third and fourth whorl, heterotopy), or at the wrong time (after they would have normally been turned off). In fact, *both* are true, since these genes do not normally function at that stage of development or in that area of the flower.

namely, the genetic basis of developmental plasticity. An example of the molecular genetics of developmental plasticity is the study by Simpson et al. (1986) on the developmental and environmentally induced expression of the *rbcS* gene. Chimeric genes obtained from peas were used as probes for the study of the behavior of the whole gene family in tobacco. The authors were able to demonstrate that the developmental stage of the plastid plays a key role in the activation of the genes. Different members of the gene family were expressed in specific tissues or responded to environmental factors, and these responses were mediated by several photoreceptors present in the plant cell. Campbell et al. (1998) found that in cultures of the cyanobacterium *Synechococcus* exposed to UV-B, there was a rapid change in expression of a family of three *psbA* genes encoding photosystem II D1 proteins. The constitutively expressed *psbAI* gene encodes D1:1; *psbAII* and *psbAIII* genes encode D1:2 and are induced by UV-B. Mutants lacking functional copies of these latter genes have significant inhibition of photosystem II function.

Emery et al. (1994) previously had revealed the role of ethylene in the plasticity of stem elongation in *Stellaria longipes*. Kathesiran et al. (1998) elucidated the mechanism of its action by examining two genes, *ACC synthase* and *ACC oxidase*. They documented genetic variability for *ACC synthase*, and demonstrated the differential regulation of this gene; transcripts under warm short day conditions are different than those produced under cold long day conditions.

The development of plant root systems is known to be sensitive to the availability and distribution of nutrients within the soil; for example, the proliferation of lateral roots within nitrate (NO_3^-) rich soil patches. Zhang and Forde (1998) characterized an NO_3^- - inducible *Arabidopsis* gene (*ANR1*) that encodes a MADS box transcription factor. Transgenic plants with repressed *ANR1* did not respond to NO_3^--rich zones by lateral root proliferation, suggesting that *ANR1* is a key component of plastic responses to soil nitrogen by *Arabidopsis* roots.

LIMITATIONS TO A "MODERN" APPROACH

Modern molecular genetics is, of course, not a panacea, and it may be therefore not be the final solution to our quest to understand epigenesis. As visualized in Figure 8.16, there are complementary approaches to the study of the relationship between genotype and phenotype, the "direct" and "reverse" genetics. Classical (direct) genetics is a "top-down" approach, in which one starts from the phenotype and attempts to identify genes somehow responsible for that phenotype. Reverse genetics is a "bottom-up" approach, in which one has a piece of genetic material and is curious to find out what phenotypic effects it has. In a sense, both direct and reverse genetics are scratching the surface of this box; they just approach the problem from distinct vantage points.

The limits that we are trying to outline here are part of the never-ending debate between reductionist and holist philosophies in the quest for an understanding of the natural world. It can be argued that the impressive progress of

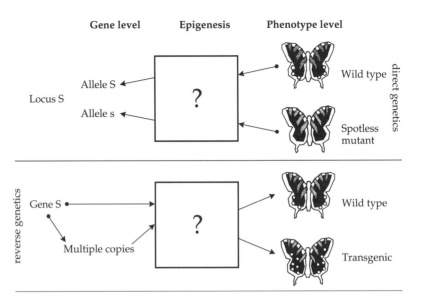

Gene level   Epigenesis   Phenotype level

Figure 8.16 Two approaches to the study of the genotype-phenotype mapping function. The upper panel depicts the classical direct genetic approach. The experimenter screens for mutations affecting the phenotype and infers the existence of one or more loci, and the relationships between two or more alleles. The lower panel shows a reverse genetic approach. For example, a known genetic element can be inserted in multiple copies as a transgene, and the effects on the phenotype of its overexpression can be studied. Note that both approaches largely leave the epigenetic process unaddressed—direct genetics scratches the outer surface of the "black box," while reverse genetics works from the core. Neither set of techniques penetrates very far toward the other.

the most reductionist of the biological sciences, molecular biology, is in fact helping to reinforce a scenario proposed long ago by the holist camp—the essence of a biological system is in the emergent properties of its interacting component parts. We can dismantle the system piece by piece, but the more we do that, the more we realize that these emergent properties can only be investigated when the parts are together. In more pragmatic terms, this is an old problem in mutagenesis studies. For example, we will never be able to uncover all the important genes contributing to the normal development of an embryo, simply because the mutations of many of these genes are lethal, precluding the study of their phenotypic effects. In other cases, redundancy of function also masks the true nature of the mutation. In a metaphorical sense, this is similar to attempting to understand how an automobile works by taking it apart, impairing one major function at a time. For example, we might surmise that the loss of directional ability caused by the "steering wheel-less" mutant is a fundamental mutation early in the guidance system pathway of the car. However, from our knowledge about automobiles, we know that the steering wheel is actually the terminal component, and we have not really untangled any of the actual complexity. And a car is orders of magnitude less complicated than even the simplest living organism . . .

In fact, the whole concept of the emergent properties of epistatic systems subverts a wholly reductionist approach to studying development. One example of the integration of holistic and reductionist approaches has been offered by Kaplan and Hagemann (1991). The cell theory is considered one of the triumphs of modern biology, and it is rooted in the mechanistic idea that by tracing in detail the movements and descendants of individual cells, we will be able to understand how the shape of that organism changes through development. Kaplan and Hagemann point out, however, that while this is all very well for animals, plant cells simply *do not move*. They refer the reader to a series of interesting examples in algae in which morphogenesis is independent of cell division (Figure 8.17). The shape of the organism is determined before cells are formed, when the whole cytoplasm and several nuclei share one common cellular environment. Clearly, the alga has to have a way of reaching its final shape without building it gradually by cell division, a notion very difficult to explain on a strictly mechanistic basis. We are, of course, not advocating any "vitalistic" principle here, but merely pointing out that dissecting a system is only one (incomplete) way of understanding it.

From the discussion and examples above, it is clear that epistatic gene action leads inevitably to biologically emergent properties. Recent empirical and theoretical literature in developmental biology has focused on alternative paradigms that acknowledge emergence as an inherent characteristic of living organisms. Nijhout et al. (1986) developed computer simulations that apply cellular automata theory to the study of development (see also Chapter 4). Their work's significance here lies in their discovery of the inability of even such a simple system to predict the "phenotypic" effect of a single mutation. Even though all the genetic rules were specified *a priori*, the system was open-ended and complex enough to exhibit emergent properties. Nijhout and coworkers described the behavior of their system in a way that could be used to summarize what we know of the development of living organisms:

1. Mutations of a single rule can affect several apparently unrelated characteristics (the cellular automaton equivalent of universal pleiotropy).

2. The pattern of cell division can be altered in similar ways by apparently unrelated mutations (genetic redundancy).

3. Canalization of a phenotype occurs whenever the "physiological" basis of that phenotype includes a threshold for the action of a "morphogen."

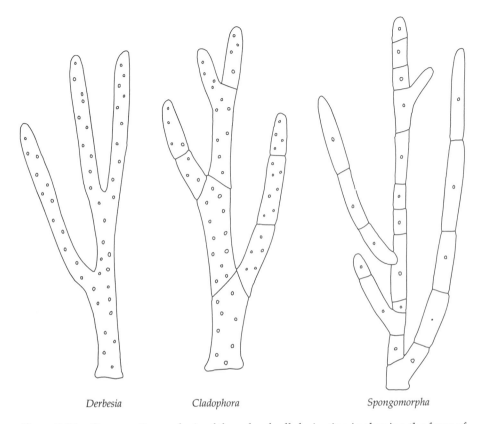

Derbesia   *Cladophora*   *Spongomorpha*

Figure 8.17   Comparative analysis of the role of cellularization in shaping the form of an organism. Some siphonous green algae clearly assume their final shape *before* cells are formed, and in coenocytic species cells are not formed at all. Left, the coenocyte *Derbesia* has multiple nuclei, but no subdivisions. Center, the partly cellular *Cladophora* is comprised of many more nuclei than cells. Right, the fully multicellular *Spongomorpha*. (From Kaplan and Hagemann, 1991, © American Institute of Biological Sciences.)

4. The effect of a mutation is entirely dependent on the developmental, physiological, and biochemical context in which it occurs.

5. The overall developmental pattern ("bauplan") is fairly resilient to changes induced by mutations, and remains recognizable despite genetic alterations.

6. Most of the phenotypic divergence caused by mutations is concentrated in later developmental stages, while the earlier ones retain a higher degree of similarity (as already pointed out by de Beer, 1930).

Analogously, molecular biology has led the way toward our acceptance that genes do not represent "blueprints" of organisms, and therefore do not constitute simply interpretable instructions of a more or less linear program. More properly, genes are molecules that code for other molecules, which in turn have only local effects. They are analogous to subroutines in a complex non-linear program, calling upon and modifying one another in a variety of ways. Consequently, there is no such a thing as a gene "for" a complex trait or behavior. All genes do is produce enzymes, receptors, and information carriers. It is from the local interactions of these simple elements, and the combinations of the products of these local programs, that complex phenotypes "emerge." *The process of development is itself a temporal reaction norm*, as cells express (or silence) genes depending on the local "environment" that they are in.

Wray (1994) suggested that there are three fundamental questions likely to continue to guide research in the field:

1. Since gene products involved in fundamental aspects of development appear to be highly conserved through evolutionary history, how can they operate in very diverse cellular and organismal "environments"?

2. Given the same premise of molecular conservatism, what causes the evolution of so many different forms and alternative developmental pathways? This is to some extent the distinction between functional genetics and the genetics of natural populations to which we were referring above.

3. Sometimes the same developmental processes are controlled by different genes ("genetic piracy": van Valen, 1982; Roth, 1988). How does that happen through evolutionary time?

Undoubtedly, satisfactory answers to these questions will go a long way to finally allowing a clearer image of the contents of the gray box of epigenetics.

CONCEPTUAL SUMMARY

1. Modern molecular biology is providing very powerful tools to peek into the "black box" of development. Unsuspected links among apparently morphologically unrelated structures are appearing, and the meaning of *homology* itself needs to be questioned.

2. The word "epigenesis" was proposed by Waddington to maintain a conceptual separation between the direct and indirect effects of gene action. The indirect effects play a role much larger and more multifarious than the direct biochemical role acted out by gene products.

3. We could not resist our own definition of epigenesis: Epigenesis is an ensemble of processes that propagate phenotypic characteristics throughout development. These processes derive from either indirect effects of gene action (emergent properties) or non-genetic phenomena (e.g., cell-cell or hormone-target communications).

4. One of the major results of modern developmental molecular genetics is the discovery and characterization of a series of important regulatory switches such as the Hox and MADS box genes. These are highly conserved across all sorts of living organisms, from animals to plants to bacteria.

5. Much of the current progress in molecular developmental genetics is being achieved by the use of a few model systems. We discuss some of the recent advances in the mechanistic understanding of three of these: the insect *Drosophila melanogaster*; the nematode *Caenorhabditis elegans*; and the flowering plant *Arabidopsis thaliana*.

6. Comparative studies of the expression of homeobox genes has fueled the hopes of understanding macroevolutionary steps within and across major groups of organisms as the result of altered timing or place of expression of a reduced number of key regulatory genes.

7. It's not all molecular. Theoretical investigations (complexity theory, cellular automata theory) and molecular biology itself are indicating that while molecular tools are and will be invaluable for our understanding of development, the whole story is bound to be much more complex. There are suites of phenomena that either have not or are very unlikely to ever receive a full mechanistic explanation. This stems in part from the complexity of gene/gene and epigenetic interactions, and from the impossibility of disrupting most of these by mutation without also killing the organism.

8. Modern studies of the genetics and molecular biology of developmental plasticity have begun to elucidate its role in varied ecological settings and to suggest its contributions to macroevolutionary trends.

9. The mechanistic and whole-organism approaches to development are delving into the black box from two sides; neither is more important or informative, and both are still quite far from each other.

10. Development and the epigenetic processes represent a temporal reaction norm as cells respond to the changing internal environment.

CHAPTER 9

Evolution of Developmental Reaction Norms and Phenotypes

A genuinely labile epigenetic system ought to be of the greatest interest to an evolutionist, and it is a far, far better thing than an occasional atavistic slip in a mature epigenetic system . . . After all, all God's hoatzins got more than wings. (Rachootin and Thomson, 1981)

INTEGRATING ALLOMETRY, ONTOGENY, AND PLASTICITY

The necessity of integrating the separate components of phenotypic expression has been recognized since very early in the ontogeny of evolutionary thought. Calls for the incorporation of development pre-date the origin of the phrase "the modern synthesis" itself (e.g., Haldane, 1932b), and continue to the present (e.g., Hall, 1992a; Wray, 1994; Coleman and McConnaughay, 1995; Galis, 1996). The recent nascence of development as a major player in evolutionary theory surely is attributable in large part to Gould, who repopularized interest in both allometry and heterochronic phenomena (Gould, 1966, 1977, 1979; Alberch et al., 1979). Phenotypic plasticity, despite the excellent treatments by Bradshaw and Levins (Bradshaw, 1965; Levins, 1968), and the efforts of Jain (Marshall and Jain, 1968; Jain, 1978, 1979), also hung around on the fringes until the independent convergence on the topic in the early 1980s by Stearns, Scheiner, Schlichting, and Via.[1]

The two-way integrations of allometry, ontogeny, and plasticity have been documented in Chapters 4, 5, and 7. Stearns has been particularly persistent and successful in promoting these mergers through both his conceptual syntheses and his own research program (e.g., Stearns, 1982, 1983; Stearns and Koella, 1986), calling for the explicit documentation of developmental processes, multivariate allometries, and reaction norms. Kaplan's students have led the movement toward putting the "development" back into developmental plasticity (Lord, 1981; Diggle, 1991, 1993, 1994; Jones, 1992, 1993). Investigations of patterns of phenotypic integration have also become more common in the past decade and are more often coupled with ontogenetic or environmental approaches (Cheverud, 1982b, 1995; Schlichting, 1986, 1989a; Kingsolver and Wiernasz, 1987, 1991; Zelditch, 1988; Zelditch and Carmichael, 1989; Wagner, 1990, 1995).

The three-way integration has been explicitly called for only recently (Coleman et al., 1994; Coleman and McConnaughay, 1995; Pigliucci and Schlichting, 1995; Pigliucci et al., 1996), and studies incorporating all three components are rare. However, those that have been done are particularly illustrative of the potentials of this research agenda.

Three-Way Integration Examples

The nonpareil of DRN studies is the comprehensive set of investigations by Brakefield and his colleagues on the seasonal polyphenism of the African satyrine butterfly, *Bicyclus anynana*. *B. anynana* occupies savannah-rainforest ecotones in wet-dry seasonal areas (Brakefield and Larsen, 1984; Roskam and Brakefield, 1996). The wet and dry season forms are phenotypically distinct, differing in the size of the eyespots (two on the forewing and seven on the hindwing) and the banding pattern on the wings (see cover illustration). The seasonal shift in rainfall is associated with changes in temperature and in the behavior of the butterflies. In the cooler dry season, butterflies rest on dried grasses or leaf litter and fly sparingly; oviposition will not occur until rains have stimulated regrowth of grasses. By contrast, in the warmer wet season, adults are active fliers, searching for mates and oviposition sites among the lush vege-

---

[1] The latter three all published their first papers on plasticity in the same year (Scheiner and Goodnight, 1984; Schlichting and Levin, 1984; Via, 1984a), Stearns several years earlier.

tation (Brakefield and Reitsma, 1991; Windig et al., 1994; Brakefield, 1997). Comparison of survival of both forms in the two seasons supports the adaptive interpretation of cryptic matching in the dry season and deception via false eyes for predator avoidance in the wet season (Reitsma, cited in Brakefield, 1997).

Various experimental studies have demonstrated that the wet season syndrome is related to faster development, whether induced by higher temperatures (Windig, 1994b; Holloway and Brakefield, 1995; Roskam and Brakefield, 1996), temperature fluctuation (Brakefield and Kesbeke, 1997), food quality (Kooi et al., 1996), or hormone treatments (ecdysteroids: Koch et al., 1996). There is substantial genetic variability for both eyespot size (Holloway et al., 1993; Monteiro et al., 1994) and the plasticity of eyespot size (Windig, 1994b; Holloway and Brakefield, 1995; Brakefield et al., 1996), as shown both by family variation in different temperature environments and by responses to direct selection. The genetic architecture of various components of the wet-dry season phenotypes has been examined as well (Holloway et al., 1993; Windig, 1994a; Brakefield et al., 1996). Monteiro et al. found that selection for size of one eyespot resulted in responses by other eyespots on the wing (Monteiro et al., 1994), as did selection for the proportion of black or gold (Monteiro et al., 1997a). Selection for elliptical eyespots also altered wing shape and scale pattern around the eyespot (Monteiro et al., 1997b).

Brakefield et al. (1996) went further, in a set of studies combining developmental and quantitative genetics, as well as phenotypic plasticity of eyespot formation. They selected lines at intermediate temperatures (20°C) for their similarity to either the wet season form (raised at 27°C) or the dry season form (raised at 17°C). Therefore selection was for increased or decreased eyespot size plasticity to temperature. Response to selection was substantial. High lines raised at 17°C had eyespots similar in size to wild-caught (23°C) wet season forms, whereas dry lines at 27°C did not approach the eyespot size of the wet season form (Figure 9.1). The family by temperature interaction, representing the genetic variability for plasticity, was significant for the unselected line, but not for either the H or L lines. The authors examined the pattern of expression of the homeobox gene *Distal-less* (*Dll*, previously implicated in control of eyespot formation; Carroll et al., 1994) in the fifth instar larvae and pupae. *Dll* expression did not differ between wet and dry season forms at the fifth instar (see cover photo), but had diverged substantially by 24 hours following pupation (cover photo; Figure 9.2), leading to the marked difference in eyespots apparent in adults.

Contrasting patterns of eyespot formation and *Distal-less* expression have been examined in mutants of *B. anynana* as well as in other butterfly species (French, 1997). Early patterns of gene expression in the butterfly *Precis coenia* are quite similar to those of *Bicyclus*, with *Distal-less* stripes forming early in each wing subdivision, but only two foci remain later, giving rise to the two eyespots (Carroll et al., 1994; Brakefield et al., 1996). The spot-less monarch butterfly

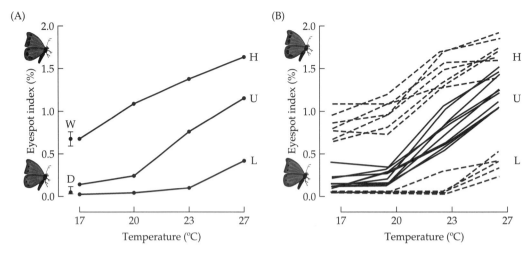

Figure 9.1   Reaction norms to temperature of eyespot size of unselected lines (U) of *Bicyclus anynana* and lines selected for larger (H) or smaller (L) eyespots at intermediate temperatures. (A) Mean reaction norms depicting the response to selection. W and D indicate the mean values for wild-collected wet and dry season forms. (B) Variation in reaction norms for female full-sib families. (From Brakefield et al., 1996.)

(*Danaus plexippus*), however, has no *Distal-less* foci (Brakefield et al., 1996). Investigation of several mutants of *Bicyclus anynana* showed varying alterations of gene expression: *Cyclops*, with a single large hindwing eyespot with a central line instead of circle, has stripes of *Dll* at the center of the focus instead of the circular spot; *Spotty*, with four eyespots on the forewing instead of two, has two additional *Dll* foci; and *Bigeye*, with much larger hindwing spots in the adult, is indistinguishable from the wild type in the late fifth instar (Brakefield et al., 1996). Thus both position and timing of gene expression can be altered to create novel phenotypic patterns (see also the spectacular work by Nijhout 1991, 1994a,b).

Gedroc, McConnaughay, and Coleman (1996) were interested in addressing the question of whether plants are able to optimally adjust resource partitioning under different environmental conditions. They examined allometric changes in root/shoot ratio as a function of age and size in two plant species, *Abutilon theophrasti* and *Chenopodium album*. Plants were started in two constant environments, low and high nutrients (an eight-fold range), and a subset of each group was later switched to the alternate nutrient condition. Three results are especially noteworthy in the context of the DRN. First, the relationship between root/shoot ratio and age differed between the low and high nutrient treatments (Figure 9.3A,B). Second, the allometric relationship between root/shoot and total size also varied with treatment (Figure 9.3C,D). Third, the responsiveness

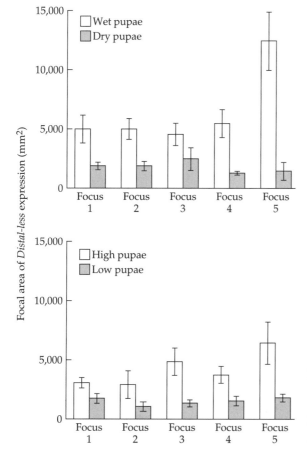

Figure 9.2 Effects of plasticity and genetic variation in plasticity on expression of *Distal-less* in *Bicyclus anynana* 24 hours after pupation. Foci 1–5 are the five anteriormost wing subdivisions, and focal area refers to the size of the *Dll* expression patch. Wet and dry (upper panel) refer to the seasonal forms of the unselected stock produced at 27°C and 17°C. High and low (lower panel) refer to lines selected at intermediate temperature (20°C) for their similarity to either wet (high) or dry (low) season forms (i.e., for more or less plasticity to temperature). (From Brakefield et al., 1996.)

to altered nutrient levels varied as a function of plant age. Plants switched to the alternate nutrient level at 3 weeks showed little modification of root/shoot relationships.[2] By conducting the ontogenetic analysis, the authors determined that, although the root/shoot ratio is always higher in low nutrient plants, the difference between treatments is largely due to a marked increase in root biomass at a very early age. Thereafter, plants in both treatments have relatively

---

[2] These authors interpreted their results as mixed support for the ability of plants to optimally alter the root/shoot ratio. The response of seedlings was taken as positive evidence; the lack of response by older plants was seen as evidence for ontogenetic constraints. We would interpret the latter result somewhat more circumspectly. The inability to respond to the shock of a rapid transition to 8× or 16× nutrient levels seems just as likely to be a function of the rarity of such a shift in nature.

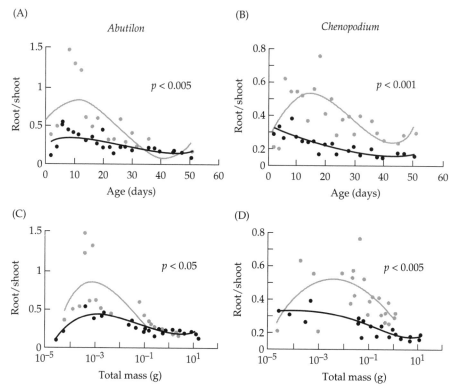

Figure 9.3 Responses of *Chenopodium album* (A and C) and *Abutilon theophrasti* (B and D) to different nutrient levels. (A and B) Relationship between root/shoot ratio and age. (C and D) Relationship between root/shoot ratio and biomass. Black circles and solid lines represent high nutrients; gray circles and lines, low nutrients. Both sets of comparisons reveal significantly different allometric relationships between nutrient treatments. (From Gedroc et al., 1996.)

higher allocation to shoots than roots, resulting in an "ontogenetic drift"[3] toward lower root/shoot ratios.

Pigliucci and coworkers (Pigliucci and Schlichting, 1995; Pigliucci et al., 1997) examined the developmental reaction norms of two species of *Lobelia* to variation in light and nutrient availability. The typical view of the end-of-ontogeny traits in one environment is shown in Figure 9.4; analysis of these data indicat-

---

[3] "Drift" is probably an unfortunate choice of term for what may be an environment-independent genetic program for a decrease in root allocation relative to shoots as plants age, a phenomenon seen in numerous plant species.

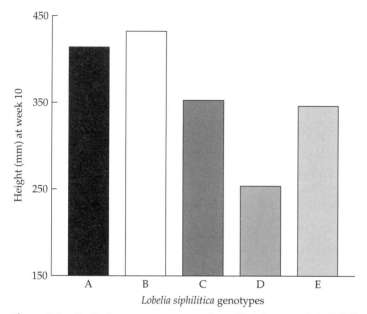

Figure 9.4   End-of-ontogeny perspective of height growth in *Lobelia siphilitica* grown in full sunlight. Genotypes A and B appear quite similar, and C and E are nearly identical. Contrast with week 10 in Figure 9.5A, showing the environment dependence of the phenotypes, and Figure 9.5B, which shows that the ontogenetic trajectories of these genotypes are distinct. (From Pigliucci and Schlichting, 1995.)

ed only three distinct phenotypes. However, the data from ontogenetic trajectories in two environments revealed that all five genotypes are distinct (Figure 9.5). They found evidence of genetic variation at the species, population, and genotype levels for the shape of the ontogenetic trajectories of traits delimiting the reproductive phase of growth in these plants, implying an ample capacity for response to selection on the shape of the trajectories. They also demonstrated environment- and population-specific phenotypic correlations between two fitness components and the adult trait values resulting from these ontogenetic trajectories, suggesting that selection could potentially shape population and species differences.

Inspection of the genetic parameters also revealed complex patterns of ontogenetic change in coefficients of genetic variation ("evolvabilities"). Note that there are almost no differences within or between species at the end of ontogeny, but substantial variation at early and middle phases (Figure 9.6). Also, at the earlier stages, the two populations of each species are quite different in their variation. Finally, the multivariate reaction norms of the two species are dis-

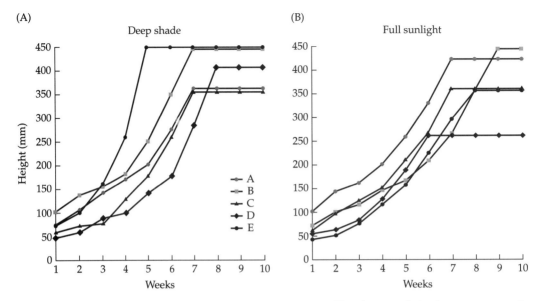

(A) Deep shade

(B) Full sunlight

Figure 9.5 Variation in ontogenetic trajectories of height growth for five genotypes of *Lobelia siphilitica* grown in shade and full sun. Note the variation not only in timing of flowering (cessation of growth), but also in genotype ranks between environments. For example, genotype E (•) is the tallest in the shade, reaching a height of 450 mm by week 5; in contrast, it is only the third largest genotype in the sun, reaching 350 mm at week 8. (From Pigliucci and Schlichting, 1995.)

Figure 9.6 Changes in the coefficient of genetic variation (CV_G) of leaf number during ontogeny for two populations of *Lobelia siphilitica* and *L. cardinalis*. (After Pigliucci et al., 1997.)

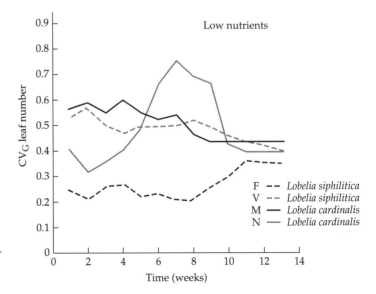

tinctive (Figure 9.7). Although both show similar changes in response to an increase in nutrients along principal component 2, differences along PC1 distinguish their plastic responses: *Lobelia cardinalis* bolts and flowers earlier and devotes more energy to vegetative reproduction (rosettes), with *L. siphilitica* increasing sexual reproduction (flowers).

Tadpoles of the gray treefrog, *Hyla chrysoscelis*, that are exposed to predators exhibit a suite of characteristics that differ from unexposed tadpoles (McCollum and Van Buskirk, 1996). Following exposure to either caged or free-ranging *Anax* dragonfly larvae, there are integrated plastic responses for behavior, tail size, and coloration. The ontogeny of these changes was followed. After 2 weeks there was reduced feeding and swimming activity and tails were larger; and after 3 weeks the tails are significantly more brightly colored with marginal dark spots. McCollum and Van Buskirk examined the issue of benefits and costs of this predator-induced phenotype by comparing the growth and survival of control tadpoles to those reared with caged or uncaged dragonflies. Growth rates of tadpoles from caged predator treatments were the same as for the controls. Tadpoles induced with caged dragonflies had higher mortality than the controls in predator-free conditions, but higher survival in trials with free-ranging dragonfly larvae present. Thus, a mismatch between phenotype and the predator environment resulted in decreased fitness. In experiments on *Pseudacris triseriata* (Van Buskirk and McCollum, 1997), they examined the strength of selection on tadpole characteristics in the presence or absence of predators, and determined that the plasticity of tail shape conformed to the directional selection produced by the dragonfly larvae (especially for tail fin depth; Figure 9.8A). However, despite apparent strong selection for decreased body depth in the presence of the predators, there was very little plasticity in this component (Figure 9.8B).

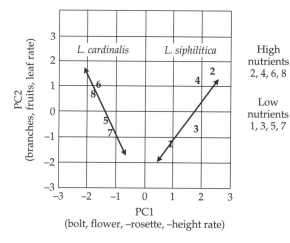

Figure 9.7 Depiction of plastic responses of two species of *Lobelia* to variation in nutrient availability. Results of a principal component analysis for seven phenological and reproductive traits (character loadings on axes). The two species have different multivariate plastic responses, indicated by the arrows. Numbers indicate different population/environment combinations. (From Pigliucci et al., 1997.)

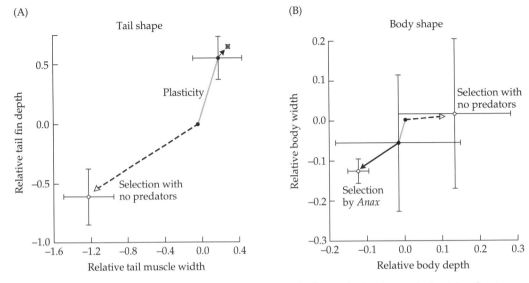

Figure 9.8 A comparison of the directional effects of selection and plasticity due to *Anax* dragonfly predation on (A) tail shape and (B) body shape of *Pseudacris triseriata* tadpoles. The phenotypic value in the no-predator treatment was set to (0,0). Changes favored by selection are indicated by the dashed line and white arrowheads for the no-predator treatment, and by solid lines and black arrowheads for the predator treatment. The bold line in each indicates the direction and magnitude of the plastic response to predation. (From Van Buskirk and McCollum, 1997.)

Yampolsky and Ebert (1994) combined environmental, ontogenetic, and allometric aspects in their investigation of the expectation of a tradeoff between growth and reproduction. They examined 56 clones from two populations of *Daphnia magna* grown with rich or poor food. They could detect no significant negative correlation between biomass allocation to growth versus reproduction in either treatment. Broad-sense heritability was higher in rich than in poor food, and there was significant among-population variation only on rich food. The genetic correlations among various traits were dependent both on the environment and the developmental stage; correlations were always higher with abundant food, and were generally lower later in development. If the correlations had been measured only at the last instar, the two treatments would not be distinct.

Nylin et al. (1996) examined reaction norms for age and size at maturity in two closely related species of *Lasiommata* butterflies. Both species are univoltine in Sweden, but *L. petropolitana* hibernates as a pupa and *L. maera* hibernates as a half-grown larva. Nylin et al. reasoned that if these species were adaptively plastic, they should increase development rate in response to cues suggesting that it is later in the season. However, they predicted that the direction of the plastic responses should be opposite in the two species, because the same day length cue would provide very different information about what time of the season it actually is—shorter day length for *L. maera* indicates that it is earlier in the spring, whereas for *L. petropolitana*, it suggests that it is later in the summer. When larvae were grown at four day lengths, results corresponded to their predictions of adaptive plasticity: larval development time was positively related to day length for *L. petropolitana*, but both larval and pupal development were negatively correlated with increasing daylength for *L. maera*. Ontogenetic sequences in different day lengths for *L. petropolitana* revealed that changes in development time were greatest in later instars. The DRN of a French population was shifted to a range of shorter photoperiods, as predicted from the shorter daylengths (Gotthard, 1998).

Perhaps the most general message to emerge from this potpourri of examples is that the discourse among genes, characters, and the environment can easily lead to complex patterns of phenotypic expression. The resulting depictions of phenotypes are much more elaborate than we hoped, although probably no more so than we expected. In the following section, we present a brief overview of some of the conceptual lines of inquiry into how organisms might deal with non-constant environments.

THEORY OF EVOLUTION IN HETEROGENEOUS ENVIRONMENTS

Early theoretical approaches to the question of adaptation to variable environments examined the potential for maintenance of genetic polymorphism as a solution (for spatial heterogeneity: Levene, 1953; for temporal heterogeneity: Haldane and Jayakar, 1963). The focus was on adaptation to subsets of the environmental range: different genotypes, specialized for the different conditions, would partition the niche space. The other side of the coin, though, was the question of whether a generalist strategy could be successful against specialists.

Levins attacked this issue from several vantage points, including his famous fitness set approach (Levins, 1968). *Convex* fitness sets are those in which the range of environments experienced is smaller than the "tolerance" of the phenotype; in a *concave* fitness set, environmental range exceeds the tolerance (Figure 9.9). Levins also introduced the concept of *environmental grain*—if an organism spends its entire life in a single environment, the grain is *coarse*; if the organism experiences more than one condition or patch, the grain is *fine*. He concluded that three optimal strategies could result. (1) For a convex fitness set, the optimum strategy is always a single specialist phenotype that, although it can survive and reproduce in each of the two conditions, is best adapted to an environment intermediate between the two. For concave fitness sets, there are two possibilities, depending on the environmental grain. (2) In a fine-grained environment, a single phenotype specialized to the most frequent environment is favored. (3) In a coarse-grained environment, both specialists will be present in frequencies proportional to the environmental frequencies.

Levins further considered optimal strategies for an organism with the ability to alter its phenotype during development, and distinguished four outcomes

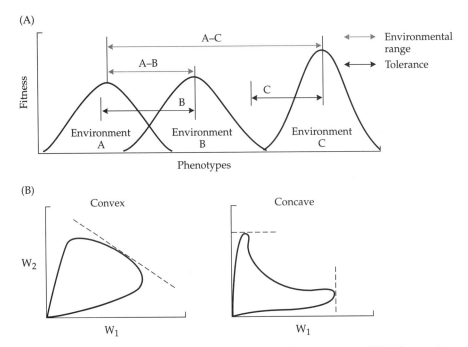

Figure 9.9 Levins' fitness set concept for a fine-grained environment. (A) Tolerance is twice the distance from the peak to the inflection point of a curve relating fitness to phenotype in a single environment, and environmental range is the difference between the phenotypic values at which fitness is maximized. (B) Each curve represents fitness values (W_1 and W_2) associated with the same phenotype in two environments. The dashed lines indicates the maximum isocline for fitness for each environment; where this line is tangential to the fitness set, there is a phenotypic optimum for that environment. Convex fitness set: there is one optimal phenotype for environments A and B, with intermediate fitness in both environments. Concave fitness set: for environments A and C there are two optima, each with high fitness in only one environment. (From Levins, 1968.)

(A–D in Table 9.1) based on environmental information, generation times, and environmental grain. Bradshaw made similar predictions (albeit without the concept of grain!), in his classic paper several years earlier (Bradshaw, 1965). Lloyd (1984) further expanded the list with his recognition that an individual plant could produce multiple structures, and we also include the concept of adaptive coin-flipping (Cooper and Kaplan, 1981).

Genetic Variation versus Plasticity, Specialists versus Generalists

Substantial discussion has focused on the question of whether genetic variation (polymorphism) can be maintained in environments that vary. Consistent spatial variation can maintain genetic variation, but the conditions for polymorphism in temporally variable environments are quite restrictive (Hedrick et al., 1976; Felsenstein, 1976; Hedrick, 1986). For fine-grained temporal variability,

Table 9.1 **Alternative phenotypic outcomes for selection in heterogeneous environments**

Phenotypes	Conditions favoring	Reference
A. Uniform	Reliable environmental cues lacking; convex fitness set	
	1. Environmental variation temporally and spatially fine-grained (environment fluctuates on a cycle much shorter than the generation time)	(Bradshaw, 1965; Levins, 1968)
	or	
	2. Environmental variation temporally fine-grained but sequence of change is predictable (fixed developmental sequence, heteroblasty)	(Bradshaw, 1965; Winn, 1996)
	or	
	3. A single optimal phenotypic state exists (stabilizing selection—canalization)	(Lloyd, 1984)
B. Plasticity: phenotypic modulation	Predictive environmental cues, convex fitness set	(Bradshaw, 1965; Levins, 1968; Lloyd, 1984)
	Environmental variation: temporal and spatial scales of fluctuation unimportant	
	Optimum is a continuous function of environment	
C. Plasticity: development conversion (conditional strategies)	Predictive environmental cues, concave fitness set	
	1. Environmental variation spatially coarse-grained	(Bradshaw, 1965; Levins, 1968; Lloyd, 1984; Lively, 1986)
	or	
	2. Environmental variation temporally fine-grained (multiple strategy—more than one phenotype on an individual, e.g., heterophylly)	(Lloyd, 1984)
	or	
	3. Environmental variation temporally coarse-grained, with predictive cues (polyphenism)	(Hazel et al., 1990; Moran, 1992; Roff, 1996)
D. Polymorphism: specialization	Concave fitness set	(Bradshaw, 1965; Levins, 1968)
	Environmental variation spatially coarse-grained, alternatives always present	
	Stochastic spatial variation over time favors evolution of habitat selection	(Maynard Smith, 1962)
E. Polymorphism: random	Reliable environmental cues lacking, concave fitness set	(Cooper and Kaplan, 1981; Bull, 1987)
	Environmental variation temporally and spatially coarse-grained (adaptive coin-flipping)	

overdominance is necessary; for coarse-grained temporal variability, polymorphism is more likely in cyclical than stochastic environments (Karlin and Levikson, 1974).

Another approach has been to model the evolution of generalist versus specialist strategies (e.g., Futuyma and Moreno, 1988; Gabriel and Lynch, 1992; Gilchrist, 1995) or phenotypic plasticity itself (e.g., Via and Lande, 1985; de Jong, 1990; Gomulkiewicz and Kirkpatrick, 1992; Gavrilets and Scheiner, 1993a). Levins (1968) found that phenotypically plastic individuals (i.e., generalists) will be favored over canalized individuals (specialists) in fine-grained environments; specialists are always favored in homogeneous environments. A recent model by Brown and Pavlovic (1992) makes similar predictions.

For quantitative traits, fine-grained environments select for phenotypic plasticity: when there are no genetic constraints the population will always evolve to a genotype that produces the optimal phenotype under each set of conditions (Via and Lande, 1985; Gomulkiewicz and Kirkpatrick, 1992). However, if the probability of encountering different environments varies, the sequence and relative frequencies of selective environments will determine the evolutionary trajectory. The form of plastic response depends on whether the fitness set is convex (favoring *phenotypic modulation*) or concave (favoring *developmental conversion*; Table 9.1). Temporally fine-grained variation favors *multiple strategies* such as heterophylly, with multiple phenotypic states on a single individual. Coarse-grained spatial variation can lead to developmental conversion early in the life cycle, as in *Nemoria arizonica*: inflorescence-feeding caterpillars develop an uncanny mimicry of the catkins, and leaf-feeding caterpillars (ingesting tannins) develop into a twiglike morph (Greene, 1989). Coarse-grained temporal variation favors the evolution of *polyphenism*, in which discrete morphs are produced in a seasonal sequence (Moran, 1992; Roff, 1996). A model by Moran (1992) shows that such conditional strategies are increasingly advantageous (1) as the accuracy of predicting the environment increases, (2) as the magnitude of the fitness cost of producing the "wrong" phenotype increases (see below), and (3) when the frequencies of the environments are equivalent.[4]

If environmental change follows a typical predictable course, organisms may adopt the option of *heteroblasty*—a developmentally fixed sequence of phenotypes during ontogeny. In an adaptive configuration this sequence would match the phenotypic progression to the sequential conditions the individual experiences. Winn (1996) examined winter-summer heteroblasty in the plant *Dicerandra linearifolia*. Through experimental manipulation she showed that, although temperature had significant effects on leaf traits, the node at which a

---

[4]Getty (1996) has shown that Moran's model is a special case of signal detection theory when the probabilities of producing the correct phenotype in each environment are equal. Signal detection theory is a form of Bayesian decision theory useful for problems of distinguishing signal from noise, and for cue reliability and response.

leaf was produced explained a far greater proportion of the variance. However, further analysis of selection differentials on leaf size, thickness, and stomatal density found no evidence for different selective optima in the two seasons (Winn, in press).[5]

Costs of Plasticity and Environmental Assessment

When the response is appropriate to the conditions, phenotypic plasticity is universally recognized as the optimum solution to the problems posed by variable environments. If organisms pay no toll for producing continuously variable phenotypes, the best phenotype should be expressed in each environment experienced (Bradshaw, 1965; Lloyd, 1984; Seger and Brockman, 1987; Schlichting, 1989a; Gomulkiewicz and Kirkpatrick, 1992; Zhivotovsky et al., 1996). However, such universal plasticity is clearly not the standard (Whitlock, 1996). Why this gap between theory and reality? The answer appears simple: plasticity has its costs! (Moran, 1992; Newman, 1992; Sultan, 1992).

Many models of adaptation to variable environments assume that there are costs to plasticity, taking to heart the old saw that "a jack-of-all-trades is a master of none." In such scenarios, adaptation to a second environment will entail a fitness loss in the first (Lynch and Gabriel, 1987; Gabriel and Lynch, 1992; Gilchrist, 1995; others reviewed in Futuyma and Moreno, 1988), and conversely, there should also be a tradeoff in performance in other environments by specialists. There is empirical support for this assumption in some studies (e.g., Berenbaum et al., 1986; Via, 1991; Groeters et al., 1994; Simms and Triplett, 1994; Mitchell-Olds and Bradley, 1996; Krebs and Feder, 1997; Zangerl et al., 1997), but in others there is not (e.g., Lenski and Bennett, 1993; Tollrian, 1995; Bennett and Lenski, 1997; and references in Fry, 1996; Mitchell-Olds and Bradley, 1996). Note that if there is a significant tradeoff in performance, there will be selection for improved performance in the poorer environment (e.g., as seen in *E. coli* and *Bacillus subtilis*; Lenski, 1988; Cohan et al., 1994), and concomitantly, less opportunity for us to observe "costs."[6]

Whitlock (1996) modeled the evolution of niche breadth without the assumption of tradeoffs and found that, generally, species with narrower niche breadths have higher probabilities of fixing advantageous alleles, and lighter drift and mutational loads. Specialists will evolve faster and will shoulder aside the generalists in their preferred habitats due to faster evolutionary rates, because the

---

[5] Winn's conclusions were based on an analysis of the data from a character state perspective. Her results indicated larger selection differentials for leaf size in the season with larger leaves. If we view these results from a reaction norm perspective, we would be inclined to interpret them as evidence that there may be selection for plasticity of leaf size.

[6] Lenski's results were counter to the expectation of *more* pronounced antagonistic pleiotropy. He argued that this prediction derives from an implicit assumption of tradeoffs due to limiting resources, and proposed an alternative hypothesis congruent with his observations, that the resistance mutation initially disrupted genomic integration.

selection intensity in that particular habitat increases with the proportion of the population residing there. We note, though, that an allele that expands niche breadth will lead to an increase in the proportion of the population in the focal habitat, and will also accrue the advantages of increased selection intensity. If overall population sizes increase and further reduce drift and mutational loads, such a mutant may gain an advantage over the specialist.

When the costs of alternative strategies are fairly high, conditional strategies will be favored over continuously labile phenotypes (Lloyd, 1984; Hazel et al., 1990); conditional strategies are also more likely in coarse-grained than in fine-grained environments (Lloyd, 1984). Van Tienderen (1991) explicitly modeled plasticity costs by incorporating a second locus. Depending on the initial conditions, selection favored either an individual whose fixed phenotype is close to the adaptive peak for one environment or an individual whose phenotype is moderately plastic and low-cost. When interactions with a second species were included in the model, costs of plasticity led to a compromise generalist phenotype; higher costs led to flatter reaction norms (van Tienderen, 1997).

The likelihood of evolving any of the different types of strategies in Table 9.1 is related to the costs of producing phenotypic alternatives (for plastic responses; Lloyd, 1984) and the capacity of the organisms to accurately assess the environmental cues (Lloyd, 1984; Getty, 1996). What are these costs of plasticity due to? Construction costs, maintenance costs, and tradeoffs with other systems? DeWitt, Sih, and Wilson (1998) have reviewed the various proposals in the literature and sorted them into two categories, "pure" costs and those factors that act to limit the benefits of plastic response.

Costs
(1) Maintenance of machinery
(2) Production cost
(3) Information acquisition
(4) Developmental instability
(5) Genetic—linkage, pleiotropy, epistasis

Limits to benefits
(6) Information reliability
(7) Lag times—response too slow
(8) Developmental range (plastic development has a more limited ability to reach extreme phenotypes)
(9) "Epiphenotype"—an "add-on" may not be as good as one integrated from the outset

Most of these are (at least theoretically) modifiable by natural selection. With enough time and the right mutations, costs might be decreased (1, 2), sensory systems fine-tuned (3), canalization enforced (4), and developmental programs modified (5, 8, 9). Reliability of information and lag times, though, would seem to be inherently more difficult for natural selection to solve. What to do about an environmental cue as fickle as rainfall (or lack thereof), or one as transient as sunflecks (e.g., Chazdon and Pearcy, 1991)? A number of investigators have begun to examine questions of sensory capability and lag times (Kuiper and

Kuiper, 1988; Burger and Lynch, 1995; Getty, 1996; Padilla and Adolph, 1996).[7] For example, Novoplansky et al. (1994) examined the role of the timing of a cue relative to the timing (rapidity) of response—is it possible for the response to effectively track environmental change on the scale of a passing cloud?

In spite of the difficulties *we* have imagining solutions, it is certainly satisfying to see that some creatures have solved both the reliability and the lag time problems. Take, for example, the dilemma of winter. For leafy deciduous plants, as the temperature drops to 0°C and below, the availability of liquid water declines precipitously, and softer tissues can suffer substantial disruption if the liquid contents of their vacuoles freeze. Both reliability and lag time are major problems here—temperatures do decline progressively, but not without considerable variation, and if the actual cue was a temperature at or near freezing, the plant would be hard-pressed to complete the necessary preparations in time. The solution to both problems is to use a more reliable cue, such as photoperiod, and to set the threshold of sensitivity so that the drought/freeze is adequately anticipated. The sophistication and speed of some systems of response is phenomenal. Ramachandran et al. (1996) studied the well-known ability of flounder to match their coloration to that of the background. They found that by adjusting the contrast of six different sets of splotches of different grain size, these fish achieved matching in 91% of the trials, and blending can be accomplished in as little as 2 to 8 seconds!

Of course, there are numerous examples in which lag time or material costs prohibit the evolution of a plastic response. We offer one example of a system—adaptation to food sources in crossbills (Benkman, 1993)—that appears to have all the hallmarks for the evolution of developmental conversion (Table 9.1C): predictable resources, fine-grained environmental variation, and concave fitness sets. A variety of conifer species produce cones that provide seed for red crossbills (*Loxia curvirostra*). Four crossbill types have morphologically distinct bill forms—variation that is not plastic in origin. Through feeding trials, Benkman tested the hypothesis that these differences represented specializations to the four key conifers found in the Pacific Northwest (North America). Fitness sets for all four types were concave, suggesting selection against intermediate morphologies. The morphologies corresponded generally to peaks in fitness surfaces (Figure 9.10). The western hemlock and ponderosa pine specialists had palate groove widths and bill depths that conformed to the optima predicted for those traits. The observed characteristics of the other two specialists were optimal for either groove width or bill depth, but not both. In this example, all the prerequisites for the evolution of plasticity are present; but instead, specialization is favored, presumably due to significant construction costs and/or lag

---

[7]In the Burger and Lynch model, the magnitude of the lag time determines susceptibility to extinction, raising the intriguing possibility of "species selection" for plasticity, if plasticity allowed significant reduction of lag times.

(A)

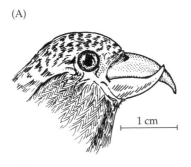

1 cm

Figure 9.10 Adaptation of red crossbills (*Loxia curvirostra*) for husking different conifer seeds. (A) Type 3 crossbill, putative Western hemlock specialist. (B) The cones and seeds of the four conifers fed on by crossbills. (C) Predicted fitness surfaces obtained from feeding trials. The observed values for ponderosa pine (PP) and western hemlock (WH) fell on the peaks; asterisks represent observed values for Douglas fir (DF) and lodgepole pine (LP). (From Benkman, 1993.)

(B)

Western hemlock

Douglas fir

Ponderosa pine

Lodgepole pine

(C)

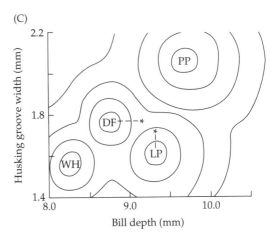

time problems. And, in this example, even behavioral modification appears not to have evolved: there was at least a 50% reduction in performance on resources other than that favored by the bill characteristics.

Getty (1996) suggested that another solution to the problem of unreliable cues is to change the sensitivity threshold, that is, change the phenotype only when the cue is very strong. This is perhaps a form of the bet-hedging proposed for situations of low information quality by Kawecki and Stearns (1993). Experimental results of Harvell (1994), and Fairbairn and Yadlowski (1997; also see discussion in Chapter 3), indicate that signal sensitivity can be a target of selection.

A few models have been constructed examining the sensitivity of threshold responses. Vidybida (1991) found that creating cooperation among receptors resulted in threshold switching; such systems had much higher selectivity and sensitivity than unconnected systems. Gibson (1996) investigated the effects of changes in (1) the number of binding sites for activator proteins, and (2) the DNA/protein and protein/protein binding energies on the stabilization of gene expression. He discovered that the minimum threshold width (i.e., maximal switch sensitivity) is a negative exponential function of both the number of binding sites and of the level of response, but is not related to binding energies. In the model, threshold widths became broader with heterozygosity, and trade-offs arose between sensitivity and the number of target genes.

One of the issues raised by a discussion of environmental signals and their reception is: How do organisms "remember?" At the level of the individual, this is the fascinating question of learning and behavior in humans and other animals (e.g., Jaenike and Papaj, 1992; Marler and Nelson, 1993), as well as in other organisms (e.g., Baldwin and Schmelz, 1996). But an equally intriguing question can be asked at another level: How does cross-generational "genetic memory" evolve (Levins, 1968; Harvell, 1990; Zhivotovsky et al., 1996)? In fact, all character states of organisms carry some memory of their evolutionary history (e.g., past selective events or genetic constraints), but it is those traits with environment-contingent development that are of the most interest. In traits such as anticipatory responses, the genetic memory is revealed in the epigenetic configurations of signals and responses. Jablonka et al. (1995) proposed that one type of short-term memory is the information transmitted as carry-over effects, that is, the persistence of current environmental effects into subsequent generations. The best known are maternal effects, but the alterations of DNA methylation patterns are another example. They modeled the performance of organisms with carry-over strategies versus fixed (no response) and immediate response plastic strategies, and found that carry-over is definitely favored in random or near-random environments (i.e., when transition probabilities between environmental states are close to 0.50), or when signal reliability is very low.

Although the preceding discussion has focused on costs of plastic responses, the basic issues are the same for any shift in gene action during development. Thus, even those events we currently observe as regular components of change in an ontogenetic trajectory (e.g., larval to adult transitions in holometabolous insects and other complex life cycles; Moran, 1994) must at some point have been integrated into a previously existing developmental program. The evolutionary "appraisal" of fitness costs and benefits necessarily occurs for changes to *any* component of the DRN, that is, plasticity, ontogeny, or allometric relationships. Although the course of evolutionary transformation is dependent on more than just the cost to benefit ratio (i.e., don't forget drift and gene flow), the product testing role of natural selection remains vital to the adaptive modification of phenotypes.

THE OBJECTS OF SELECTION

The two extreme perspectives on the objects of selection are (1) that all traits can be selected independently (Weber, 1996), and (2) that all traits are subsets of functional complexes, which are in turn the objects of selection. The former represents the Fisherian view of particulate (mosaic) evolution; the latter mirrors the Wrightian paradigm emphasizing the importance of epistasis. As an example, interpretations of phenotypic evolution in humans often broadly correspond to the aforementioned perspectives: changes in particular traits are either interpreted as an indication of mosaic evolution, with individual traits responding to selection on particular behavioral attributes, or alternatively as correlated responses to selection on integrated systems related to broader aspects of form. Churchill (1996) compared the postcranial skeletons of Eurasian Neandertal, early modern fossils, and recent humans, and found that a model incorporating morphological integration fits the data better. However, this integration accounts for less than half of the variance in upper limb traits, suggesting that there has been substantial particulate evolution in the upper body.

From the perspective of the DRN we obviously predict that selection generally operates on traits in combination rather than in isolation. The fact that fitness integrates the ups and downs experienced during the life span suggests to us that the DRN is itself the object of selection—it is the aggregation of trait values into a functional whole that is of primary importance. As we have pointed out before, this is by no means a novel or even unorthodox view;[8] its significance, however, lies in the recognition and emphasis that *context* can be everything (Stearns and Kaiser, 1996). Thus, not only the intensity of selection, but also the extent of correlated selection—that is, whether single traits (e.g., an enzyme that permits detoxification of insecticides; see Thompson and Burdon, 1992; McKenzie and Batterham, 1994), or suites of integrated characters are affected (e.g., feeding, see Lauder, 1989; or locomotion, see Gatesy and Dial, 1996)—depends on the environment (biotic or abiotic selective agents), the ontogenetic stage, and the consequent genetic architecture. We have cited illustrations of the importance of context in various chapters throughout, but here we will bring examples together to focus the issue.

Simultaneous Selection on Multiple Traits

Armbruster has, in a series of studies, focused on selection on the integration of traits critical to pollination success in the plant genus *Dalechampia* (Armbruster, 1988, 1990, 1991). He analyzed various traits of the inflorescence and their relative positions, and found that the probability that the pollinators (bees of different sizes) would actually come in contact with the stigma was related most closely to the distance between the resin gland (the bee's food source) and the stigma. The

---

[8]Clearly not new ideas; see Wright, 1931, p. 147. "Individual adaptability is, in fact, distinctly a factor of evolutionary poise. It is not only of the greatest significance as a factor of evolution in damping the effects of selection . . . but is itself perhaps the chief object of selection." And Wright, 1935, p. 264: "harmonious adjustment of all characteristics of the organism as a whole that is the object of selection, not the separate metrical characters." And, of course, Schmalhausen in his book endorsed similar notions.

largest bees can contact all stigmas, but concentrated their visits on species with large glands (positively correlated with the gland-stigma distance: Armbruster, 1990). The adaptive surface predicted from these data, however, indicated that for optimal pollen receipt, the gland-stigma distance should be short and the resin gland should be large (Figure 9.11). Notice, however, that the positions of populations are often far below the predicted optimal values. Armbruster (1990) proposed that the lack of inclusion of information on costs and benefits might affect the degree of conformation to the adaptive surface, and suggested that including such factors might alter our depiction of the surface to an adaptive "ridge" running along the diagonal rather than a single peak.

Brodie (1992) detected correlational selection in garter snakes that varied in color pattern and antipredator behavior. Measurements on neonates revealed a negative correlation between stripedness and the number of evasive turns (reversals) during escape. These individuals were marked and released, and the characteristics of those recaptured was recorded. The selection surface showed two distinct peaks: survivorship was highest for striped individuals employing straight line escape and for spotted snakes that zigzagged (Figure 9.12). Brodie interprets these results in light of the optical illusions that these behaviors produce—the apparent speed of striped snakes moving in a straight line is much less than their actual speed. Spots, on the other hand, tend to provide fixed points of reference for a visual predator, and reversals of direction would tend to disrupt this effect.

Carriere, Masaki, and Roff (1997) examined the joint evolution of ovipositor length and egg size in 40 cricket species. Crickets in harsh habitats tend to lay

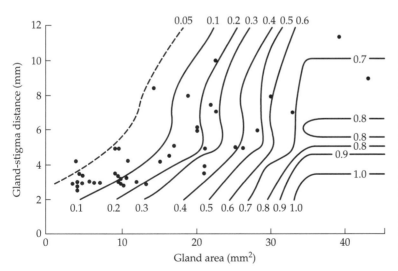

Figure 9.11 The adaptive surface predicted from data on floral morphology and pollinator visitation in *Dalechampia* (Euphorbiaceae). The points represent the actual values of gland area and gland-stigma distances for 43 populations of 25 species. Note that none of the populations are near the predicted peak fitness of 1.0. (From Armbruster, 1990.)

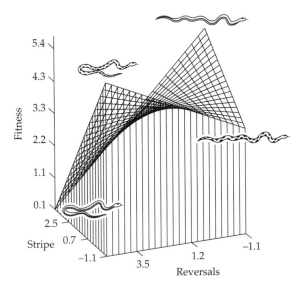

Figure 9.12 Selective topography of the relationship between antipredator behavior (reversals) and color patterns (stripe) in garter snakes (*Thamnophis ordinoides*). The fitness peaks are for two particular combinations: striped pattern with straight line movement (top right), and spotted pattern with multiple reversals (top left). Striped snakes with zigzag behavior, or spotted snakes with straight line movement, are bird food. (From Brodie, 1992.)

eggs deeper in the soil to ameliorate drying, but deeper hatchlings have lower survivorship. After removing phylogenetic signal and body size effects, the analysis revealed a significant relation between ovipositor length and egg size, suggesting that increased egg size has evolved specifically in taxa that have evolved longer ovipositors. Similar specificity of the relationship between timing of diapause and duration of development has been documented for the pitcher plant mosquito, *Wyeomyia smithii*, by Bradshaw, Hard, and coworkers (Hard et al., 1993; Bradshaw and Holzapfel, 1996).

Also in the category of simultaneously selected traits are those that we associate with the classic tradeoffs: size versus number, growth versus reproduction, longevity versus fecundity, etc. (Stearns, 1989b, 1992; Roff, 1992). For example, Sinervo et al. (1992) showed that fecundity selection favoring production of large clutches of smaller eggs was balanced by survival selection favoring large offspring and thus fewer, larger eggs. Mole and Zera (1993) dissected the differential allocation of resources underlying the dispersal/reproduction tradeoff in the wing-dimorphic cricket *Gryllus rubens*. They found that the short-winged form allotted 172 mg to ovaries and only 31 mg to wing muscle, whereas the long-winged form allotted 105 mg to reproduction, but 55 mg to wing muscle (i.e., dispersal). (See Chapter 7 for discussion of allocation/acquisition models and examples.)

That selection operates at various stages and on different traits at those stages has long been recognized (Buchholz, 1922; Haldane, 1932b), and a variety of studies have found variation in selection intensity on different aspects of life cycles (Christiansen and Frydenberg, 1973; Clegg and Allard, 1973; Clegg et al., 1978; Brassard and Schoen, 1990; Jordan, 1991; Searcy, 1992). For example, Grace (1985) discovered that the competitive abilities of two species of cattails (*Typha*) varied with age and habitat. *T. latifolia* was better both above and below water as adults, but as seedlings *T. domingensis* excelled below the water table. King (1993) showed that selection on color patterns of Lake Erie water snakes differs between juveniles and adults because of their differential susceptibility to predation.

Variation in selection pressure, both temporal and spatial, has been repeatedly documented (Schmidt and Levin, 1985; Hairston and Dillon, 1990; Reznick et al., 1990; Lechowicz and Bell, 1991; Stratton, 1992, 1994; Hutchings, 1993; Stibor and Luning, 1994; Bell, 1997). Benkman and Miller (1996) have shown that the selection pressure on bill depth in crossbills varied seasonally. In the summer there was directional selection for smaller bills that could extract seeds from spruce cones, but in the winter stabilizing selection operated for seed removal from lodgepole pine cones. The observed bill size was that predicted by stabilizing selection, and the authors concluded that selection for the scarcer resource, that is, winter food, has dominated the morphological evolution. When multiple years of data have been collected, the fluctuating nature of selection intensities is often revealed (Figure 9.13).[9] And, even if selection on a focal trait is the same, the environment dependence of character correlations can have profound effects on the correlated responses to selection (see Chapter 7).

In an inverted approach to this question, several investigators have looked at responses to the same selection applied at different stages (Robertson, 1964; Danjon, 1995; Loeschcke and Krebs, 1996). Zera and coworkers investigated the direct and correlated responses to selection on juvenile hormone esterase activity in juvenile and adult crickets (*Gryllus assimilis*). Selection on juveniles was much more effective than on adults, and there were strong correlated responses within stages. However, there was little or no correlated response across stages, that is, between juvenile and adult traits (Zera and Zhang, 1995; Zera et al., 1997). These results are congruent with findings that the genetic architecture changes during the life cycle (see Chapter 4 and Chapter 7, e.g., Figure 7.12).

To conclude with a counterexample, Weber (1992), with monkey wrench in hand, weighed in on this issue of the objects of selection with an experiment to selectively modify very small regions of *Drosophila* wings. Asking "How small are the smallest selectable domains of form?" the resounding answer was pretty damn small (literally a fly "speck of tissue" < 0.2 mm)! Of course, these results don't address the issue of what the actual objects of selection have been, but they certainly do give pause to those of us who ardently advocate that selection operates largely on coordinated systems!

---

[9] Reznick et al. (1997) have argued that such variation may be at the root of some cases of apparent stasis, rather than constant stabilizing selection.

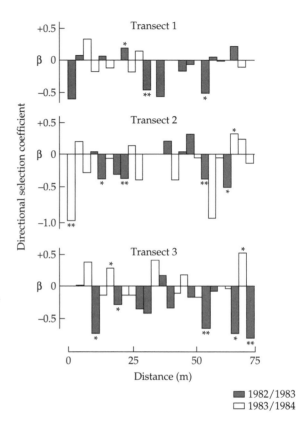

Figure 9.13 Variation in selection intensities (β) on timing of seed germination in the annual plant *Collinsia verna*. Data are for two overwinter periods and for 25 quadrats along three transects. Asterisks indicate coefficients significantly different from zero. (From Kalisz, 1986.)

BUILDING A NORM OF REACTION

Bottom-Up Evolution of Genetic Systems

The key sequence of events that we seek to characterize is the mapping of genotype to phenotype (G→P) (Lewontin, 1974b; Scharloo, 1987). Deciphering the arrow in this transformation is problematic for even the simplest of extant organisms (as we have seen in detail in Chapter 8), as it represents the epigenetic translation of genetic information by way of the internal and external environmental milieu into the expressed phenotype (Feder and Watt, 1992; Rollo, 1994). The complexity of such transformations is daunting even in systems that appear simple, such as the genetic control of bristle production in *Drosophila* (Ghysen and Dambly-Chaudiere, 1988). In the achaete-scute complex there are eleven bristles (macrochaetes); *achaete* is responsible for three bristles and *scute* for nine (both are required for one bristle). There is no bristle production following inactivation of these two genes, leading to the easy interpretation that these are *the* genes for bristle number. However, this clear picture is clouded by the discovery that the *scute* gene can be subdivided into smaller regions, each apparently directing the development of one or a few bristles. The authors conclude: "... this dilemma reflects the inherent duality of the eukaryotic genes—their molecular functions, encoded in the translated sequence, and their regulation, which involves a far larger segment of the chromosome ... it seems clear that it is the combination of the two that matters for the animal ..." (Ghysen and Dambly-Chaudiere, 1988).

Going back to the earliest stages of life when there were few genes is a useful aid in thinking about how genetic systems and their attendant phenotypes may arise. Phenotypes came into existence as soon as there was storage and transmission of more information than was contained in the genetic sequence alone (Jablonka and Szathmary, 1995). For example, the formation of secondary and tertiary structures of molecules such as RNA would represent such information, and may have been among the very first phenotypes. At this level there seems to be little in the way of epigenetics—the genotype directly determines the phenotype, and selection can sort among variations in sequence that result in variations in structure and function (Huynen et al., 1996; Nakaya et al., 1997). However, environmental modification would occur when other molecules were present to bind with the RNA, whether such binding hindered or promoted the function. Genetic interaction would arise when the products or structures resulting from different genetic sequences were coupled. It is easy to envision the origin of epigenetic systems at this point—if the products of two genes combined to form a new molecule with features that differed from those of either initial product, an emergent property would arise.

We can imagine, even at this primeval stage when cells first arose, that selection would already be strongly favoring those same attributes that we immediately think of today as targets of selection; namely, survival (production and modification of the protective membrane surrounding cell contents), growth (transfer of substances in/out of the cell, efficiency of metabolism), and reproduction (mechanisms for replication/reproduction and its accuracy) (de Duve, 1987). Clearly, selection for all of these will also implicitly favor evolution of homeostasis to minimize any disruptive effects of external environmental conditions.

The further evolution of metabolic complexity is an issue of considerable interest, because therein lies the origin of capabilities of responses to changes in both the internal and external environments. Several simulations suggest that the evolution of complexity appears to be an inevitable outcome of selection for efficiency or growth rate. Kacser and Beeby (1984) started with a precursor cell containing only a few relatively inefficient, multifunctional enzymes, and, allowing only mutation and natural selection for faster growth, showed that the descendants had a large number of specialized enzymes with high catalytic efficiencies. De Duve (1987) found that an increasingly complex network of metabolic pathways would evolve, and that the catalysts that would arise were selected both for modifications that enhanced their efficiency and for their role in extending the metabolic pathway.

In an interesting simulation, Beeby and Kacser (1990) examined the evolution of metabolic enzymes in an "organism" with a simple metabolic pathway of eight reaction steps. Initially, a single enzyme catalyzed all transitions, but the efficiency (complementarity) with which it could carry these out varied. The authors allowed gene duplication and mutation of the enzymes and imposed selection for population growth rate. In each of three experiments, genome size (total number of genes) increased quickly, peaking at 6 in two experiments, but rising to 65 in the third! Final genome sizes were 2, 3, and 23 respectively (Figure 9.14). Complexity (number of different gene sequences) also rose quickly, peaking at 4 in the first two experiments, but rising to 30 in the third. In two experiments, final complexity was equal to genome size (i.e., there was a single copy of each enzyme); however, in the third, there were 18 different gene sequences present in the population, representing five different functional enzymes (all but one with multiple copies). The salient points are: (1) in each case there was rapid evolution of more complex metabolisms; (2) each had a different outcome despite starting from the same conditions,[10] (3) genetic redundancy (see below) was common—enzymes retained capabilities of catalyzing multiple reactions. All three experiments eventually achieved equilibria, but these steady states are probably due to two factors, a constant environment (no abiotic or biotic change) and the immutability of the pathway itself; that is, the eight reactions were not allowed to change.

The inclusion of gene duplication in such models is indicative of the acceptance of its role in generating new genetic material and permitting the evolution of new functions (Ohta, 1991; Iwabe et al., 1996; Lupski et al., 1996). In those

Figure 9.14   Details of the simulation of metabolic evolution. (A) The eight transitions ▶ that are the target of selection. Each block represents the optimal shape of the enzyme catalyzing the reactions in the pathway. (B) The initial conformation of the ancestral enzyme, and its efficiency at catalyzing each of the eight reactions. (C–E) Three simulated runs in which duplication and mutation of the ancestral enzyme allowed selection for increased metabolic efficiency, with different evolutionary outcomes. The scale of the enzyme activities in C–E is 100× the scale in B. C, D, and E show the evolution of new specialized enzyme forms and a marked increase in the efficiency of the metabolic transitions. D also shows the evolution of multiple copies of the same enzyme. This example shows the complexity of the evolutionary process, with a mixture of historical accidents, constraints, and selection determining outcomes. (From Beeby and Kacser, 1990.)

---

[10] Actually, experiment 3 had slightly less intense truncation selection applied initially; however, the authors report that higher complexities can be derived in replications of the conditions of experiments 1 and 2.

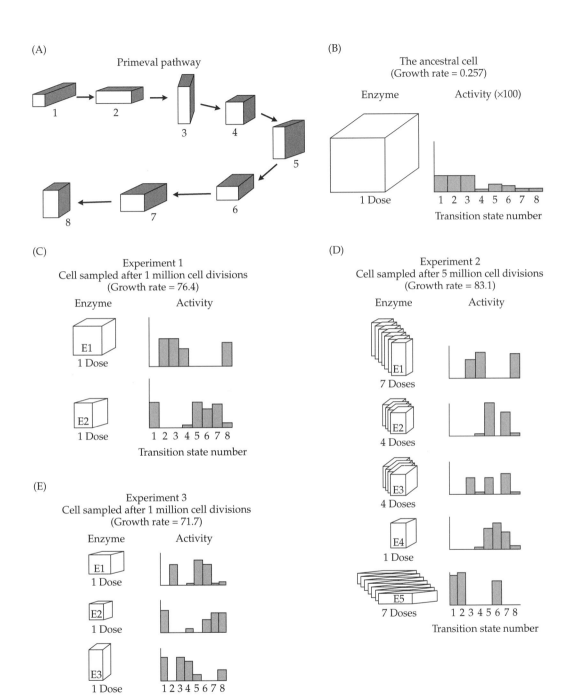

(A) Primeval pathway

(B) The ancestral cell
(Growth rate = 0.257)
Enzyme   Activity (×100)
1 Dose
1 2 3 4 5 6 7 8
Transition state number

(C) Experiment 1
Cell sampled after 1 million cell divisions
(Growth rate = 76.4)
Enzyme   Activity
E1
1 Dose
E2
1 Dose
1 2 3 4 5 6 7 8
Transition state number

(D) Experiment 2
Cell sampled after 5 million cell divisions
(Growth rate = 83.1)
Enzyme   Activity
E1
7 Doses
E2
4 Doses
E3
4 Doses
E4
1 Dose
E5
7 Doses
1 2 3 4 5 6 7 8
Transition state number

(E) Experiment 3
Cell sampled after 1 million cell divisions
(Growth rate = 71.7)
Enzyme   Activity
E1
1 Dose
E2
1 Dose
E3
1 Dose
1 2 3 4 5 6 7 8
Transition state number

organisms for which much or all of the genome has been sequenced, estimates of the minimum fraction of genes that have originated by duplication range from 30% of the 1680 proteins in *H. influenzae* to 45% in *E. coli* (Brenner et al., 1995). Even the tens of thousands of genes in vertebrate genomes are estimated to be composed of derivatives of only about 1000 gene families (Chothia, 1994). Additionally, there are several deep lineages that have their origins in polyploidy (many flowering plants, for example: Levin, 1983; Masterson, 1994; and some fish groups: Bailey et al., 1978).

Duplication would seem to lead inevitably to epistasis, initially, because dosage effects need to be ameliorated. Later, those genes not silenced will generally continue to interact until either divergence allows the uncoupling of pathways, or the patterns of expression (timing or location) are changed. These are not mere speculations, because investigations of multigene families have abundantly demonstrated that the duplication of genetic material has allowed divergence of both the gene's function and the specificity of control. This diversification has been especially well documented in the immunoglobins (see, e.g., Hunkapillar and Hood, 1986; Rast et al., 1996; Hughes and Yeager, 1997), ribulose-1,5-bisphosphate carboxylase (*rbcS*), chalcone synthase (*Chs*), the alcohol dehydrogenases (*Adh*) (Clegg et al., 1997), the MADS homeotic gene family (Doyle, 1994; Purugganan et al., 1995; Munster et al., 1997) and the Hox homeotic gene family (e.g., Holland and Garcia-Fernandez, 1996; Zhang and Nei, 1996; Averof and Patel, 1997; Grenier et al., 1997; and discussion in Chapter 8).

The superlative example of convergent evolution of functional antifreeze (AFP) and antifreeze glycoproteins (AFGP) in cold-water fishes (with at least four unrelated gene groups differing in their structure and amino acid composition; Logsdon and Doolittle, 1997; see also Chapter 3) also offers an extraordinary example of duplication and divergence. The "normal" *AFGP* gene from the Arctic cod, *Boreogadus saida*, is nearly identical in amino acid composition to that of the unrelated Antarctic notothenioid *Dissostichus* (mainly Thr-Ala-Ala repeats). However, the flanking coding regions and gene structure of the Arctic cod are substantially different, with distinct numbers and locations of introns (Chen et al., 1997a). In fact, the progenitor of the notothenioid *AFGP* gene appears to be a *trypsinogen* gene, which has nearly identical sequences at both ends (Chen et al., 1997b).

The standard idea about how gene duplication contributed to functional divergence was that with the origin of a second identical copy, one copy was released from selection pressure and free to vary (Ohno, 1970). However, several theoretical analyses have suggested that things are not quite so simple—the new copy must be adaptive (i.e., either contributing to the original function through genetic redundancy, or gaining a new one), with rapid loss or silencing if it is not (Clark, 1994; Hughes, 1994; Walsh, 1995). These results would appear to be at odds with our observations of rampant duplication: Nadeau and Sankoff (1997) examined empirical distributions of copies in multigene families and reported that nearly as many genes had acquired new functions as had been lost. There are several possible ways around this dilemma. A. Wagner (1994) found that either single genes or large complexes of genes are more likely to be duplicated and retained. Averof et al. (1996) have suggested a scenario that allows for simultaneous duplication and divergence. Tandem duplication can result in production of copies of an entire coding sequence, without copying all of its regulatory elements (Figure 9.15). The duplicated coding regions will still carry out the same function overall, but they will have divided the labor, operating somewhat independently in space or time. This not only creates genetic redundancy, but also sets the stage for further relatively rapid adaptive divergence.

Genetic redundancy (GR) is of interest to us at several hierarchical levels. At the organismal level, redundancy has been used to refer to the production of the

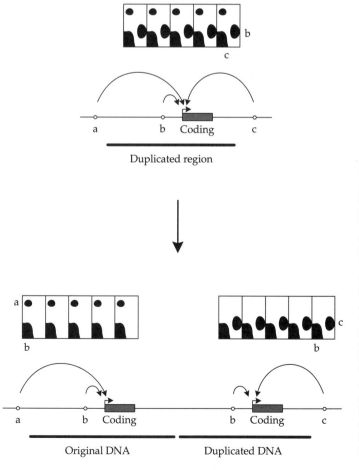

Figure 9.15 A model for the evolution of gene function based on a duplication-divergence scenario. Upper panel: The ancestral gene consists of a coding region and three regulatory sites; a, b and c. The regulators produce three products (differently shaped spots) in the cells (blocks). A tandem duplication of the coding and b regulatory regions results in (lower panel) two genes with partially overlapping functions. The original sequence now codes for a and b, while the duplicate codes for b and c. Regulatory site b can now evolve in one or the other locus, due to its redundancy. (After Averof et al., 1996.)

same gross phenotype by different constellations of genes—to avoid confusion we will designate this second phenomenon as *phenotypic convergence* (see below). At the molecular level, redundancy refers to multiple genes contributing the same or similar products to a process or metabolic pathway, a topic we will discuss here. GR has been known for a long time (e.g., when loss-of-function mutations of individual genes produce weak or no phenotypic effects in the homozygous state), but the recognition of its scope is relatively recent. GR is not an all-or-none phenomenon (Thomas, 1993; Pickett and Meeks-Wagner, 1995). It may range from complete identity of function when a simple quantitative increase is desirable (e.g., multiple copies of "housekeeping" genes) to quite low levels of functional overlap; for example, when a primary enzyme is knocked out and a second enzyme with lower efficiency takes over its function.

Because complete overlap is the expectation for a gene immediately following duplication, and because of the theory that suggests that any changes in function need to be adaptive, both the maintenance and the reduction of GR are of evolutionary interest (Cooke et al., 1997). Thomas (1993) proposed four categories in which GR may be selectively advantageous: (1) increased production of a particular gene product; (2) direct selection for homeostasis favoring overlap in function, with more fine-tuned control from multiple genes (Gralla, 1991); (3) the new copy is selected for a divergent function; (4) genes whose functions singly are overlapping may perform a different function in concert (emergence). A complete loss of function indicates active or passive silencing. A fifth advantage might be to maintain functionality in the face of mutation (A. Wagner, 1996b).

Nowak et al. (1997) formulated models of the evolution of homeostatic GR in cases in which there is a of loss of function of the primary gene, either due to mutation or due to developmental instability. For a two-gene system, they found that partial overlap of function can be maintained (1) if gene *A* has a higher functional efficiency (>) than gene *B* for function X, and gene *B* > *A* for function Y; (2) if gene *A* > *B* for function X, but *A* also has a higher mutation rate, which favors *aB* heterozygotes; (3) if the rate of developmental errors is greater than the mutation rates of the two genes. These authors favored the third scenario, arguing that the rate of developmental errors is likely to be much higher than typical mutation rates. They extended their two-gene model of pleiotropic effects to multiple genes, and found that situations with pervasive GR evolved, particularly when the effectiveness of genes at performing other functions was high. Note that these results are similar not only to those of the models of Beeby and Kacser (1990, see above), but also to findings in real systems (see, for example, Davis and Capecchi, 1996; Bayer et al., 1997). A. Wagner (1996b) also modeled the evolution of GR, with results implying that in some cases it is favored in the context of the efficiency of the entire metabolic network rather than for the individual genes in that pathway, and also suggesting that GR will remain higher in pathways with lower epistasis.

As genome size increases linearly, and redundancy of function diminishes, there is an exponential increase in the number of possible epistatic interactions.

Phenotypes inevitably become more complex (one's particular conception of complexity notwithstanding; McShea, 1996), and epigenetic systems assume greater responsibility for determining the phenotype. Several authors have suggested that, in addition to this general increase in genetic information, crossings of certain thresholds of genome size or organization have been catalysts for macroevolutionary change (Holland and Garcia-Fernandez, 1996). Bird (1995) suggested that an explosion in gene number accompanied the transition from the prokaryotes (~4000 genes) to the early eukaryotes (~7000 to 20,000 genes) and later to the vertebrates (50,000 to 100,000 genes). These increases were in turn made possible by innovations enabling the biochemical machinery to distinguish expressed gene signals from background noise; namely, the evolution of the nuclear membrane, histones, and methylation of DNA. A similar theme has been echoed by Jablonka and Szathmary (1995) through their suggestion that evolutionary transitions coincide with major changes in hereditary systems; for example, steady-state inheritance (autoregulatory—positive feedback), methylation, and the evolution of behavior and learning. The recent discovery that eukaryotic cells can begin protein folding while translation is proceeding (co-translationally as opposed to only post-translationally in prokaryotes), has been suggested as yet another important step toward cellularly more complex organisms (Netzer and Hartl, 1997).

Although there is a tendency to overlook prokaryotes as "primitive," they are in fact amazingly complicated, with intricate patterns of gene expression, regulation, and feedback. Just one example will suffice here. Grossman (1995) reviewed the basis of sporulation in *Bacillus subtilis*. During endospore formation two cells are produced, with the mother cell engulfing the forespore cell. This process is tightly regulated. Different sets of genes (100+) in the two cells are expressed in a distinct temporal order, with feedback between changes in gene expression modifying morphology, and *vice versa*. A single transcription factor, Spo0A, starts the cascade. External environmental factors have substantial influence on this process as well—it can be induced by increased density, and increased nutrient levels can release the spore's dormancy.

Ontogeny and Epigenetics

ONTOGENY IS CREATED, BUT WHY? Following the evolution of the eukaryotic cell, major breakthroughs arose in the establishment of colonial, and ultimately multicellular, organisms and sexual systems of reproduction. The origin of multicellularity is seen as a pivotal event, opening the door for differentiation and functional specialization of cells (Bell and Mooers, 1997)[11] and forming the basis for the evolution of increasingly complex ontogenies. From a DRN perspective,

---

[11] The increased efficiency derived from cellular division of labor is presumed to be the major advantage of specialization. Niu and Chen (1997), however, have proposed that the expression of fewer genes in individual cells, and methylation of the others, allows many genes to be protected from damage by mutagens.

though, the origins of ontogeny were probably purely *passive*, in that the simple addition of more cells creates an interior versus exterior and establishes position effects and a gradient of different exposures to the outside environment. The cells in the interior are now in a *new* environment that could result in the expression of previously hidden portions of their reaction norms. This scenario would also obtain any time there was growth. Mutations that led to differentiation or specialization of particular cells could have subsequently been directly selected.

As more details of regulation and cell fate specification are revealed, we are constantly amazed by their astonishing complexity. Salser and Kenyon (1996) reported that in the six lateral ectodermal (V) cells of *C. elegans*, the *Antennapedia* homologue *mab-5* regulates at least five cell fate choices at five different times. It first switches on to stimulate cell proliferation, then off to specify epidermal structures, on again in a single lineage branch for neuroblast formation, and off again to allow proper sense organ morphology (Figure 9.16)! And there is no particular reason to imagine that this is an extreme case. Among the handful of genes that have been well characterized, there are already examples in *Drosophila*, echinoderms, and mammals of genes with 30 or more *cis*-regulatory binding sites, and there may be tens to hundreds of targets that interact with each regulatory element (see Mastick et al., 1995; Arnone and Davidson, 1997; Chapter 8). The further evolution of this *"developmental choreography"* (Atchley

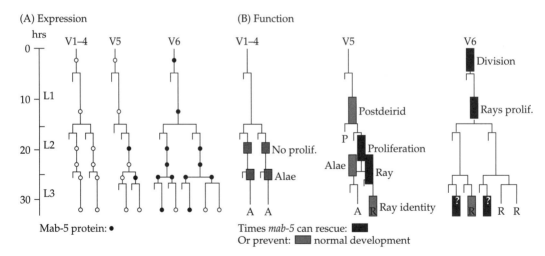

Figure 9.16 Multiple activations of the polyfunctional gene *mab-5* during *Caenorhabditis* development. (A) The temporal sequence of *mab-5* expression in the lateral ectodermal V-cell lineages. Black circles represent gene on, white circles gene off. (B) *mab-5* expression can rescue or prevent normal development when it is activated at various times in a *mab-5* background using a heat shock promoter. A, alae cells; R, ray cells. (From Salser and Kenyon, 1996, © The Company of Biologists.)

et al., 1990) would have paralleled the evolution of metabolism, discussed above. The control of timing and location of gene actions, signals, and responses would have become increasingly important for specifying particular cell identities and functions. Innovations such as methylation and chromatin structuring appear to have been vital in this progression (Finnegan et al., 1996; Michelotti et al., 1997).

THE EVOLUTION OF GENETIC ARCHITECTURE   The pioneering works of Olson and Miller (1958) and Berg (1960) nicely demonstrated the quantification of functional relationships among traits. Here we are interested in how such correlation networks can be built at the genetic level, with the goal of forging links between the evolution of genetic architecture and phenotypic integration (Chapter 7). These elaborations of gene pathways, cascades, and networks; negative and positive feedback loops; and signal reception and transduction (i.e., the evolution of epigenetic systems), clearly represent solutions to achieving the coupling and uncoupling of traits given the challenges of novel biotic (internal and external) and abiotic environments. We have covered some of the mechanistic details of such systems in Chapter 8; here we take a conceptual approach to the general issue of how such architectural plans are devised.

How do pleiotropy, epistasis, and genetic correlation evolve? Again, we can step back and examine how they arose initially. Both the incorporation of new genes and the duplication of existing genetic material will inevitably result in epistasis, as the products of new and preexisting genes interact. These new associations of gene products will generate new characters via the direct formation of new molecules, or by creating new "environments" for the functioning of the other components. The functional divergence of those genes will begin to alter the epistatic relationships, and will lead to initially very high genetic correlations while their functions overlap substantially.

Pleiotropy originates with the "emergence" of these new phenotypic characteristics. A single gene cannot influence multiple phenotypic traits until those traits actually exist. Nascent phenotypic evolution will *de facto* propagate the pleiotropic effects of genes. For example, cells may expose unexpressed portions of their reaction norms in the "new" internal environments created by growth: the genes responsible for the basic traits of a single cell are now also associated with the new traits in this developmental reaction norm. Likewise, the expression of these genes will come to be influenced as well by the environmentally mediated expression of other genes. Although there is the potential for endless feedback, the pattern of gene expression will eventually stabilize, perhaps through selection for a loss of sensitivity to signals (thresholds). An excellent example of the role of the internal environment in gene expression is seen in the comparison of *in vivo* versus *in vitro* binding of Even-skipped protein to gene promoter regions (the pattern of cross-linking to genes in the embryo is markedly different than *in vitro* (Figure 9.17).

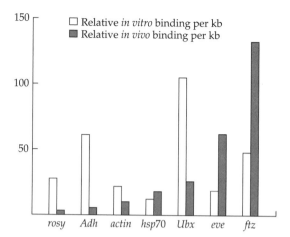

Figure 9.17 The importance of the internal environment for determining gene expression for the segmentation gene *even-skipped* (*eve*). The graph contrasts the differential cross-linking of Eve protein *in vivo* to the promoter regions of various genes with the *in vitro* binding of Eve protein to DNA fragments containing promoter regions. Note, for example, that *Adh* is bound more strongly *in vitro* than *eve*, while the reverse holds true *in vivo*. (From Biggin and McGinnis, 1997, © The Company of Biologists.)

If early organisms were starting from this foundation of high genetic correlation and increasing pleiotropy, the most interesting question about the early evolution of genetic architecture is not "How did traits come to be put together?" but rather "How did they come apart?" (Riska, 1986). It is only later in organismal evolution that the phenomenon of character coupling achieves as much importance as that of character uncoupling. Efforts at understanding how genetic architecture is constructed have taken two partially different approaches: a quantitative genetic approach focusing on the evolution of genetic correlation structure, spearheaded by the work of Cheverud and Atchley; and an approach from the perspective of the mapping of genotype to phenotype (G→P), advanced by Günter Wagner.

Quantitative genetic assessments of pathways for evolutionary change of genetic architecture use as a plynth Lande's model of the evolution of quantitative traits (Lande, 1982; Lande and Arnold, 1983). The focus is on the genetic and phenotypic variance/covariance matrices of traits and the effects of selection on these.[12] Cheverud, Atchley, and coworkers have explicitly taken an ontogenetic perspective in their empirical and theoretical investigations, examining the magnitudes and signs of phenotypic and genetic correlations of traits within and between particular age classes.

Atchley's seminal work was first encapsulated in papers in 1984 and 1987, in which he discussed how organisms solve the problems associated with getting from the zygote to the reproductive stage (Atchley, 1984, 1987). Minimizing the

---

[12] As always, the pertinence of quantitative genetic studies depends on the degree to which the quantitative genetic parameters reflect the biologically relevant genetic architecture (Chapter 3).

inconveniences posed by this inescapable transition phase represents a serious challenge to organisms, and Atchley recognized that quantitative genetic parameters at the adult stage were rooted in the previous ontogeny of the organism. Most importantly, he underscored the essential point that selection at different ages or stages may produce distinct responses based on the quantitative genetic parameters obtaining at that time (Atchley, 1984). Unraveling the evolution of complex structures or suites of functionally related traits requires an understanding of (1) the ontogenetic trajectory of the structures; (2) how modification of one component (or trajectory) affects the others (i.e., some knowledge of the scope and details of the genetic architecture leading to developmental integration); and (3) the target of selection, that is, the individual parts or some composite structure/function (Atchley, 1987).

Subsequent investigations have amplified these themes. For example, selection for fat or lean mice resulted in correlated responses of several characteristics of the mandible. The differences between lines were multidimensional in nature and also age-dependent (Atchley et al., 1990). Cowley and Atchley (1990) investigated patterns of correlation for 17 head and thorax traits of *Drosophila melanogaster*, and related these patterns to the origin of the traits in different imaginal discs. Correlations among traits were generally higher within than between disc origins, although selection on individual disc products generated relatively strong correlated responses by products of other discs. They concluded that selection may have a greater effect if it acts on functional units than if it acts only on the components derived from a single imaginal disc. Atchley and coworkers have recently reported on the results of an ingenious study of 14 generations of selection for early versus late development in mice (Atchley et al., 1997). Directional selection on one phase was indexed in such a way as to hold the other phase constant (e.g., selection for early development was done in concert with stabilizing selection on late weight gain). Again, the results demonstrated that selection "can occur at different points in ontogeny with different consequences." Lines selected for early development showed hyperplasia (a change in cell number), whereas the late lines were hypertrophied (changes in cell size). The early lines diverged from the control before 10 days of age, while the late lines diverged only after 28 days.

Whereas Atchley has hewn closely to the path of basic quantitative genetics (Atchley, 1990; Atchley and Hall, 1991; Atchley et al., 1994), Cheverud has explored a bit farther afield. Riedl (1978) and Frazzetta (1975) recognized that organisms with traits under separate genetic control would require multiple mutations to bring the entire system to a new coordinated state. They argued that the evolution of pleiotropy/epistasis is advantageous because it enables the consolidated genetic control of those traits involved in a single functional complex. The signature of such advantages for trait coupling should be legible as a correspondence between their patterns of genetic control (i.e., degree of epistasis and pleiotropy) and the necessity of their functional integration.

Elaborating on these ideas, Cheverud (1984, 1988) argued that epigenetic concepts should be coupled to the predictive power of quantitative genetics. If internal stabilizing selection for functional integration and the distribution of mutational effects on phenotypic coordination are important forces, then patterns of genetic covariance should conform to the adaptive topography for those traits. The empirical work by Cheverud and others on morphological integration supports this prediction (see Cheverud 1995 for an overview; Chapter 7). A study of the effects of 37 gene loci (QTL chromosomal regions) on mouse mandible integration found 26 loci that affected more than two traits (Cheverud et al., 1997). Approximately 80% of these loci have effects only on traits within a functional complex versus 20% with pleiotropic effects across complexes.

Waxman and Peck (1998) developed a mathematical model that makes some interesting predictions about the evolution of pleiotropy. They constructed a viability selection scenario in which the effects of individual mutations can be pleiotropic, and each of the phenotypic characters affected is under stabilizing selection. Their analysis shows that when three or more characters are affected by each mutation, a single genetic sequence may become optimal, and other genetic variation will be largely dissipated. Although new mutations may arise that optimize an individual character, the likelihood that such a mutation will improve all of the pleiotropically affected traits is very small. They relate this result to the observed low levels of variation and substitution rates at some protein-coding loci. Wagner (1998) more broadly interprets these results as a potential explanation for the "crystallization" of baupläne—the probability that random mutation will significantly improve a majority of the integrated traits diminishes as systems become more complex (see also Duboule and Wilkins, 1998). However, if Waxman and Peck's finding is a significant solution to the conundrum of baupläne, the obverse of the coin—the profusion of quantitative genetic variability for genetically correlated traits—may need a new explanation as well.

A slightly different perspective comes from quantitative genetic models of phenotypic evolution that depart from the standard assumptions of Gaussian fitness surfaces. Models that have combined both directional and stabilizing forces (corridor models) have demonstrated that stabilizing selection on increased variation orthogonal to the axis of directional selection will inhibit phenotypic evolution. It was found, however, that this resistance could be effectively sidestepped through differential genetic coupling or uncoupling of traits (Burger, 1986; Wagner, 1988a,b, 1996).

These studies and approaches have as their focus some aspect of integration of traits, with the attendant expectations associated with such genetic architecture—constraints on, or correlated responses to, selection. These expectations are often realized (e.g., see Scheiner and Istock, 1991), but this is not always true. Fink and Zelditch (1996) compared the spatial and temporal integration of allometric ontogenies among species of piranha. Taking a phylogenetic perspective, they concluded that developmental integration does *not* act as an unequivocal constraint. In fact, patterns of integration appear to be evolutionarily quite labile, changing at nearly every speciation event.

Wagner's quantitative genetic modeling and interests in homology led him to a complementary approach—envisioning phenotypic evolution from the perspective of the mapping of genotype to phenotype. He has focused on the construction of *modules*, that is, complexes of functionally related characters that are integrated via pleiotropy, and that evolve relatively independently of other such complexes (Wagner, 1996). He proposed that there are two general processes that can alter the current status of the suite of characters in a module. *Integration* enhances pleiotropy and epistasis within modules, whereas *parcellation* suppresses the coupling of characters between modules (Figure 9.18; Wagner and Altenberg, 1996).[13] In the case of integration, genes are assumed to initially have little or no pleiotropic effect, but selection that favors coupling for functional purposes will lead to an increase in the level of pleiotropy (Cheverud, 1996). Evolution of modularity by parcellation occurs when selection for reduced pleiotropy operates on a set of genes with widespread pleiotropic effects (i.e., many characters are strongly genetically correlated). Character coupling and decoupling have often been associated with the evolution of new functions or behaviors (see Schaefer and Lauder, 1996 for a review).

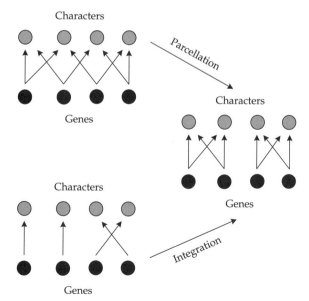

Figure 9.18   Scenario for the evolution of pleiotropy. Characters can be decoupled through parcellation, a process diminishing the pleiotropic effects of single genes. Conversely, an increase in pleiotropy leads to character coupling (integration). (From Wagner and Altenberg, 1996.)

---

[13]Bonner (1988) referred to these as integration and isolation.

A superb example of the evolution of modular organization is the work of Gatesy and Dial (1996) examining the transition to flight by early birds. In their study they proposed that the repatterning of the skeletomuscular system involved *both* parcellation and integration. They defined a "locomotor module" (LM) as an integrated subregion of the skeletomuscular system. Bipedal basal dinosaurs and theropods had a LM comprising the hind legs and the tail (Figure 9.19; Gatesy and Dial, 1996). Gatesy and Dial postulated that bird flight evolved by superposition of a new LM for forelimb-powered flapping flight. The theropod LM was then anatomically and functionally subdivided (*parcellation*). This

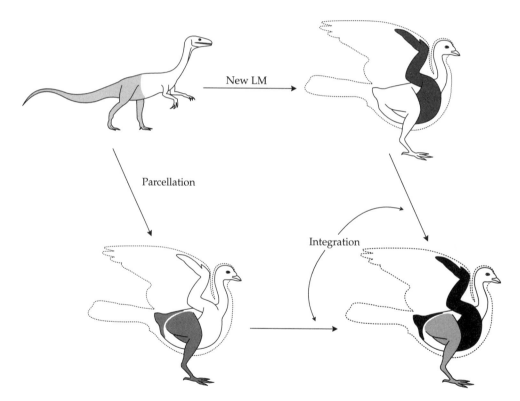

Figure 9.19 Episodes of parcellation and integration hypothesized for the evolution of avian flight. There are two evolutionary events in the lineage leading from dinosaurs to birds. On one hand, the ancestral locomotor module (LM) in theropod dinosaurs (upper left) is parcellated into tail and hind limb components (lower left). On the other hand, a new forelimb LM arises for wing flapping (upper right). Subsequently, the tail and forelimb LMs are integrated into a single module (lower right), with the leg LM now independent. (After Gatesy and Dial, 1996.)

initially decoupled the tail, allowing it subsequently to be newly coupled with the forelimbs (*integration*; Figure 9.19).

Galis and Drucker (1996) examined the extent of structural coupling between upper and lower pharyngeal jaws of centrarchid and cichlid fishes. Cichlids have much more independent jaw movement due to the decoupling of epibranchial four from the upper pharyngeal jaws. The authors suggest that a cascade of subsequent changes was facilitated by this alteration, including the shift in insertion of m. levator externus four, which had previously been proposed as a primary innovation.

A concept related to integration is *genetic piracy* (Roth, 1984; or *epigenetic trapping*, Wagner, 1989; Chapter 6). As these authors delved into the murkiness of homology, each noted that extremely problematic interpretations would arise if genes (or pathways) are co-opted to participate in new functions (Wray, 1994; Lowe and Wray, 1997). Problems for assigning homology are clearly daunting in instances in which old genes are doing new jobs. However, as in the case of gene duplication, in which new genes may do new jobs, the possibilities for evolutionary innovation are enormous.

Perhaps the most prominently discussed example comes again from vertebrate locomotor systems—the evolution of the similarity of the vertebrate fore- and hindlimbs (Tabin and Laufer, 1993). As noted by many previous workers, both fore- and hindlimbs often express comparable evolutionary modifications of locomotion within lineages. More recently, the homeobox detective story revealed that in vertebrates, *Hox* genes 1–13 are expressed in a spatial sequence along the body axis (Duboule, 1994), and genes 9–13 are expressed in the hindlimb. Correspondingly, one would expect *Hox* 2–7 to be expressed in the forelimb, with an attendant different morphology from the hindlimbs. However, it is again the expression of *Hox* genes 9–13 that is observed. It has been hypothesized that, early in the evolution of paired appendages, gene expression in forelimbs was brought under the same regulatory control as that of hindlimbs (a good case of heterotopy). This coupling would explain not only the overall morphological similarities among subcomponents in these structures, but the correlated evolution of both fore- and hindlimb features observed in many groups (Shubin et al., 1997).[14]

This train of thought can be extended to incorporate *phenotypic convergence*, in which the same gross phenotype can be achieved by the actions of different constellations of genes, and *iterative evolution* (Rensch, 1959, p. 97), in which there has been repeated evolution of the "same" phenotypic features (cf. the experiment of Atchley et al., 1997, discussed above). In these cases, there has been no

---

[14] The question remains: What benefit of bringing these structures under common gene regulation has so obviously outweighed the potentials for divergence in structure that separate regulation might allow? for example, pelvic fins and tail fin as propellers/rudder coupled with grasping appendages in front. Perhaps the advantage arose from coordinating the nervous control of their activities.

"piracy" in an evolutionary context; but the action of selection on a stage set by historical contingency and mutation may effect different solutions to an equivalent problem. It is obvious from artificial selection experiments that the route to a particular phenotype need not be the same in all replicates (e.g., Robertson, 1964; other references in Bell, 1997). It is much less clear, however, what the frequency of iterative evolution of phenotypes within species is (e.g., Riska, 1985; Mauthe et al., 1984; Schluter and McPhail, 1992, 1993; Harris et al., 1995; McPeek, 1995; Taylor et al., 1996; Walker, 1997), and what fraction of those events share a common genetic basis.[15]

Perhaps the highest occurrence (and broadest distribution) of iterative evolutionary events is that documented for the evolution of pelvic reduction in stickleback fishes. Cases are recorded from Alaska, British Columbia, California, Nevada, Quebec, Norway, and the Outer Hebrides; and there is evidence for multiple origins even among Alaskan populations (Bell, 1987, 1988; Bell and Orti, 1994). Another excellent example comes from studies on blind cave fish, in which there appears to be repeated loss of vision or even eyes (Wilkens, 1988). Crosses between some blind populations resulted in the production of offspring with eyes. A similar tale is revealed from studies of paedomorphosis in *Ambystoma*. Paedomorphosis may be obligate or facultative (Harris et al., 1990), and crosses between obligate paedomorphic individuals of different origin resulted in percentages of metamorphs ranging from 0 to 100 (Voss and Shaffer, 1996).

In plants, there are also numerous examples of apparent iterative evolution. Sinha and Kellogg (1996) document the multiple origins of alternative (C_4) photosynthetic pathways in the grass family.[16] Their survey of the diversity of C_4 taxa revealed that although the general feature of spatially separated light and dark reactions was maintained, the details of the divergence from the typical C_3 pathway varied broadly among the taxa investigated. In fact, only two of six enzymes examined had the same patterns of expression conserved across the three subfamilies examined. Paterson et al. (1995) made the striking observation that the evolution of several characters in three species of domesticated grains is determined by changes in the same small set of corresponding QTLs, despite 65 million years since a common ancestor! The conservative nature of these parallel changes might suggest that we would find the same pattern in other instances of iterative evolution. However, although there has not been extensive work on the comparative genetics of natural populations, examples of divergent genetic

---

[15] Operationally, the two are effectively equivalent: detection of genetic divergence is the basic evidence for iterative evolution.

[16] C_4 is the name given to pathways that spatially separate the light reactions (light capture and ATP synthesis) from the dark reactions (carbon fixation and CO_2 uptake). This separation greatly increases photosynthetic efficiency in environments in which the stomata must remain closed during the day to conserve water.

bases for convergent phenotypes exist for a variety of traits: morphology (Hobbs and Burnett 1982; Shaffer and Voss, 1996), behavior (Joshi and Thompson, 1995, on competitive ability), and life history (Reznick and Bryga, 1996).

Burton's (1987, 1990a,b; Burton and Lee, 1994) work on the evolution of physiological tolerance in tidal pool copepods represents a model for such studies. Despite evidence that each population was capable of living in the hypersaline conditions that develop in these tide pools, Burton (1990a) found hybrid breakdown in the offspring of crosses between individuals from populations of copepods separated by only 15 km (Figure 9.20). Individuals from genetically differentiated populations were also found to have quite similar patterns of free amino acid accumulation under hyperosmotic stress, but their F_2 offspring showed substantial variation (Burton, 1990b).

Of particular fascination in understanding parcellation and integration is the evolution of complex life cycles. For algae and plants this involves the alternation of haploid gametophyte and diploid sporophyte generations. The morphological similarity of these phases ranges from nearly identical (e.g., in the sea lettuce *Ulva*, Chlorophyta) to vastly different in groups such as brown algae and seed plants. There are distinct larval and adult phases in many animal phyla as well. Perhaps the most remarkable transition is that of the holometabolous insects, in which a functional ambulatory larva is transformed to a distinct functional ambulatory adult, via the bag of goo familiar as the pupal stage. The reasons for adopting such complexity are doubtless diverse (Moran, 1994), but considerations of the mechanisms underlying the evolutionary transitions are few (Davidson et al., 1995; Nijhout and Wheeler, 1996; Bayer et al., 1997; Kondrashov, 1997). Processes akin to integration and parcellation must be happening

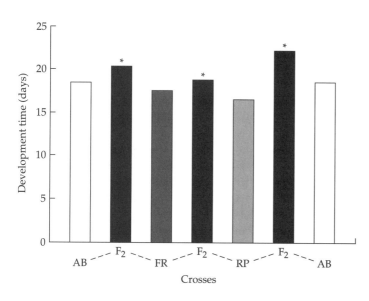

Figure 9.20 Hybrid breakdown in interpopulation crosses of the copepod *Tigriopus californicus*. Crosses were made between three populations in the Los Angeles area (AB, FR, RP). Each set of F_2 offspring had significantly delayed development times relative to the parentals. This suggests that the similarities among parents are not due to identical genetic systems. (Burton, 1990a.)

within such life cycles, with oscillations of shifts in regulatory control between phases. Equally fascinating are the cases in which the complexity has been decreased. For example, Hanken et al. (1997) discuss the manifold changes observed in embryonic structure and timing of events resulting from the evolution of direct development in anurans.

Although parcellation, integration, and genetic piracy are satisfying conceptualizations, giving them form does not necessarily lead to an understanding of the mechanisms that permit or drive such transitions. However, they do lay the groundwork for meshing the discoveries of molecular developmental genetics with experimental and comparative studies of phenotypes. Clearly, though, we have substantial work ahead of us to dissect even simple examples of the evolution of integrated character complexes.

Context-Dependent Gene Regulation: Plasticity and Canalization of Development

A DRN perspective leads us to envision the evolution of development as proceeding largely through changes in sensitivity to internal and external environmental signals by devising genetic and epigenetic machinery that will either *inhibit* or *enable* responses. In this context, then, much of what occurs can be discussed in terms of plasticity and canalization (and combinations of the two).

Despite the seemingly incredible potential for environmentally induced variation, few organisms produce offspring that develop radically differently from themselves (and even those that do, as in cases of polyphenism, typically produce a consistent progeny). The coordination of the unfolding of a complex developmental reaction norm requires the capacity to exert accurate and precise regulatory control over the expression of the genetic template (Wilson, 1976; Smith, 1990; Davidson et al., 1995; Jablonka and Szathmary, 1995; Pigliucci et al., 1996). Regardless of whether the "environmental" stimulus lies on the internal or external portion of the continuum, selection will favor coordinated systems of "signal→receptor→response" that are appropriate for the conditions encountered . . . *and* that do not disrupt the functioning of other systems (i.e., benefits must outweigh costs). Such considerations must favor substantial "coevolution" among genes involved in the developmental process (Bonneton et al., 1997); either through coadaptation of alleles via selection for their joint effects on fitness (Fenster et al., 1997), or through the evolution of specific homeostatic systems (see below). If coadaptation or the assembly of functional groups is the goal, the processes of developmental and genetic integration may themselves be somewhat uncoupled (Muller and Wagner, 1996). In this case the desired phenotypic outcome may be achieved before the genetic linkages are fully formed (perhaps by means of the expression of hidden portions of the reaction norm).

Stabilizing selection on the phenotype, whether physiological, morphological, or behavioral, should lead to canalization of expression, but the means for achieving it are less than transparent. One commonly invoked solution is the maintenance of redundant gene expression pathways (Stearns, 1994; Davis and Capecchi, 1996; Bayer et al., 1997; Wilkins, 1997; Nowak et al., 1997). Another is the selection of changes in the oft-invoked "modifier" loci. An example for antibiotic resistance was documented by Schrag and Perrot (1996), who showed that, despite initially large fitness "costs" to mutations that confer resistance, natural selection can rapidly fix second-site compensatory mutations (i.e., modifiers) in *E. coli* that restore impaired physiological functions; for example, peptide elongation rates are returned to normal.

Another well-documented example is seen in the ongoing investigation of the relationship between asymmetry and resistance to organophosphorus (OP) insecticides in the Australian sheep blowfly, *Lucilia cuprina*. After 20 years, resistance to the OP diazinon was nearly fixed in *L. cuprina* populations. Resistance was accompanied, however, by a significant increase in asymmetry of bristle number and a decrease in fitness (relative to OP-susceptible flies). The identified resistance gene, *Rop-1*, encodes a carboxylesterase enzyme, and its action results in a significant increase in asymmetry (McKenzie, 1993). Over time, however, the degree of asymmetry declined in natural populations even with continued pesticide application (Clarke and McKenzie, 1987). The candidate gene for the suppressor of asymmetry appears to be a dominant mutation involving the gene *Scalloped wings*. This is an apparent *L. cuprina* homologue to the *Drosophila* gene *Notch*, which is important in determining fates of sensory organ precursor cells (Batterham et al., 1996; Clarke, 1997). The precise mechanism by which *Scalloped wings* might canalize *Rop-1* is under investigation.

It has also been argued that the mere fact of "connectivity" of biochemical networks leads to the emergence of stability or canalization (e.g., Kauffman, 1993; Hunding and Engelhardt, 1995; Barkal and Leibler, 1997). This explanation, although appealing as a historical explanation for the general pervasiveness of connected systems, does not readily explain the existence of differences in developmental homeostasis of genotypes that differ at one or a few loci! For example, the number of second vibrissae in wild type house mice is 19, and practically invariant. *Tabby* mutants have significantly fewer vibrissae (males about 9, females 15), but with greatly increased variability (Dun and Fraser, 1959). *Tabby* lines responded to up- and down-selection for vibrissae number (h^2 males = 0.44). Dun and Fraser found that they could produce mice that expressed *tabby*-like variation even in the absence of the *tabby* allele.

Evolutionary prospects for the importance of canalization have been discussed sporadically for years (Chapter 3), but recently there has been renewed interest in constructing a conceptual framework for understanding when canalization will be favored or possible (Stearns, 1994; Stearns and Kawecki, 1994; Stearns et al., 1995; Higgins and Rankin, 1996; Vogl, 1996; A. Wagner, 1996a; Wagner et al., 1997).

Wagner et al. (1997) have distinguished environmental canalization (insensitivity of a trait to external sources of variation) and genetic canalization (insensitivity of a trait to mutational effects; Wilkins, 1997). Their models showed, as expected, that environmental canalization should become stronger as stabilizing selection intensity increases. This accords with the work of Stearns and coworkers (Stearns and Kawecki, 1994; Stearns et al., 1995), who predicted and found a positive relationship between the amount of canalization and the contribution of a trait to fitness. However, Wagner et al. point out that, over time, even weak stabilizing selection should be effective in reducing environmental variance to zero, a situation that has *not* been found for any traits. They suggest that a correlation between the strength of stabilizing selection and fitness is expected only if the genes responsible for the canalization also have negative pleiotropic effects on traits other than the focal trait.

The results for genetic canalization are based on models that alter epistatic relationships among genes. An increase in genetic canalization is favored by increases in population size, mutation rates, and the number of genes affecting the trait. Stabilizing selection favors genetic canalization as well, but only at relatively weak levels: strong stabilizing selection that purges deleterious alleles simultaneously reduces the need for genetic canalization. Wagner et al. argue that in the case of genetic variation maintained by mutation/selection balance, there will be no gain in mean fitness through genetic canalization, because the benefits of modulating the effects of individual mutations will be offset by an increase in the overall number of mutations in the population. A subsequent shift from stabilizing to directional selection, however, can lead to a release of this cryptic genetic variation. As Wagner et al. (1997) point out, there is a dearth of empirical work, especially for the distribution of mutational effects on genetic canalization.

The alternative developmental choice to canalization is to *encourage* a response to environmental stimuli—namely, plasticity in one form or another. Although whatever advantages accrue to homeostasis of development in one environment will probably obtain in another, it is the getting there that is of interest here. How is the genetic architecture of plasticity constructed? The more complex the response, the greater the necessity for regulatory control of some form. And, rather than building numerous new pathways to other genes, altered regulation of existing gene cascades is much more likely, especially given the penchant for evolution to use the available raw materials. Remember that although it may be easier to envision this scenario for the evolution of developmental pathways under internal conditions, it is equally applicable to the evolution of phenotypic plasticity to external environments.

The survey of regulatory epigenetics in Chapter 8 surely will suffice to convince any possible skeptics about the role of regulatory interactions in development *per se*; we have also previously reviewed the considerable evidence linking regulatory genetic control and adaptive plastic responses (Schlichting and Pigliucci, 1995a; Pigliucci, 1996a; Callahan et al., 1997), and will not belabor it here (see our favorite example, Figure 9.21). Smith (1990) as well as Schlichting and Pigliucci (1993; 1995a) have argued that adaptive phenotypic plasticity can

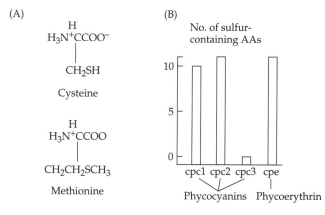

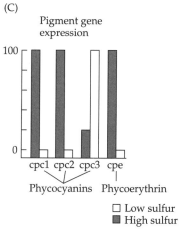

(C)

Pigment gene expression

☐ Low sulfur
■ High sulfur

Figure 9.21 Regulatory control of plasticity of pigment gene expression in the cyanobacterium *Calothrix*. (A) The two sulfur-containing amino acids, cysteine and methionine. (B) Three of the four pigment molecules used in photosynthesis have these sulfur-containing amino acids; the phycocyanin cpc3 does not. (C) Under sulfur-limited conditions, the expression of cpc3 is boosted and the other three genes are turned off. (Mazel and Marliere, 1989.)

evolve by duplication of gene functions. Smith, in particular, reviewed some of the evidence that the differential regulation of members of multigene families represents the molecular basis of plastic responses. Several further examples have been presented since then that support his contention of differential gene regulation for some plastic responses.

Nordin et al. (1991), investigated the environmental induction of expression of a cold acclimation polypeptide in *Arabidopsis thaliana,* and found that two additional stimuli, abscisic acid and water stress, triggered separate signal pathways that converged on production of the same polypeptide. Also in *A. thaliana,* Dolferus et al. (1994a,b) studied the induction of the *Adh* gene by several different environmental factors. By characterizing deletions in several regions upstream of the gene, they localized one promoter that responded only to conditions of low oxygen, and another that catalyzed gene transcription only after dehydration or cold (Figure 9.22). However, a third promoter with a GT/GC-rich motif was found that triggered a response to *any* stress. It is unclear how the action of the different promoters is balanced within the cell and at the organismal level, although the hormone abscisic acid appears to be a common regulator/repressor. Even more elaborate regulatory machinery might be expected in cases of anticipatory plasticity, in which the inducing environmental cue is only a proxy for the conditions that are to be ameliorated.

Khavkin and Coe (1997) reported redundant clusters of genes that are distributed non-randomly on the chromosomes of maize. These clusters included a wide variety of genes important for basic processes of plant growth and development: environmental and hormonal sensors, enzymes for hormone synthesis, sporophyte and gametophyte development, programmed cell death, and transcription factors. They hypothesized that these clusters represent functional sets of genes that regulate plant development and responses to environmental factors; furthermore, the grouping of these genes would facilitate the regulatory control of diverse genetic elements. These sets of results, taken together with the elucidation of large banks of regulatory elements (e.g., Figure 8.10), clearly indicate a possible route by which the sensitivity of responses to internal or external environments could be modified. The maintenance of largely parallel genetic pathways that differ in the environmental sensitivity of their controls may go a long way toward explaining apparent genetic redundancy.

Environmental specificity of gene action may also help to explain the presence of some of the numerous mystery genes currently residing in GenBank. Oliver (1996; Goffeau et al., 1996) made this proposal for the nearly 3000 "functionless" genes of *Saccharomyces cerevisiae* after noting the following: (1) most of the yeast genome is transcribed into RNA; (2) 70% of these genes were *not* necessary on a rich medium; and (3) many of the unknown genes appear to be

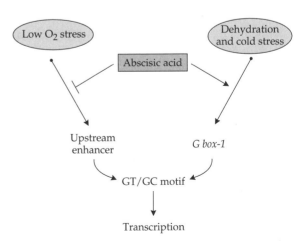

Figure 9.22 Proposed multiple signal transduction pathways in the environment-dependent regulation of *Adh* in *Arabidopsis*. Two different factors trigger *Adh* production, low O_2 or dehydration/cold. Abscisic acid may be involved in regulating both by activating one or failing to deactivate the other. (After Dolferus et al., 1994a.)

expressed at low levels or only transiently, as might be expected of regulatory elements. The findings of Smith et al. (1996) support this perspective: 11 novel genes have growth advantages or defects *only* in the particular environments (e.g., caffeine, high temperature, hypersaline) in which the yeast were grown.[17] Similar suggestions have been advanced by *E. coli* researchers for the unexplained functions of 30–40% of that genome (Moxon and Higgins, 1997). If these suppositions are confirmed, this would indicate that substantial fractions of these genomes are devoted to the specification of often *unexpressed* portions of the developmental reaction norm!

The examples we have supplied, and the conceptual meanderings we have offered, probably only scratch the surface of the intricacy of the generation of organismal form. We have found the DRN perspective to be extremely useful for focusing attention on the details of the developmental process as well as on the mechanisms responsible for the evolution of phenotypes. In Chapter 10 we present two further examples of how this perspective led us to other deviant views on the primary mode of adaptive phenotypic change, and on the capabilities of organisms to move around in morphospace.

CONCEPTUAL SUMMARY

1. We review a variety of empirical studies in which the integration of all three components of phenotypes (allometric, ontogenetic, and plastic) has been achieved. Such studies have varied in their success at unraveling some of the causes of variation in the form of organisms, but in each example the information derived from a DRN perspective has substantially enriched comprehension of the system. One common result of investigating even one more component of the phenotype is the observation of "unexpected" results, or previously hidden variation.

2. Theoretical models of evolution in heterogeneous environments nearly always predict the evolution of an appropriate plastic response to environmental change. If adaptive plasticity is not possible, other alternatives can be adopted. We summarize the conditions favoring genetic variation or plasticity, and generalist versus specialist strategies, in Table 9.1.

3. Evolution of adaptive plasticity may be precluded by a variety of factors, collectively referred to as costs. Chief among these are features such as

---

[17] Thatcher et al. (1998) also found that most yeast genes are non-essential under laboratory conditions. They also found, however, that many mutant strains without obvious defects still have substantially lower fitness in competition with wild type on standard lab medium. They interpret these results as favoring a hypothesis of the "marginal benefit" of these genes—very small, not readily detectable increments to fitness.

construction costs, as well as obstacles such as predictability of the environmental change. The reliability of cues and the problem of the lag time between the signal and the response would appear to be especially difficult barriers to surmount.

4. Although there are many instances in which selection appears to be operating on one or a few traits, we examine several examples in which the selection pressure is on combinations of traits, or on the capabilities of functional groups of characters. We conclude that, given that fitness integrates events throughout the entire developmental sequence, and may be affected by changes in the environment, it is the developmental reaction norm itself that is the chief object of selection on the phenotype.

5. The roles of gene duplication in creating new genetic material and in the production of genetic redundancy are examined. Although models of duplication suggest that neutral evolution of duplicated genes is likely to be of quite short duration (thus unlikely to allow new gene functions to arise), the evidence for duplication/divergence events is overwhelming. We discuss a model by Averof et al. that can circumvent this incongruity between theory and reality. Genetic redundancy has been proposed as a fail-safe system to counteract developmental or mutational instability.

6. The evolution of ontogeny is examined, and it is observed that the evolution of multicellularity will result in the creation of different environments as new cell layers are produced, opening the possibility that these cells may express new reaction norms.

7. The process of development, even in a constant external environment, can be envisioned as a reaction norm itself, with cells and tissues forming as a result of their interactions with internal environmental conditions produced by themselves and their neighbors.

8. We take a giant step backward to very simple organisms to envision the evolution of simple genetic architecture (i.e., pleiotropy, epistasis, genetic correlation). The major question remains: How do genotypes map to phenotypes? The key to answering this lies in an understanding of epigenetic processes.

9. Two approaches to investigating the evolution of genetic architecture are discussed. Quantitative genetics provides statistical power and quantification of patterns. Application of the concept of genotype→phenotype mapping to the evolution of modular organization emphasizes the mechanics and function of genetic architecture. Each method has its attractions and distractions; conceivably, the best of both can be merged.

10. Iterative evolution, the independent evolution of the "same" phenotype, is potentially widespread. These replicate "experiments" offer substantial potential for understanding selection and constraint on genetic architecture.

11. The process of development and the extent of internal and external environmental influences are depicted as a tension zone between the need for canalization and plasticity. Selection operating on phenotypes must feed back through the genetic architecture to find solutions that incorporate homeostatic mechanisms as well as the ability to shift to alternative phenotypes when appropriate.

CHAPTER 10

Implications
and Projections

"Some books are lies frae end to end." Robert Burns

In the preceding chapters we have developed our ideas on how the developmental reaction norm perspective has informed our understanding of organismal integration and constraints and of the mechanisms of constructing the genetic architecture that underlies complex phenotypes. The DRN has also led us to reexamine the framework that we use on a daily basis (largely unconsciously) to understand or explain evolutionary patterns and processes. In the bulk of this chapter we will examine two issues that the DRN directly impinges upon: modes of evolutionary change and the origin of evolutionary novelties. In the final section we will present a synopsis of the major concepts and implications of the DRN perspective in the form of a "research program" (*sensu* Lakatos, 1974).

MODES OF EVOLUTIONARY CHANGE

The paradigmatic mode of evolutionary response of a population to environmental challenges is *allelic substitution*. This is the well-known process (taught in all biology classes) by which a common allele is replaced over time by a mutant allele that is present in the population, but that has been lurking at very low frequencies. The rate of replacement depends on the selective advantage of the mutant over the wild type allele under the new environmental regime. A lack of genetic variability, of course, prohibits evolutionary change, and if the environmental change is drastic enough, may lead to the extinction of that population.

In Figure 10.1 we depict this process from a developmental perspective, in which genotypes give rise to ontogenetic sequences that are represented by the phenotypes produced at different developmental stages (e.g., A→B→E). The developmental reaction norm is controlled polygenically, and the characteristic ontogeny is produced epigenetically from the pleiotropic and epistatic effects of gene expression. Note that here, as well as in Figures 10.2 and 10.3, we have chosen to indicate a change in the terminal phenotypic state only for the sake of simplicity: initial or intermediate states may also be altered, depending on the target of selection and the available genetic variation. In this temporal scheme genotype *X* represents the initial ontogenetic phenotype sequence in environment 1,

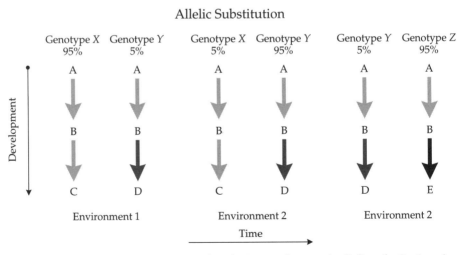

Figure 10.1 The standard depiction of evolutionary change via allelic substitution. A population is composed largely of one genotype, *X*, which produces an adaptive phenotype sequence A→B→C for environment 1. New conditions arise in which the optimum phenotype would be A→B→E. Genotype *Y*, previously present only in very low frequencies, is now favored because its phenotype A→B→D is closer to the environment 2 optimum. Following an appropriate mutation to genotype *Z*, A→B→E phenotypes would come to predominate.

Evolution of Phenotypic Plasticity

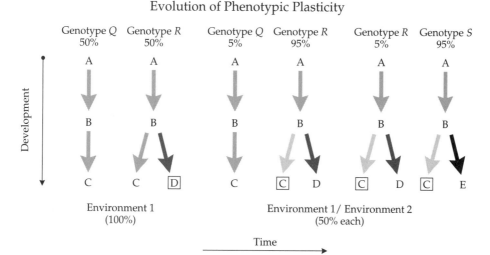

Figure 10.2 Scenario for the evolution of phenotypic plasticity. An adapted population contains genetically different individuals Q and R that each express phenotype
A→B→C in environment 1. An increasing frequency of a new environment, with an
optimum of A→B→E, selects for the plastic genotype R that expresses a hidden part of
the reaction norm, A→B→(C)D, when environmental condition 2 occurs. Subsequent
mutation could result in A→B→(C)E. Boxed letters indicate unexpressed portion of
reaction norm.

A→B→C. There is also a lower-fitness genotype Y present at low frequency,
which produces terminal state D. With the advent of new environmental condition 2, there is a new phenotypic optimum, terminal phenotype state E. Because
D is more fit than C under these conditions, genotype Y eventually supplants
genotype X. Over time, as new mutations arise, a new genotype Z with the optimal ontogenetic sequence A→B→E will be favored over genotype Y.

A second mode of phenotypic evolution that we recognize is the *evolution of
phenotypic plasticity* (Figure 10.2). Although this process also involves the
replacement of one genotype by another, it differs from allelic substitution in
several key respects. First, there is explicit recognition of the alteration of phenotypic characteristics in response to changes in environmental stimuli. This is
depicted in the figure by the bifurcated ontogenetic sequence, with the unexpressed pathway represented in parentheses, for example, A→B→C(D) or
A→B→(C)D. Second, it injects some variety into the life of the organism. Instead
of the either/or scenario of coarse-grained environmental alternatives of the
allelic substitution model, it incorporates the realities of temporally or spatially
fine-grained environments, as well as the likelihood that appropriate plastic
responses will be favored in such situations. In the depicted evolutionary
sequence, the population is initially composed of two genotypes (Q and R) that
produce the same phenotype, A→B→C, in environment 1. These genotypes dif-

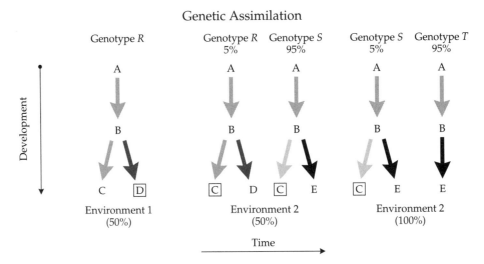

Figure 10.3 Evolution of an adapted phenotype via genetic assimilation. In this process genotype R's phenotypic plasticity, A→B→C(D), initially offers an advantage in a changing or variable environment. Subsequently, a better-adapted plastic genotype S, with phenotype A→B→(C)E, replaces R. If environment 2 supplants environment 1, *and* there is a cost to plasticity, a new mutation to genotype T will be favored and the ability to respond plastically will be lost. Now even with a return of environment 1 conditions, phenotype A→B→E will be expressed. Both the evolution of plasticity and allelic substitution (yielding canalization) occur in this sequence. Boxed letters indicate unexpressed portion of reaction norm.

fer with respect to their plasticity—genotype Q is unresponsive to changes in environmental conditions, whereas genotype R reveals the hidden portion of its reaction norm in environment 2, the better-adapted terminal phenotype D.

Note that this scenario makes implicit assumptions about costs of plasticity: by allowing the initial coexistence of the two genotypes, it is specifying that there are no such costs. Any appreciable cost to plasticity would lead to primacy of genotype Q. However, an increase in frequency of environment 2 will begin to balance the costs with the benefits of expression of A→B→(C)D, and once the environmental frequency at which costs are equal to benefits is surpassed, replacement of genotype Q by R will begin. The more appropriate plastic response of genotype S would be favored, again assuming that costs are equal for the two genotypes.

The third mode of phenotypic evolution is the generally ignored proposition of *genetic assimilation*. The concept of genetic assimilation has an extended history, dating at least to the writings of Baldwin (1896), Osborn (1897), and Lloyd Morgan (1896) on what they referred to as "organic selection." Its present incarnation emerged from the work of Waddington, who coined the phrase, and Schmalhausen, who referred to it as "stabilizing selection" (see Chapter 2.)

Genetic assimilation involves both phenotypic plasticity and allelic substitution (Figure 10.3). A population of genotype *R*, experiencing environmental change, responds initially via a plastic reaction that enables the genotype to persist. Note that there is *no necessity for genetic variation* at this point. The plastic response that permits survival, being previously unselected, is unlikely to produce the optimum phenotype E. Subsequent improvements in the reaction norm toward phenotype A→B→(C)E (of genotype *S*), due to new mutations or recombination, will be favored by directional selection. In this scenario the environmental variable of importance is fixed at condition 2, over the long term. In the absence of a cost to plasticity (or a new environmental shift) this will be the end of the adaptive phenotypic evolution. Mutations altering the expressed reaction norm in environment 2 will be selected against, while those altering only unexposed portions of the reaction norm may accumulate.

However, if there is a *net* cost to plasticity, selection will favor a new reaction norm that lacks the machinery to produce phenotype C if the organism again encounters environment 1. Mutations that eliminate the genetically controlled plastic response are favored, replacing the phenotype A→B→(C)E of genotype *S* with A→B→E of genotype *T*. This step is the culmination of Waddington's genetic assimilation: a process initiated by a purely environmentally induced phenotypic change subsequently becomes fixed in the genotype by means of conventional natural selection. After this modification, reversion of the environment to condition 1 will no longer result in the production of terminal state C.

This is precisely the scenario obtained in the experiments on assimilation by Waddington (Waddington, 1942, 1952, 1953a, 1956, 1959a, 1961) and others (e.g., Dun and Fraser, 1959; Robertson, 1964; Scharloo, 1964; Anderson, 1973; te Velde et al., 1987; te Velde and Scharloo, 1988). Waddington's experiments on the bithorax phenotype are representative. Phenocopies of the bithorax mutation (Figure 10.4) appeared at appreciable frequency when eggs of *Drosophila melanogaster* were treated with ether. Artificial selection for and against this phenotypic response was quite successful (Figure 10.5). Following selection for ether-induced production of the bithorax phenocopy, many flies subsequently were constitutively bithorax; that is, they retained the bithorax morphology in the absence of the previously required ether treatment.

Both Waddington and Schmalhausen considered "new" phenotypes to be typically developmentally unstable, with attendant fitness disadvantages. This final phase would then also characteristically contain a bout of selection for canalization of the phenotype. Developmentally this would be observed as a reduction of variation around the ontogenetic trajectory in environment 2. Canalization is, as discussed before, a phenomenon that strictly applies only to single genotypes. At the population level, stabilizing selection on the new norm will cause a convergence of individual reaction norms toward the average optimal norm, *for that portion of the norm exposed to active selection.*

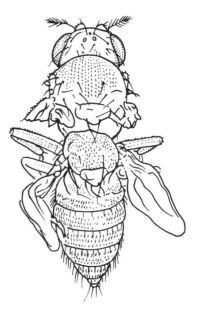

Figure 10.4 The bithorax phenotype of *Droso-phila melanogaster* originally produced as a pheno-copy following treatment of eggs with ether. This particular fly is from the genetically assimilated selection line, and developed from eggs that were not subjected to ether. (From Waddington, 1959b.)

Perhaps the most important consequence of the genetic assimilation scenario is the persistence of some genotypes even in the absence of appropriate genetic variation for evolutionary change. Given that this first step does not require a genetic change, the presence of a plastic reaction norm ameliorates the intensity of selection, creating a lag time between the environmental change and the necessity of genetic adjustment to such change.

This scenario has historically been relegated to the status of an oddity. Simpson (1953), for example, stated: "the Baldwin effect . . . is an interesting but, I would judge, relatively minor outcome of the [synthetic] theory," and Arthur, in his book *A Theory of the Evolution of Development* (1988, p. 56), concluded that there was no particular importance for genetic assimilation. We, however, believe that genetic assimilation represents a common and important mode of phenotypic evolution, a view embraced (to varying extents) independently by a number of others in the context of the evolution of phenotypes (Matsuda, 1982; Wallace, 1990; Hall, 1992a; Roth, 1992; Harvell, 1994; Rollo, 1994) and behavior (Hinton and Nowlan, 1987; Gottlieb, 1992; Mayley, 1996).

The assignment of genetic assimilation to a role as a subsidiary factor appears in most cases to arise from an inappropriate application of Occam's razor.[1] The reasoning is well represented by a passage from Simpson (1953, p. 113):

---

[1] Not an isolated example, this is also seen in some arguments that pit sympatric against allopatric speciation; and in discussions of the irrelevance of Wright's shifting balance theory (see below).

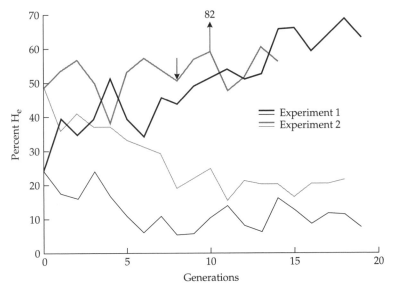

Figure 10.5   Results of artificial selection for and against expression of the bithorax phenocopy in *Drosophila melanogaster*. After 20 generations bithorax developed at a high frequency in the absence of the ether treatment. (From Waddington, 1959b.)

> . . . when the characters in question are demonstrated to be hereditary there is no evidence whatever that they had occurred as accommodations [plastic responses] before they became hereditary.

While this represents a fine example of parsimonious logic, it unfortunately dismisses the impact of several lines of evidence. First, it underestimates the pervasiveness of plastic responses. It is the rule rather than the exception for traits to exhibit some, or even substantial, environmental modification. Second, it ignores the reality of environmental heterogeneity. The vast majority of organisms do not live in chemostat-like environments; fine- or coarse-grained environmental variation is ubiquitous in natural habitats. Third, it overestimates the probability that allelic substitution by itself has been the predominant evolutionary mode by downplaying (1) the low likelihood of having just the right mutation hanging out in one's gene pool, or (2) the reduced expectation of population persistence with neither an appropriate mutation or plasticity.

Taken together, these observations seem to us to elevate genetic assimilation to a position where it is the *most likely* mode of evolutionary change for many characters (see also Rollo, 1994). As environmental signals are received and processed, there will be changes in gene expression (leading to changes in observable genetic variance) and changes in the phenotype mediated through the epigenetic system. If, as we proposed in Chapter 9, a major catalyst in the evolution of the phenotype has been the creation of new internal environments, then the evolution of development itself has been a continual process of genetic assimilation, with oscillations between the phases in which plasticity is favored and those in which canalization is favored. Selection acting on the phenotypic expression of previously hidden portions of reaction norms should lead to the genetic stabilization of a fit phenotype. The resulting "optimal" genotype, though, may again just be the tip of another unexposed iceberg of reaction norm potential.

THE ORIGIN OF PHENOTYPIC NOVELTY

Phagocytosis, photosynthesis, shell, bone, jaws, xylem and phloem, legs, wings, amniotic eggs, pollen, flowers . . . where would the descendants of apterygote insects, jawless fishes, and non-vascular plants be without their Eurekas? Although the influence of the gravitational pull of historical contingency as an overriding factor in the origin of novelty is undeniable, the question persists: are there any patterns to be discerned in the inception of novelty that might serve as predictors of future events?

It has been drummed into all of us by now that evolution is pragmatically opportunistic: it uses that which is available as the framework for new structures and pathways (Jacob, 1977; Duboule and Wilkins, 1998). The combination of chance and natural selection has produced any number of solutions to problems at which biochemists and engineers can only wonder (e.g., the human knee). Take, for example, the enzyme rubisco (ribulose-bisphosphate carboxylase). Rubisco is the linchpin in the Calvin cycle, responsible for facilitating the uptake of CO_2 from the atmosphere during photosynthesis. One subunit of this enzyme is encoded in the chloroplast genome, while the other is nuclear-encoded (as are most other Calvin cycle enzyme genes). On top of this odd division is the unfortunate propensity of rubisco to operate as an oxygenase as CO_2 concentrations decline, which leads to a disruption of the cycle and a drastic loss in photosynthetic efficiency for many plants in hot, dry conditions. Unfortunate indeed, because the Calvin cycle is the only eukaryotic pathway of CO_2 fixation during photosynthesis.

We can easily imagine a higher adaptive peak in this system: Why not an enzyme that is only a carboxylase? Many authors have asked questions such as this in attempts to understand the process of adaptive phenotypic evolution. Although these conjectures span the range from microevolutionary to macroevolutionary scales, they are part and parcel of the same general issue: how do we get from our current phenotype to something that is better?

Sewall Wright formulated the shifting balance theory (SBT) as a description of the microevolutionary process of achieving a new phenotype that incorporates both selection and drift. The SBT has the unfortunate distinction of simultaneously being conceptually alluring and mathematically refractory. Wright conceived a three-phase process by which a species can move among adaptive peaks from lower to higher fitness. Phase I posits that genetic drift displaces subpopulations from a local peak to a valley, which allows them, in Phase II, to move to other, perhaps higher, local peaks. Phase III occurs with selection among populations on the disparate peaks; the advantage lies with populations on the higher peaks that can produce more offspring, and whose migrants will invade other gene pools.

The SBT has been subject to intense scrutiny regarding (1) the parameter space that it can theoretically occur in, and (2) the likelihood that such conditions actually occur frequently enough to account for a significant fraction of adaptive phenotypic change (Crow et al., 1990; Barton, 1992; Phillips, 1993; Moore and Tonsor, 1994; Goodnight, 1995; Gavrilets, 1996; Weber, 1996). Coyne et al. (1997) comprehensively reviewed both the theoretical and empirical examinations of the SBT. They concluded that although Phases I and II are theoretically plausible, Phase III would appear to be possible in only a very restrictive parameter space.

In addition, they point out that there are few empirical examples conforming to the three-phase process, and none that cannot be accounted for more easily by Fisherian processes of mass selection (Coyne et al., 1997). As we suggested above, though, this may be another case of shaving a bit too close with Occam's razor. The fact that a Fisherian model *can* explain putative examples of the phases of the shifting balance is essentially irrelevant. The important issue is: Are the conditions (e.g., the nature of the fitness landscape, the pervasiveness of genetic correlations and their resistance to change, and the effective sizes of natural populations of organisms) that favor one model more prevalent? The empirical data are scant.

The macroevolutionary perspective on phenotypic change has typically been discussed in terms of novelty. Mayr (1960) presented an in-depth appraisal of the origin of evolutionary novelty. His discussion focused on the origin and subsequent gradual modification of new features via three mechanisms: (1) as a pleiotropic by-product of some other change; (2) selection for an intensification of function; and (3) selection for a new function. Mayr's roster of ideas has been expanded and fine-tuned by various subsequent authors. For example, Galis (1996) proposed four categories within Mayr's category 3: (1) excess structural capacity, due to bet hedging or as a result of multiple functions; (2) decoupling of developmental pathways (e.g., heteromorphic life histories); (3) variable developmental pathways; (4) phenotypic plasticity—accommodating structural changes (learning).

One main thrust of Mayr's analysis was a firm rejection of the potential for the macromutational genesis of such features, a perspective that is still widely held (e.g., Charlesworth et al., 1982; Charlesworth, 1990). The lack of evidence for actual macromutational divergence notwithstanding, both empirical and theoretical examples of the potential for discontinuous evolutionary change are accumulating from various directions, with the instances of homeotic transformations produced by Hox gene mutations merely the most dramatic.

An interesting investigation of the nature of phenotypic change was devised by Huynen et al. (1996). They analyzed perhaps the most direct G→P mapping relationship available, the connection between genotype and phenotype for RNA secondary structure. A sequence of a given length has a potential phenotype space determined by the number of possible permutations of the four bases at each site in the sequence. Huynen et al. identified a set of RNA sequences that shared an identical structure, but differed from each other via connections of one point mutation (a neural network). When these sequences were allowed to mutate, some point mutations resulted in no structural change, while others led to large changes (elements of a different neural net). Because of the redundancy of the mapping function, that is, many G to each P, the resulting populations occupied a much broader expanse of sequence space than structure space.

Huynen et al. also allowed an initially uniform population of sequences to evolve via mutation. The fitness of a sequence (its replication rate) was computed as a function of the similarity of its structural phenotype to an optimum. The evolutionary trend was punctuational in form (Figure 10.6; see also Agur and Kerszberg, 1987 for a similar system). The resulting population at various times consisted of a collection of different "quasi-species," connected in genotypic space via their interwoven neural networks. Exploration of new phenotypic space was not uniform: redundant point mutations did not depart from a

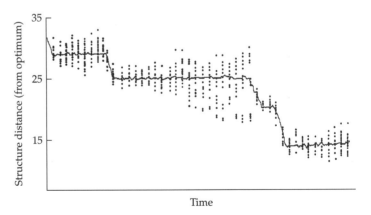

Figure 10.6 Evolutionary trajectory of a population of RNA sequences toward an optimum. A single initial sequence of 76 bases was allowed to mutate, and evolution proceeded with the optimum set as the tRNA[Phe] cloverleaf. The solid line represents the mean distance of a structure from the optimum, and the points represent the variety of structures present in the population at time intervals. (After Huynen et al., 1996.)

network, other mutations moved quasi-species only slightly off of a network, and a few resulted in abrupt transitions of structure or fitness.

Complexity theory, of course, is notable for its predictions of the prevalence of such results, but other theoretical investigations have found similar behavior. Kitchell (1990) presented models that allowed joint evolution of the timing of reproduction and the allocation of resources to growth versus reproduction for a prey species in response to changes in predator size and selection intensity. The results revealed: (1) non-linear trajectories of prey phenotypes in response to changes in predator size (Figure 10.7); (2) the potential for quantum changes in the joint character states characterizing the fitness peaks, even with small changes in predator size; and (3) the existence of regions of "dynamic stasis" in phenotypic space where large drop-offs in predator fitness will result in oscillating selection (dotted regions in Figure 10.7).

The phenotypic adaptive landscape for Kitchell's prey species would appear to be quite rugged. The frequency of smooth versus rugged landscapes is central to debates about the relevance of complexity theory and the shifting balance to evolutionary theory. The issues involved with landscapes are complicated, and the distinction between phenotypic and genotypic landscapes is one of the more challenging questions (Weber, 1996).

Gavrilets (1997) argued that to some extent our fascination with the metaphor of a fitness landscape is due to our restricted experience in a 3-D world. He pointed out that the potential dimensionality of the genetic adaptive landscape is on the order of the number of loci. The significance of this lies in the fact that the *connectedness* of the genetic system (the degree to which phenotypes are linked via point mutations, e.g., Figure 10.8) is much greater on higher dimensional landscapes. In low dimensionality systems, regions of genotype space

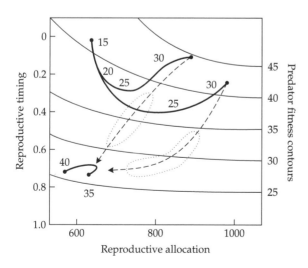

Figure 10.7 Results of simulation of predator/prey coevolution. Solid lines represent joint changes in values of timing and allocation that maximize prey fitness as predator size increases from 15 to 40 mm. Dashed lines represent abrupt transitions between character combinations on fitness peaks. Areas enclosed by dots represent regions of predator/prey "stasis." The contours are the relative fitnesses of predators in the prey morphospace. (From Kitchell, 1990.)

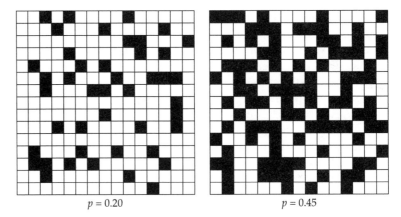

$p = 0.20$ $p = 0.45$

Figure 10.8 A two-dimensional representation of the concept of mutational connectedness. Black squares represent the possible phenotypic states; if another black square is adjacent, it can be reached with a single mutation. At low levels of connectedness ($p = 0.20$), the ability to move from one phenotype to others via single mutations is limited. As connectedness increases to $p = 0.45$, the pathways between different phenotypic states become increasingly linked.

with very low fitness represent adaptive "valleys," and crossing these is a daunting task. In a high dimensionality system, there are many more direct evolutionary (mutational) pathways between combinations of alleles that have higher fitness values. The route may be circuitous, but there is a continuous "ridge" of relatively high fitnesses connecting the peaks (Gavrilets, 1997).

There would appear to be two key issues determining the applicability of such a view. (1) The true dimensionality of the surface will to some extent be determined by the strength of selection on the components. If there is intense selection on a few components and weak selection on most, then the effective dimensionality will be dominated by the few. Although we would argue on the one hand that the importance of the integration of the phenotype will subject many of its facets to simultaneous selection, we also expect that the strength of selection on various components will vary considerably. (2) The connectedness of genotypes/phenotypes on the landscape will be strongly influenced by the nature of the genetic architecture. Extensive genetic redundancy and phenotypic convergence will tend to increase connectedness, while modularity, coadaptation, and regulatory plasticity will tend to decrease it.

A population yearning for higher fitness on a rugged landscape is not out of luck if the conditions of the shifting balance do not obtain. As various authors have pointed out, there is more than one way to hop a peak (Price et al., 1993; Whitlock, 1995). Coyne et al. (1997) review these as well as other reasons populations may be on separate peaks; for example, because of different mutations in the two populations that lack connectivity (no transition between them via mutation; Johnson et al., 1995), or due to stabilizing selection on coadapted allele combinations. The importance of environmental changes has been demonstrated by Price et al. (1993), who showed that changes in the selection intensity on correlated traits could result in peak shifts. Whitlock (1997) demonstrated that adaptive depressions could disappear due to environmentally driven changes in the intensity of stabilizing selection or due to changes in the relative heights of the peaks (Figure 10.9; *cf*. environmentally altered character correlations, Chapter 7).

Our intuition on peak shifts from a reaction norm perspective suggests to us that the changes in character correlations that are induced by environmental change represent a prime route for either (1) moving between peaks without traversing valleys, or (2) altering the landscape itself. As we have documented in Chapter 7, there is substantial evidence that correlations between traits are malleable, including not only the phenotypic correlations that influence the intensity of selection on characters, but also the genetic correlations that determine

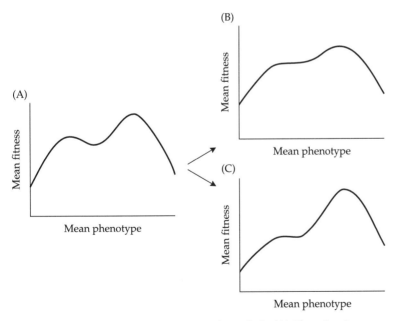

Figure 10.9   Facilitating peak shifts without drift. (A) The adaptive landscape initially has a valley separating the two peaks. (B) Changes in environmental conditions that reduce the intensity of stabilizing selection can remove the valley (24% decrease in this example). (C) Alternatively, the heights of the peaks themselves can be modified (the height of the lower peak is reduced by 57%). (From Whitlock, 1997.)

responses to selection. Although an expectation of no change is a convenient null hypothesis, the observation that correlations should have reaction norms of their own is not surprising given the component variances and covariances (see Equation 7.2). A lack of environmentally mediated change in correlation is only possible in two rather unlikely scenarios: (1) there is no change in any of the components, or (2) there are simultaneous and proportional changes in the variances and covariance. Our expectation, then, is that correlations among traits are likely to change along with the environment.

Given this expectation, the coordination of these changes is of vital importance, and the ability to achieve smooth transitions (via the DRN) to new or different peaks will be under intense selection. In fact, this is precisely what happens in those cases of phenotypic plasticity that we refer to as developmental conversion (i.e., threshold plasticity). A regulatory switch leads to the production of a radically different phenotype that is superior in performance under those conditions that triggered its formation. There is no transition through the phenotypic intermediates (with the potential for drastically lowered fitness) between forms A and B.

We have previously called attention to the role of such regulatory control in the calibration of multiple plastic responses (see also West-Eberhard, 1992). However, it has been difficult to imagine how the fine-tuning of these to match several (or many) different environmental regimes could be accomplished. Nonetheless, a likely mechanism may be discerned from the recent results that indicate a multiplicity of *cis*-regulatory sites for individual genes (see, e.g., Yuh and Davidson, 1996, and Chapter 9).

Furthermore, the DRN also contains the raw material for additional shifts in its unexpressed portions, thus new phenotypic combinations will be exposed as internal and external environments change.[2] The notion that organisms occupying marginal environments may have a significant role in the genesis of diversity has been around a long time (Simpson, 1944; Mayr, 1963).[3] Levin (1970) proposed that, in addition to the variability exposed by homozygosity and recombination in small populations, there are other sources of phenotypic variation in peripheral isolates. The joint effects of inbreeding, directional selection, and drift can lead to the formation of developmentally unstable genotypes that leak previously masked variability. He also pointed out that the environmental stresses more likely to be encountered on the periphery could also lead to the exposure of novel phenotypic variation (Service and Rose, 1985; Holloway et

---

[2] Including even the alterations brought about by changes in the mating environment, e.g., the effects of bottlenecking on genetic variances (Bryant and Meffert, 1995; Cheverud and Routman, 1996).

[3] Although Parsons (1993) argued that just the opposite would occur—because the cost of accommodating environmental stress is high, evolutionary change may be deterred at species boundaries that occur in novel environments.

al., 1990). This second proposal is very similar to Schmalhausen's (1949) concept of the expression and subsequent stabilizing selection of a previously unexposed portion of a reaction norm. Although developmental instability may play a significant role in such peripheral populations, it seems likely that the exposure of uncharted reaction norms and morphospace is more important in releasing hidden variability in a novel environment.

West-Eberhard (1989) vigorously promoted phenotypic plasticity not just as an adaptation to varying environments, but as a stimulus for evolutionary diversification as well. She argued that more plastic traits (and especially behavior) may evolve more rapidly because (1) a more labile trait may have more optional forms, leading to a higher overall probability of one being associated with higher fitness; (2) there are more cues available to initiate adaptive behavioral plasticity; and (3) unlike morphological responses, behavioral innovations may have only ephemeral effects on the organism. She emphasized that phenotypic change through plasticity is facilitated by the capacity for correlated shifts in related traits, and by the condition-sensitive phenotypic expression that allows for the somewhat independent evolution of alternative forms. Although directly opposed to the notion that plasticity should inhibit evolutionary change (see Chapter 3), this proposition is clearly in synch with Waddington's genetic assimilation and Schmalhausen's stabilizing selection (see above). Furthermore, it should be directly testable through experiments that are coupled with comparative phylogenetic analyses.

Doebley et al. (1995, 1997) have illuminated a genetic transition in the evolution of maize from teosinte that appears to represent an example of West-Eberhard's proposal. Maize (*Zea mays* spp. *mays*) has substantially greater apical dominance than its putative wild ancestor, teosinte (*Zea mays* ssp. *parviglumis*). The pattern of expression of the gene *teosinte branched1* (*tb1*) and the morphology of *tb1* mutant plants suggest that this gene acts to repress the growth of axillary organs and to promote the formation of female inflorescences. Furthermore, this gene is involved in the response to environmental conditions to produce either long or short branches. The *tb1* allele of maize is expressed at twice the level of that of teosinte, indicating changes in gene regulation. Doebley et al. suggested that the evolution of the regulation of this plastic response led to a morphological change whereby short branches are produced under all environmental conditions (i.e., the maize phenotype).

Several authors have observed that epigenetic systems are actually primed for the genesis of novelty due to the DRN and emergent properties. Rachootin and Thomson (1981) argued that the features of self-assembly, feedback, and alternative pathways that compensate for irregularities make the stabilization of new developmental patterns relatively straightforward. For example, Brylski and Patton (1988) speculated that the unusual external cheek pouches of geomyoid rodents originated as a small change during development that led to a marked morphological shift.

Müller (1990) extensively discussed the role of epigenetics in the formation of novelty. He argued for important roles for plasticity and regulatory feedbacks, and emphasized again the capacity of developmental systems to "integrate" changes. Müller pointed out two common outcomes in developmental systems. (1) The reaction of tissues and organs to changing conditions during development produces "intermediate" structures. Even though Müller was referring to changes in the internal environment during development, his reasoning can be extended to external environmental changes as well. (2) Threshold properties of developmental systems easily explain discontinuity in phenotypes. Threshold characters are far from uncommon in biological systems. In quantitative genetics, thresholds provide a working hypothesis to connect observable phenotypic discontinuities with the assumed underlying genetic continuity of information (i.e., it allows for the reconciliation of the existence of discontinuous traits with the genetic model based on many genes with small additive effects; Roff, 1996). Müller pointed out two consequences of these features. (1) The evolution of phenotypic novelties does not require the evolution of new genes or gene functions, but only a combination of modified regulation and epigenetic adjustments (Schlichting and Pigliucci, 1995a). (2) The resulting pattern of morphological evolution fits well with observations of "macroevolutionary" stasis and punctuated equilibria (Eldredge and Gould, 1972).

We propose that it is these general tendencies for redundancy and compensation that allow the raw materials for peak shifts (contained in the unexpressed portions of the DRN) to be utilized. The modular nature of organisms can accommodate changes in components of one module via adjustments of other components, without dire consequences for the functional integrity of the whole (e.g., the case of the fibular crest in birds compared with theropod dinosaurs; Müller and Wagner, 1996). Furthermore, the origin of novelties can be achieved without new genes, via repatterning of existing genetic architecture (Müller, 1990; Schlichting and Pigliucci, 1995a).

EVOLUTIONARY FUTURES

For a truly predictive theory of phenotypic evolution, we would dearly love to have techniques that enable us to determine where lineages can travel in morphospace, and whether certain evolutionary outcomes are more likely than others. Many evolutionists might argue that such forecasting may be forever beyond our reach. Others, however, continue to look for clues to the *evolvability* (e.g., Houle, 1992; Jones and Price, 1996; Wagner and Altenberg, 1996) of organisms, that is, their (latent) potential for evolutionary change. In the following sections we will describe four complementary methods for investigating the limits of phenotypic space. None of these methods will provide definitive answers to the evolutionary trajectories that will be followed by species and their descendants, but they may furnish us with possible solutions to the riddle.

Selection Experiments

The most direct approach to discover whether a lineage can attain particular character combinations is to try to move it there. This is an extremely valuable approach, with innumerable applications (see Chapter 1, and recent reviews by Reznick and Travis, 1996; Rose et al., 1996; Bell, 1997). The drawbacks to selection experiments lie in the brevity of the time scale, and in the strong likelihood that the sampled populations will not be representative or that they will contain only a fraction of the available mutational variation.

The Past as a Template for the Future?

The exploration of phenotypic space has traditionally been carried out as a means of understanding "constraints" on organismal form (Raup, 1966; McGhee, 1980; Gould, 1984; Barel, 1993; Stone, 1996). The power and interpretation of such investigations, however, requires considerable care in the choice of parameters (Schindel, 1990; Stone, 1996).

We will highlight two recent approaches to examining morphospace. Thomas and Reif (1993) proposed the concept of the *skeleton space*, representing the possible combinations of states of various characteristics of the skeletons of organisms. The possibilities were delimited based on geometric and physical rules and included such aspects as location of skeleton (internal versus external), mode of growth, number of skeletal parts, rigidity versus flexibility. Because the number of possible combinations of the various architectural features is enormous (greater than 1500), they chose to simplify their analyses by looking for the presence or absence of only the pairwise combinations of traits. They found that, of the 178 possible pairwise combinations, 172 were observed. From this, they argued that the skeleton morphospace had been nearly fully exploited. However, there are several unexplored higher dimensional levels (i.e., three-way and higher combinations of traits) that may or may not have similar fractions of occupancy.

Dommergues et al. (1996) presented a detailed analysis of the changes in morphospace occupancy for 436 species in 156 genera of early Jurassic ammonites (Figure 10.10). They used 18 parameters to define their morphospace, without regard to taxonomy. Overlaying the putative taxonomic relationships on this morphospace revealed two major groups with four or five subgroups each. An interesting facet of their results is their analysis of changes through time. This revealed numerous shifts in the amount of variation among those ammonites extant at any given time, and also in the amount of morphospace that was occupied during particular time periods (Figure 10.11).

Foote (1997) reviewed the literature in this area and identified a variety of areas in need of further work: (1) expanding the taxonomic scope of morphospace investigation; (2) developing methods of comparing occupancy of morphospace; (3) assessing the reality of the unoccupied portions (for fossil taxa); and (4) evaluating the contribution of environmental effects to differences in morphology. At present, the number of studies is small, and the analysis and

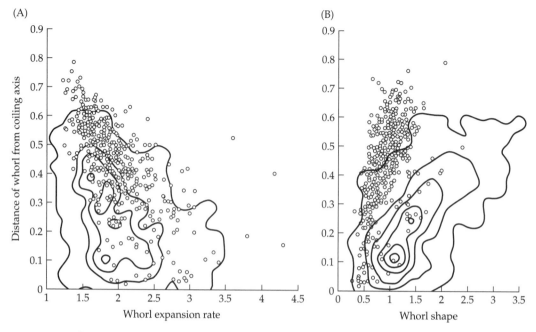

Figure 10.10 Differential occupation of morphospace over time for ammonoids. The contours represent the distribution of forms in the Nammurian (mid-Carboniferous), and points of the scattergram represent taxa in the Hettangian-Pliensbachian (early Jurassic). Note the major displacement of the whorl-coiling axis distance. (From Dommergues et al., 1996.)

interpretation of such data are in their adolescence. There is room for more investigation of extant organisms and fossil groups, and, perhaps, for some modeling (e.g., Niklas, 1994, 1997).

Developmental Reaction Norms as Evidence of Evolvability

We believe that the investigation of reaction norms offers an important source of insights into the potential for phenotypic evolution. Various authors have examined reaction norms for evidence of latent evolutionary potentials or constraints (Partridge and Harvey, 1988; Futuyma et al., 1995; and references in Sinervo and Basolo, 1996). Others, however, have disputed the utility of reaction norm studies for evolutionary prediction. Reznick (1985; 1992) has argued that manipulation experiments only measure plasticity and expose the existence of genotype by environment interactions, rather than uncover the true genetic architecture of tradeoffs. Rose et al. (1996) dismiss phenotypic manipulation experiments because their results "may be a poor guide to the response to selection," but soften that stance (Chippindale et al., 1997) to admit the possibility that environmental responses may provide some "valuable information."

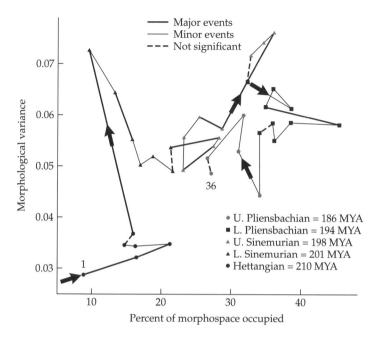

Figure 10.11 Changes in morphospace occupancy by ammonoids. The arrows represent direction of change through time; the symbols indicate the different time periods during the early Jurassic. (From Dommergues et al., 1996.)

In the figure legend:

- Major events
- Minor events
- Not significant

• U. Pliensbachian = 186 MYA
■ L. Pliensbachian = 194 MYA
▲ U. Sinemurian = 198 MYA
▲ L. Sinemurian = 201 MYA
• Hettangian = 210 MYA

Morphological variance (y-axis): 0.03, 0.04, 0.05, 0.06, 0.07

Percent of morphospace occupied (x-axis): 10, 20, 30, 40

We see both of these arguments as a bit myopic. They appear to assume that phenotypic responses are somehow non-genetic, and as such are less interesting than the "real" genetic architecture; nothing could be further from the truth. Plastic responses are direct manifestations of the underlying genetic architecture, and genotype by environment interaction represents genetic *variation* for the responses to manipulation; by its very nature genotype by environment interaction complicates interpretation of tradeoffs. Such variation is clearly indicative of a genetic capacity to produce a phenotype. Granted, this is just one step in adaptation. The other necessary step is to develop a system of signal perception and response that links particular environmental conditions to that appropriate phenotype.

One possible explanation for an apparent lack of correspondence between plastic and selection responses may be that researchers have not taken the appropriateness of the plastic response into account when evaluating their results. Only in the case of an *adaptive* plastic response will the responses to selection be in the same direction as the plasticity. If the plastic response is maladaptive, then selection on the trait will naturally be opposed to the direction of plasticity (and a *decrease* in plasticity will be favored).

We can recast this distinction in terms of co-gradient and counter-gradient variation (Conover and Schultz, 1995). A co-gradient pattern occurs when plastic responses are in the same direction as the geographic variation; for example, if organisms in cooler climates are genetically determined to take longer to develop, and they take longer still when subjected to cooler temperatures (Figure 10.12). Counter-gradient variation refers to cases in which genetic variation and plasticity show opposite patterns; organisms that are genetically slow to develop respond to decreasing temperatures by accelerating development. Counter-gradient patterns are relatively common for life history traits, while cogradient variation is more often found for morphological traits (Berven et al., 1979; Conover and Schultz, 1995). If the pattern of trait mean and plasticity is counter-gradient, then plastic responses will be in the opposite direction to any selection aimed at altering the trait mean.

An understanding of the elements of the DRN is clearly useful for assessing short-term adaptability of organisms to environmental change (and we would suggest that it is vastly more important for conservation biology than is cataloguing genetic variation or fluctuating asymmetry). Because these phenotypes are environment-dependent, they will not necessarily be evoked under conditions in which they would be most advantageous. However, because the DRN is a manifestation of the flexibility of the underlying genetic architecture, it would appear to be a prime resource for identifying potential phenotypic states of characters, as well as for appraising the likelihood that particular combinations of character states are achievable (e.g., Stebbins and Hartl, 1988).

Mutational Variability and Latent Potentials

The fourth means of assessing evolutionary potentials is by examining the range of mutational effects on the phenotype. As an analogue to the **G** (genetic variance/covariance) matrix, the mapping of mutational effects on various traits has

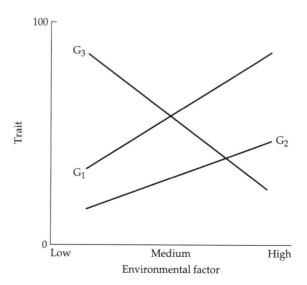

Figure 10.12 Co-gradient and counter-gradient patterns of variation. Genotype 2 is found naturally at high values of the environmental factor, whereas genotypes 1 and 3 are found at low values. Exposure of each of these to the alternative environment results in a plastic response. Genotypes 1 and 2 exhibit co-gradient variation: their plastic responses parallel their genetic differences. Genotype 3 exhibits a counter-gradient pattern, with plastic responses opposite to genetic differences.

been referred to as the **M** matrix. Several approaches have been taken, but perhaps the most famous is the saturation method pioneered by Nüsslein-Volhard and applied to *Drosophila* and zebrafish (*Danio*; Nüsslein-Volhard, 1994). Some studies have examined the effects of spontaneous mutation on phenotypic trait distributions (Mackay et al., 1992b; Houle et al., 1994) and even to some extent on reaction norms (Fry et al., 1996) in *Drosophila*. Another approach has been to insert P-elements into *Drosophila* chromosomes (Mackay et al., 1992a; Stearns and Kaiser, 1996). For example, Clark and Wang (1997) constructed all possible two-locus genotypes for each of eight pairs of P-element insertions, and analyzed the metabolic phenotypes (fat and glycogen contents, enzyme activities, total protein, and body weight). Their results indicated that there were significant epistatic effects in 27% of the trait by P-element combinations, and that these mutations had epistatic effects of the same order of magnitude as additive and dominance effects.

Although there has been this variety of studies that utilize mutagenesis, few have been concerted efforts to examine mutational effects from an evolutionary perspective. The questions of interest are: What are the limits to mutation, and are there inherent biases favoring particular classes of mutation? Perhaps the first study to examine these issues is that performed by Koufopanou and Bell (1991), who examined patterns of variation in the colonial green alga *Volvox*. Both spontaneous and chemically or UV-induced (macro)mutants of *Volvox carteri* were catalogued and compared with patterns of variation within the genus and among genera within the family Volvocaceae. Mutational variances were found not to be distributed evenly around the wild type mean, nor were they the same for different strains (Figure 10.13). Furthermore, the covariance generated via mutation between traits differed in direction from the wild type covariance for two of three trait pairs. When these covariances are contrasted with the covariation patterns among species (Figure 10.13), the patterns are concordant for the relationship between somatic and germ line tissues, discordant for the correlation between number and volume of somatic cells, and mixed for number and volume of germ cells (only mutational and among species patterns were concordant). The mutational and "evolutionary" variances generated varied among the different traits measured (Figure 10.14). For the number and volume of somatic cells, mutation generated nearly as much or more variation than is observed among higher taxa. For germ tissues, mutational variance was greater only for cell number, and less for cell volume and total germ cell volume.

Schluter (1996) examined patterns of morphological differentiation in a variety of vertebrates. He found a strong correlation between directions of morphological change and the predominant vector representing the multivariate genetic variation and covariation within populations. He interpreted this as evidence for the strong impact of genetic constraints (availability of genetic variation or restrictions posed via covariation) on evolutionary change. He points out, however, that his interpretation cannot be distinguished from the

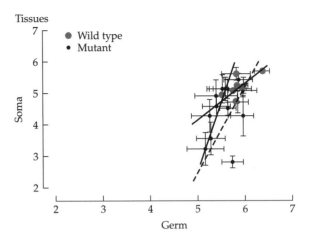

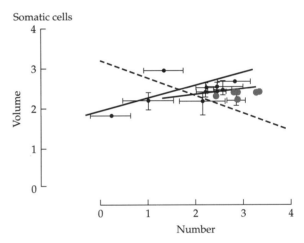

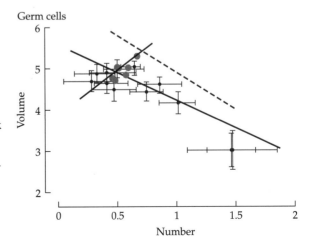

Figure 10.13 Variances and covariances of traits in wild type and mutant *Volvox*. Black circles represent the different mutant strains, shaded circles are wild type strains. The dotted lines represent the covariances between traits when calculated across the different species examined. (From Koufanopou and Bell, 1991.)

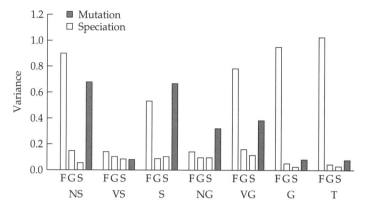

Figure 10.14 A comparison of the variances due to mutation with those observed among species of colonial green algae. For the open bars, F represents variation within the family Volvocaceae, G the genus *Volvox*, and S a section within *Volvox*. The black bars are the mutational variance. The variances among genera were, as might be expected, considerably higher in most cases than the variances at the lower taxonomic levels. NS, number of somatic cells; VS, volume of somatic cells; S, somatic tissue; NG, number of germ cells; VG, volume of germ cells; G, germ tissue; T, total tissue. (From Koufanopou and Bell, 1991.)

possibility that genetic variance/covariance structures are themselves the objects of natural selection. This second possibility is what we would predict from a DRN perspective.

We have previously mentioned studies by Markwell and Osterman (1992) and Pigliucci and colleagues (Pigliucci et al., unpublished; Camara and Pigliucci, unpublished; Chapter 6) that have addressed the nature of mutationally produced variation and covariation. Many more efforts along the lines of the studies surveyed above will be necessary before we can assess the utility of information about **M** for the prediction of evolutionary trajectories.

Studies of aspects of evolvability are in their infancy, but we are excited about the prospects of learning more about evolution's past and future from these different approaches. It may be that the best they can offer is only a rough template, but as we learn more, it is also possible that the combinations of different types of information about the mapping of genotype to phenotype will present new insights into patterns of evolutionary change and stasis.

A RESEARCH PROGRAM

In this section we introduce an outline for an integrated research program on phenotypic evolution from the perspective of the developmental reaction norm. Our objective is to present a set of structured ideas, to facilitate discussion, and to promote empirical and conceptual research on the issues raised.

In the philosophy of science, several schemes for the organization of a framework for the study of natural phenomena have been proposed (Chalmers, 1976). We have chosen Lakatos' (1974) idea of a *research program* for our description of the body of concepts dealing with phenotypic evolution. Lakatos' research program consists of three major parts: (1) the core concepts (the hard core), (2) the auxiliary concepts (the protective belt), and (3) the research projections (the positive heuristic).

The core concepts represent a set of statements that define the central ideas of the new framework. They cannot be changed without recasting the whole approach. In the theory of punctuated equilibria (Eldredge and Gould, 1972), for example, the core concepts may be considered to be (1) that most morphological changes are associated with speciation events, and (2) that for most of the time species are in a state of morphological stasis. Refutation or removal of either effectively terminates that particular research program.

The auxiliary concepts are complementary to the core concepts, but not vital to the development of the theory. They can be modified, or even dropped, without endangering the identity of the research program. For punctuated equilibria, an example would be the concept of species selection: it was derived in association with ideas stemming from the central proposal, but its rejection would not invalidate the core concept of an alternation of stasis and punctuation.

Research projections are suggestions about the direction of future investigations that are a direct consequence of embracing the new framework. Their purpose is to indicate ways to either confirm predictions or falsify the core concepts. An example in the case of the punctuation versus gradualism debate is the search for appropriate statistical tests to distinguish between the two alternative modes of evolution.

Core Concepts

THE DEVELOPMENTAL REACTION NORM IS THE OBJECT OF SELECTION   The phenotype of an organism is characterized by the potential for its genotype to express a series of developmental trajectories, depending on the particular set of environmental conditions to which the individual is exposed. The view of the DRN as an object of selection arises from two basic considerations. First, the common focus on adult stages may lead to a restricted understanding of evolutionary potentials. The origin of the observed differences in adult phenotypes can be understood only by tracing the ontogenetic trajectories. In some cases, similarities in adult phenotypes may be derived from different developmental pathways. Differences in ontogenies may originate from a variety of sources, with both heterochronic and sequence changes playing major roles.

Second, organisms cannot be considered independently from their environments. To discuss phenotypes without the context of particular environments neglects a crucial aspect of phenotypic complexity. We explicitly integrate these two components of the phenotype into our concept of the developmental reaction norm. As emphasized by Levins and Lewontin (1985), evolution results from a dialectical interaction between organisms and environments through ontogeny.

The DRN also leads us to eliminate the dichotomy between internal and external environments. In fact, we view the developmental process itself as a reaction norm. Gene expression may be constitutive or determined by the cellular environments to which the genes are exposed. The ensuing phenotypes will be dependent on these patterns of gene expression as well as on the interactions of the gene products mediated by the other features of the environment. We offered as an example the change in the environment of cells and gene products during the transition to three-dimensionality—previously all cells were in contact with the external environment, now some are never exposed to those conditions. Any new phenotypic characteristics of this internal cell population will be observed as pleiotropic effects of the original genetic complement.

THE INNATE COMPLEXITY OF GENETIC SYSTEMS NECESSARILY LEADS TO EMERGENT PROPERTIES, ARISING FROM THE EPIGENETIC PROCESSES GOVERNING THE FORMATION OF PHENOTYPES VIA GENE BY ENVIRONMENT INTERACTION   The common metaphor of genes encoding a "blueprint" for the phenotype leads to a distorted view of the process of phenotype formation. More appropriately, genes represent a series of subroutines that interact with one another, to initiate and locally control the events that unfold during ontogeny. In principle, even knowing the mechanistic details of what each subroutine does and how it interacts with the others does not allow direct prediction of what the final phenotype will be. A consequence of this is that the phenotypic effects of altering the rules (i.e., mutation) are also virtually unpredictable; conversely, it is not possible to derive the genetic basis of a particular phenotypic change by examining the phenotypic effects of a mutation. This "undecidability" (Langton, 1986) eliminates the possibility of computing *a priori* what the output will be under the range of possible circumstances. The only way to determine the outcome is to let the program run.

Epigenesis, then, is the deterministic, yet "undecidable," series of ontogenetic events resulting from the interaction of numerous local genetic systems (as opposed to a global central control) with the prevailing "internal" and "external" environments.

BOTH THE CONTROL OF THE DEVELOPMENT OF COMPLEX PHENOTYPES AND THE CAPABILITY OF THESE SYSTEMS TO RESPOND TO ENVIRONMENTAL VARIABILITY (INTERNAL OR EXTERNAL) REQUIRE A SYSTEM OF BALANCED REGULATORY INTERACTIONS   There are theoretically three types of control systems for a complex phenotype, given N genes and an average of K interactions among genes (Kauffman and Levin, 1987; Bak and Chen, 1991). At one extreme, there is no interaction among the parts ($K = 0$; a type I or subcritical system). The phenotype results from the simple addition of the individual effects (a Fisherian model of genetic evolution), and will evolve until the system is at a local maximum. Evolution will occur on a correlated adaptive landscape: the fitness value at any given point is positively correlated with the fitness of the points around it. Given the lack of connectivity within the system, a phenotype will be unable to leave the local peak even if much higher peaks are available in the landscape. This is basically the equivalent of the predictions of Fisher's fundamental theorem (Fisher, 1930). At the other extreme, each part is interconnected with all the others ($K = N - 1$; a type III or super critical system): the system is totally integrated (a Wrightian model with truly universal epistasis). In this case evolution occurs on a completely uncorrelated adaptive landscape: the fitness value of neighboring points can be high, intermediate, or low with no predictable relationship. Given the very high connectivity of the parts, any local change will cause a major rearrangement through cascade effects. Type III systems tend to behave chaotically (that is, deterministically unpredictable), and their fitness tends toward the mean fitness of the landscape.

Finally, in intermediate situations (type II or critical systems) genes interact with a limited number of other genes in a loosely connected network. Such systems are characterized by an intermediate level of connectivity and evolve on a partially correlated landscape. This means that they are capable of a balance between the two extreme behaviors described above. Most of the time, change will be gradual, with only local movements on the landscape; from time to time, however, the same small change will have long-range repercussions, resulting in a major alteration of the phenotype (a "jump" to a different portion of the landscape). We conceive the majority of biological systems to be of the critical or type II kind. Langton (1986) demonstrated that type II cellular automata are characterized by the same fundamental properties possessed by biological macromolecules: they have the ability to transport, and can catalyze or regulate reactions; and they can be used for transmission and storage of information, as building blocks for more complex structures, or for defensive purposes.

A major problem remains, however: When exactly does a system qualify as "intermediate" on the scale of connectivity? According to Kauffman (1988), type II properties hold whenever the number of connections among parts (genes) is much lower than the total number of parts, and the number of connections does not increase proportionally with the number of parts. In biological terms, this means that a developmental system will behave as a critical one when the number of gene-gene epistatic interactions is much less than the total number of genes, and when the level of epistasis is less than directly proportional to the number of genes.

HIERARCHICAL GENE REGULATION IS FUNDAMENTAL TO THE CONTROL OF ADAPTIVE PLASTICITY IN PARTICULAR, AND DEVELOPMENT IN GENERAL   We hold that most adaptive plasticity manifests itself in the form of developmental conversion, because of the advantage of environmentally cued discrete phenotypes. We also propose that the major genetic mechanism leading to this is hierarchical gene regulation, given the benefits of anticipatory and coordinated plastic responses (Schlichting and Pigliucci, 1995a). In our view of the ontogenetic trajectory as a reaction norm, the process of development consists of a sequence of environmentally mediated conversions resulting from the differential control of gene expression at different levels in a regulatory cascade. This point seems to be widely accepted as far as development goes, but it is still rather controversial when it comes to phenotypic plasticity.

GENETIC ASSIMILATION IS A SIGNIFICANT MODE OF PHENOTYPIC EVOLUTION   Allelic substitution, phenotypic plasticity, and genetic assimilation are three modes of phenotypic evolution. In genetic assimilation, phenotypic plasticity to a novel environmental condition results in the production of a new phenotype, resulting from the (genetically determined) existing reaction norm, that enables the organisms to survive. Given that this step does not require a genetic change, the presence of a plastic reaction norm ameliorates the intensity of selection, creating a lag time between the environmental change and the requirement for a genetic adjustment to such change. An important consequence of this scenario is the persistence of some populations even if they do not harbor genetic variation immediately available for appropriate evolutionary change. Subsequent genetic change (allelic substitution) can alter the reaction norm, allowing the new phenotype to be produced even without the novel environmental stimulus. The evolutionary process of canalization of the new developmental pathway is made possible by the evolution of a series of homeostatic mechanisms buffering the phenotype against minor environmental fluctuations (canalizing selection).

Given the ubiquity of both the plasticity of the phenotypes of organisms and of environmental variation, we argue that not only does genetic assimilation occur more frequently than has been assumed, but that it represents a predominant mode of effecting phenotypic evolution in heterogeneous environments.

PLASTICITY CAN EVOLVE BOTH AS AN INTERMEDIATE AND AN END-POINT IN PHENOTYPIC EVOLUTION   As shown above, phenotypic plasticity may play an important role in the intermediate stage of genetic assimilation. If a new environment permanently replaces the old one, then the second step of genetic assimilation (above) leads to canalization and homeostasis, and the reaction norm at equilibrium may not be plastic. Phenotypic plasticity *per se*, however, could also be the outcome of selection. If the two (or more) environments co-occur in some form of dynamic equilibrium, then the evolution of a plastic reaction norm should be favored. In this case there would be "incomplete" assimilation, when the favored genetic change is for controlling the switch between phenotypes instead of fixing one of the alternatives. Canalizing selection would still be expected to fine-tune the plastic responses.

PARAMETERS DESCRIBING THE GENETIC ARCHITECTURE ARE NOT CONSTANT   Changes in the genetic constitution of a population, due to either selective or non-selective forces, will alter genetic variation and covariation. Genetic variances and covariances are also labile during ontogeny and sensitive to environmental change.

This occurs due to alterations in gene expression or, more generally, because of epistatic effects altering the contribution of multiple loci to the phenotype. These alterations may represent the inescapable outcome of biochemical properties of molecules, but, more likely, are in many cases features of adaptation to varying environments.

Auxiliary Concepts

APPARENTLY CONTRASTING MECHANISMS CAN ACT SIMULTANEOUSLY AT DIFFERENT LEVELS OF BIOLOGICAL ORGANIZATION   As a consequence of epigenesis and genetic regulation of development, there is a hierarchy of biological phenomena during ontogeny, leading to a partial decoupling of levels of organization. For example, genetic heterochrony does not necessarily correspond to morphological heterochrony (or *vice versa*). In fact, Nijhout et al. (1986) presented a biologically relevant model for ontogenetic trajectories using cellular automata. They found that some heterochronic mutations did not result in heterochronic phenotypes; conversely, heterochronic phenotypes were not necessarily the result of heterochronic mutations. As another example, morphological homeostasis may be derived from physiological plasticity during development; conversely, non-plastic physiological mechanisms can result in morphological plasticity under conditions of environmental heterogeneity.

SOFT SELECTION AND FREQUENCY-DEPENDENT SELECTION ARE THE PREDOMINANT REGIMES OF NATURAL SELECTION   Although we have not explicitly discussed this before, we agree with Wallace (1982), who convincingly argued that soft selection (i.e., selection based on the relative rank of the phenotypes) and frequency-dependent selection (i.e., based on the relative abundance of phenotypes) are the most common types of selective regimes (compared with hard, i.e., threshold, with selection and density-dependent selection). These forms of selection could work synergistically with the previously mentioned moderating effect that plasticity can have on the intensity of selection. The net result would be to further increase the time lag between environmental change and the necessity of a genetic change to improve or fine-tune adaptation.

DEVELOPMENTAL CONVERSION TENDS TO EVOLVE FROM PHENOTYPIC MODULATION Developmental conversion is a more sophisticated and complex mode of plastic response compared with phenotypic modulation. As such, it cannot be easily conceived as the original state. The necessity of its evolution lies in the increased control over the developing phenotype, with the consequent elimination of intermediate or extreme, and possibly maladaptive, phenotypes.

CANALIZATION IS AN IMPORTANT EVOLUTIONARY PROCESS LEADING TO THE PHYSIOLOGICAL PHENOMENON OF HOMEOSTASIS   Canalization leads to the production of more stable developmental trajectories, and appears to play a strong role in the evolution of ontogenies. The result of canalization is a decline in the production of "errors" around a norm of reaction; we observe this as homeostasis. The tendency toward canalization may be balanced by advantages accruing to plasticity.

EVOLUTIONARY "CONSTRAINTS" ARE ULTIMATELY GENETIC IN NATURE Any limitation in the degree or direction of phenotypic change can be due to selection or constraint. However, the multifarious adjectives attached to the term constraint nearly all describe the effects of selection on alternative phenotypes. For example, "developmental" constraints can be seen to be due to stabilizing selection on genetic rules of phenotype construction. Ultimately, true constraints are genetically based, arising either from a lack of appropriate genetic variability (*genostasis*; Bradshaw, 1991) or from features of the genetic architecture (e.g., pleiotropy, epistasis).

QUANTITATIVE GENETICS WILL GENERALLY BE INADEQUATE AS A PREDICTOR OF LONG-TERM EVOLUTIONARY CHANGE This stems from three points: (1) the statistical nature of quantitative genetics models. Quantitative genetic prediction of evolution is a form of statistical extrapolation, and as such it should not be considered with confidence beyond a few generations from the actual data set. (2) Given the environmental/evolutionary lability of genetic variances and covariances, any model that assumes a constancy of such parameters is likely to give only approximate predictions for the near future (i.e., until the magnitude of the change in the "constants" becomes measurable). (3) If evolutionary dynamics are highly non-linear or dependent on initial conditions (Van Tienderen, 1991; Kirkpatrick and Lofsvold, 1992), then long-range predictions are intrinsically impossible due to the so-called "butterfly effect" (Lorenz, 1963), the exponential amplification of very small initial differences in two otherwise indistinguishable systems.

COMPLEX TRAITS ARE THE RESULT OF EPIGENESIS, AND NOT ATTRIBUTABLE TO THE DIRECT ACTIONS OF INDIVIDUAL GENES A proper consideration of epigenesis leads to a revision of our concept of "genetic" control of complex phenotypes and of behavioral traits. The perception of traits such as size, shape, or aggression as "polygenic" characters is too narrow. On the one hand, these traits are not directly controlled by genes, in the sense that they are the result of complex epigenetic interactions and emergent outcomes of developmental systems, only remotely connected with the DNA level. On the other hand, they are controlled to some extent by genes, and it is possible to elicit a response to selection, because those very same epigenetic phenomena are the result of localized and heritable gene action.

This view of polygenic control accommodates many controversial data in evolutionary genetics: for example, Gottlieb (1984) pointed out that the same trait (seed size) appears to be controlled by many genes if we are measuring it at the adult stage, but it obeys simple two-factorial Mendelian inheritance if it is measured as an ontogenetic parameter (the slope of the growth curve)!

Analogously, the endless debate between "deterministic" and "holistic" approaches to the description of variation in natural and especially in human populations is put in a different perspective by the above reasoning.

RESEARCH PROJECTIONS

1. One reason for the universally recognized exclusion of developmental biology from the present incarnation of the neo-Darwinian synthesis is the limitation of currently existing models of ontogenetic change. Some attempts are under way to include ontogenetic and epigenetic effects in quantitative genetic models (Atchley and Hall, 1991; Kirkpatrick and Lofsvold, 1992). Although interesting, such attempts are destined to be limited by the general drawbacks of quantitative genetics. Nijhout et al.'s (1986) model of ontogeny based on cellular automata theory represents an example of a mechanistically oriented approach, and includes for the first time elementary information about gene action and ontogenetic processes to study the relationships between genes and phenotypes.

2. Also as far as plasticity is concerned, modeling has generally been confined within a quantitative genetic perspective (Via and Lande, 1985; Via 1987; van Tienderen, 1991, 1997; Gomulkiewicz and Kirkpatrick, 1992; Gavrilets and Scheiner, 1993a,b), although there have been some optimality models (Kozlowski and Wiegert, 1987; Houston and McNamara, 1992; McNamara, 1994; McNamara and Houston, 1996). There appears to be ample information available in the molecular literature on mechanistic details of plastic responses mediated by regulatory genes (Schlichting and Pigliucci, 1995a; Pigliucci, 1996a) to allow development of a new generation of models using a "bottom-up" approach instead of (or along with) a "top-down" one.

3. At the same time, we need a more detailed mechanistic description of the action of plasticity regulatory genes. Model organisms such as *Arabidopsis thaliana*, *Caenorhabditis elegans*, *Drosophila melanogaster*, and others can serve as exemplars for the exploration and mapping of the metabolic pathways that mediate phenotypic responses to changing external conditions. Modern molecular techniques allow us to identify genes controlling such responses and to study their interactions within the normal developmental pathways of the organism (Prandl et al., 1995; Smith, 1995; Nijhout, 1996).

4. In the study of selection and constraints, it is necessary to include explicit and specific null hypotheses (Antonovics and van Tienderen, 1991). Given that constraints are relative to the particular system under study and to its past evolutionary history, any discussion of constraints should start from a specification of what a comparable "unconstrained" system would look like.

5. Phenotypes are complex, integrated assemblages of character sets, and they should be studied as such (Gould, 1984). The multivariate techniques of the "new" morphometrics (Marcus et al., 1996; Myers et al., 1996) allow the investigator to mathematically describe the changes in coordinates of morphologically homologous landmarks in a way that was intuited by D'Arcy Thompson (1942). The major advantage of these new techniques is their potential for retaining meaningful biological information in multi-variate space. However, there is still a need for other classical multivariate methods whenever the new morphometrics approach cannot be applied (see Chapter 6).

6. The concept of heterochrony, although generally quite useful, lends itself to misapplication and incongruities. As discussed, heterochrony can be recognized at distinct hierarchical levels. However, changes in timing or rate of processes at one level do not necessarily translate into analogous changes at another.

7. The other forms of alteration of ontogenetic trajectories (heterotopy, changes in the intensity of gene action, etc.) need to be highlighted and their relative roles in phenotypic evolution assessed. Because little is known about the genetic bases of developmental changes in natural populations, there is a need for baseline surveys of genetic variation in developmental parameters, for example, distribution within and among species.

8. Along the same line of reasoning, we also need more basic information about natural variation for the genetic control of phenotypic plasticity. Although research on this aspect is more common than on ontogenies, there has still been no systematic approach to the problem comparable to electrophoretic surveys for enzyme polymorphism. There have been few phylogenetic approaches to the evolution of plasticity using comparative methods to investigate the evolution of patterns and amounts of phenotypic plasticity.

9. The genetic bases of plasticity have been examined in a few selection experiments (Scheiner and Lyman, 1991; Leroi et al., 1994d; Partridge et al., 1995; Chippindale et al., 1997). However, we need results from a broader survey of organisms, as well as experiments that incorporate more than two environments, a design that yields only limited information on the nature of the genetic control of plasticity (Falconer, 1990; Schlichting and Pigliucci, 1993).

10. Plasticity being not just an attribute of the adult phenotype, but a developmental phenomenon, more information on the ontogenetic features of reaction norms is necessary. The response to environmental factors may be stage-specific, and the dynamics of genetic and interaction variances

through ontogeny might play a key role in our understanding of evolutionary change. Few studies have addressed this aspect of the problem (Novoplansky et al., 1994; Cheplick, 1995; Pigliucci and Schlichting, 1995; Pigliucci et al., 1997).

11. Our interest lies in adaptive reaction norms. This is the territory of ecological genetics, where a better mapping of reaction norms onto fitness functions is necessary to understand the evolution of adaptive plasticity. Weis and Gorman (1990) have shown how this problem can be approached by combining data from controlled experiments with field data. Transplant experiments are another approach to the same question. Once again, there currently is a paucity of data on this aspect of the problem.

12. The issue of evolvability has been garnering increased attention. We present several complementary approaches to examine the propensity for evolutionary change: selection experiments, analysis of fossil and extant morphospaces, investigations of mutation matrices, and more thorough examinations of unexposed portions of developmental reaction norms.

13. Given the limitations of the classical approach of evolutionary theory outlined so far, an expansion along new routes and a search for new metaphors needs to be undertaken. Non-linear dynamics in general (Brock and Potter, 1992; Ascioti et al., 1993; Stone, 1993; Ellner and Turchin, 1995; Ferriere and Fox, 1995), and complexity theory in particular (Kauffman, 1993; Perry, 1995) currently seem to be promising research fields. Within quantitative genetics, researchers are already carefully scrutinizing assumptions and ways to relax them (Turelli, 1988; Barton and Turelli, 1989). Our use of vector space to describe the interplay of genetic constraints and natural selection, or the application of transition probability matrices to specify evolutionary null hypothesis (Janson, 1992), are other possible examples.

14. There is clearly a need for a more detailed understanding of what epigenesis is and how it emerges from local gene action. One area of research could be devoted to the question of how to recognize emergent properties. For example, the demonstration of undecidability (in the sense of complexity theory) leads to the expectation of emergent properties. Therefore, whenever a portion of a developmental program can be demonstrated to be undecidable, we should observe characteristics not directly derivable from knowledge of gene action, and *vice versa*.

APPLYING THE RESEARCH PROGRAM

For a detailed discussion of the idea and use of research programs, the interested reader should consult Lakatos (1974) and Chalmers (1976). We wish only to summarize a few important points.

1. The outline of a research program is not seen by Lakatos as an immutable entity, but as a set of ideas that evolves and interacts with experimental evidence. However, there are strict rules on how the program can be changed. In particular, there can be no alteration of the core concepts: these are the defining intellectual blocks, and any change in them amounts to a redefinition of the program. Conversely, the auxiliary concepts can be altered as new evidence is gathered, with one major constraint: each change has to be independently testable (i.e., no *ad hoc* hypotheses).

2. Lakatos acknowledged the importance of both major types of tests of scientific theories, confirmation and falsification. The two, however, play very different roles in a research program. Although confirmation of a basic statement does add to our knowledge, confirmation of a bold prediction has to be regarded as of primary importance. On the other hand, falsification of a bold prediction does not necessarily endanger the theoretical framework, whereas falsification of a basic premise must lead to a critical re-examination of the entire edifice. From these considerations, it follows that falsification is of primary importance in examining the core concepts; confirmation plays a central role for the auxiliary concepts.

3. It is obviously difficult to evaluate a research program with respect to alternative models by following objective criteria. Two useful yardsticks are proposed by Lakatos. (1) Coherence: A good research program is an integrated unit of ideas from different fields, not just a collection of disconnected hypotheses. (2) Degeneracy: A healthy program is one that generates, at least from time to time, novel predictions. If a program fails to do so, it is called "degenerate," and it should be abandoned in favor of appropriate alternatives.

CONCLUSIONS

The evolution of phenotypes is an incredibly complex but fascinating aspect of evolutionary biology. For a true understanding of its intricacy, the integration of information from previously disparate fields (such as systematics and molecular developmental genetics) will be essential. Central to our view of phenotypic evolution is the concept of the developmental norm of reaction of a genotype, explicitly integrating the undeniable influence of "environment" on the mapping of genotype to phenotype. What results is a view of the expressed phenotype as a sort of iceberg: at any given instant we observe only the tip, but there are other hidden dimensions. There is the temporal dimension, to be investigated by examining ontogenetic trajectories; and the environmental dimension, to be explored as conditions change predictably, or in new ways. We can also expand the dimensionality of our snapshot by incorporating more traits into our focal "phenotype." The DRN is intended to supplant any vestiges of the particulate view of an organism somehow independent of its environment. Given the Mobius strip connecting the genetic and environmental components of phenotypes, both of these must be considered to be important engines driving the evolution of reaction norms.

Many of the ideas that we have attempted to tie together here were proposed initially by other researchers. The terrain investigated by these innovators was to a large extent unincorporated in the "map" drawn for the "evolutionary synthesis." We have highlighted components due especially to C. H. Waddington, I. I. Schmalhausen, Jens Clausen, David Keck, and William Hiesey, and, to a lesser extent, Richard Goldschmidt. Our conception of the developmental reaction norm owes much to their insights. We hope that our elaborations of their ideas and our proposed innovations pay proper homage to the extent of their contributions.

Literature Cited

Aastveit, A. H., and K. Aastveit. 1993. Effects of geno-type-environment interactions on genetic correlations. *Theor. Appl. Genet.* 86:1007–1013.

Abouheif, E. 1997. Developmental genetics and homology: A hierarchical approach. *Trends Ecol. Evol.* 12:405–408.

Abrams, P. A., Y. Harada, and H. Matsuda. 1993. On the relationship between quantitative genetic and ESS models. *Evolution* 47:982–985.

Ackerly, D. D., and M. J. Donoghue. 1995. Phylogeny and ecology reconsidered. *J. Ecol.* 83:730–733.

Adams, M. B. 1980. Severtsov and Schmalhausen: Russian morphology and the evolutionary synthesis. In E. Mayr and W. B. Provine (eds.), *The Evolutionary Synthesis*, pp. 193–225. Harvard University Press, Cambridge, MA.

Agrawal, A. A. 1998. Induced responses to herbivory and increased plant performance. *Science* 279: 1201–1202.

Agur, Z., and M. Kerszberg. 1987. The emergence of phenotypic novelties through progressive genetic change. *Am. Nat.* 129:862–875.

Akam, M., M. Averof, J. Castelli-Gair, R. Dawes, F. Falciani, and D. Ferrier. 1994a. The evolving roles of Hox genes in arthropods. *Development Suppl.*: 209–215.

Akam, M., P. Holland, P. Ingham, and G. Wray. 1994b. *The Evolution of Developmental Mechanisms*. The Company of Biologists, Cambridge.

Alberch, P., and M. J. Blanco. 1996. Evolutionary patterns in ontogenetic transformation: From laws to regularities. *Int. J. Dev. Biol.* 40:845–858.

Alberch, P., S. J. Gould, G. Oster, and D. Wake. 1979. Size and shape in ontogeny and phylogeny. *Paleobiology* 5:296–317.

Allard, R. W., G. R. Babbel, M. T. Clegg, and A. L. Kahler. 1972. Evidence for coadaptation in *Avena barbata*. *Proc. Natl. Acad. Sci. USA* 69:3043–3048.

Almeida, J., M. Rocheta, and L. Galego. 1997. Genetic control of flower shape in *Antirrhinum majus*. *Development* 124:1387–1392.

Altesor, A., C. Silva, and E. Ezcurra. 1994. Allometric neoteny and the evolution of succulence in cacti. *Bot. J. Linn. Soc.* 114:283–292.

Ambros, V., and E. G. Moss. 1994. Heterochronic genes and the temporal control of *C. elegans* development. *Trends Genet.* 10:123–127.

Anderson, E. A. 1949. *Introgressive Hybridization*. John Wiley & Sons, New York.

Anderson, W. W. 1973. Genetic divergence in body size among experimental populations of *Drosophila pseudoobscura* kept at different temperatures. *Evolution* 27:278–284.

Andersson, S. 1989. Phenotypic plasticity in *Crepis tectorum* (Asteraceae). *Plant Syst. Evol.* 168:19–38.

Andersson, S. 1996. Flower-fruit size allometry at three taxonomic levels in *Crepis* (Asteraceae). *Biol. J. Linn. Soc.* 58:401–407.

Andersson, S., and R. G. Shaw. 1994. Phenotypic plasticity in *Crepis tectorum* (Asteraceae): Genetic correlations across light regimens. *Heredity* 72:113–125.

Andruss, B. F., A. Qing Lu, and K. Beckingham. 1997. Expression of calmodulin in *Drosophila* is highly regulated in a stage- and tissue-specific manner. *Dev. Genes Evol.* 206:541–545.

Antonovics, J., and P. H. van Tienderen. 1991. Ontoecogenophyloconstraints? The chaos of constraint terminology. *Trends Ecol. Evol.* 6:166–168.

Aphalo, P. J., and C. L. Ballarè. 1995. On the importance of information-acquiring systems in plant-plant interactions. *Func. Ecol.* 9:5–14.

Arboleda-Rivera, F., and W. A. Compton. 1974. Differential response of maize (*Zea mays* L) to mass selection in diverse selection environments. *Theor. Evol.* 45:1229–1244.

Armbruster, W. S., and K. E. Schwaegerle. 1996. Causes of covariation of phenotypic traits among populations. *J. Evol. Biol.* 9:261–276.

Arnold, M. L. 1995. Are natural hybrids fit or unfit to their parents? *Trends Ecol. Evol.* 10:67–70.

Arnold, S. J. 1989. Group report: How do complex organisms evolve? In D. B. Wake and G. Roth (eds.), *Complex Organismal Functions: Integration and Evolution in Vertebrates*, pp. 403–433. John Wiley & Sons, Chichester, UK.

Arnold, S. J. 1992. Constraints on phenotypic evolution. *Am. Nat.* 140, *Suppl.*:S85-S107.

Arnone, M. I., and E. H. Davidson. 1997. The hardwiring of development: Organization and function of genomic regulatory systems. *Development* 124: 1851–1864.

Arthur, W. 1988. *A Theory of the Evolution of Development*. John Wiley & Sons, New York.

Arthur, W. B. 1988. Self-reinforcing mechanisms in economics. In P. W. Anderson, K. J. Arrow and D. Pines (eds.), *The Economy as an Evolving System*, pp. 9–31. Addison-Wesley, Redwood City, CA.

Ascioti, F. A., E. Beltrami, T. O. Carroll, and C. Wirick. 1993. Is there chaos in plankton dynamics? *J. Plankt. Res.* 15:603–617.

Ashman, T.-L., and D. J. Schoen. 1994. How long should flowers live? *Nature* 371:788–791.

Asins, M. J., P. Mestre, J. E. Garcia, F. Dicenta, and E. A. Carbonell. 1994. Genotype × environment interaction in QTL analysis of an intervarietal almond cross by means of genetic markers. *Theor. Appl. Genet.* 89:358–364.

Atchley, W. R. 1984. Ontogeny, timing of development, and genetic variance-covariance structure. *Am. Nat.* 123:519–540.

Atchley, W. R. 1987. Developmental quantitative genetics and the evolution of ontogenies. *Evolution* 41: 316–330.

Atchley, W. R. 1990. Heterochrony and morphological change: A quantitative genetic perspective. *Sem. Dev. Biol.* 1:289–297.

Atchley, W. R., and B. K. Hall. 1991. A model for development and evolution of complex morphological structures. *Biol. Rev.* 66:101–157.

Atchley, W. R., and J. J. Rutledge. 1980. Genetic components of size and shape. I. Dynamics of components of phenotypic variability and covariability during ontogeny in the laboratory rat. *Evolution* 34: 1161–1173.

Atchley, W. R., D. E. Cowley, E. J. Eisen, H. Prasetyo, and D. Hawkins-Brown. 1990. Correlated response in the developmental choreographies of the mouse mandible to selection for body composition. *Evolution* 44:669–688.

Atchley, W. R., S. Xu, and C. Vogl. 1994. Developmental quantitative genetic models of evolutionary change. *Dev. Genet.* 15:92–103.

Atchley, W. R., S. Xu, and D. E. Cowley. 1997. Altering developmental trajectories in mice by restricted index selection. *Genetics* 146:629–640.

Averof, M., and S. M. Cohen. 1997. Evolutionary origin of insect wings from ancestral gills. *Nature* 385: 627–630.

Averof, M., and N. H. Patel. 1997. Crustacean appendage evolution associated with changes in Hox gene expression. *Nature* 385:627–630.

Averof, M., R. Dawes, and D. Ferrier. 1996. Diversification of arthropod Hox genes as a paradigm for the evolution of gene functions. *Sem. Cell Dev. Biol.* 7:539–551.

Bagchi, S., and S. Iyama. 1983. Radiation induced developmental instability in *Arabidopsis thaliana*. *Theor. Appl. Genet.* 65:85–92.

Bagnall, D. J. 1992. Control of flowering in *Arabidopsis thaliana* by light, vernalisation and gibberellins. *Aust. J. Plant Physiol.* 19:401–409.

Bagnall, D. J. 1993. Light quality and vernalization interact in controlling late flowering in *Arabidopsis* ecotypes and mutants. *Ann. Bot.* 71:75–83.

Bailey, G. S., R. T. M. Poulter, and P. A. Stockwell. 1978. Gene duplication in tetraploid fish: Model for gene silencing at unlinked duplicated loci. *Proc. Natl. Acad. Sci. USA* 75:5575–5579.

Bak, P., and K. Chen. 1991. Self-organized criticality. *Sci. Am.* January:46–53.

Baldwin, I. T., and E. A. Schmelz. 1996. Immunological "memory" in the induced accumulation of nicotine in wild tobacco. *Ecology* 77:236–246.

Baldwin, J. M. 1896. A new factor in evolution. *Am. Nat.* 30:441–451; 536–553.

Baldwin, J. M. 1902. *Development and Evolution*. Macmillan, New York.

Bales, G. S. 1997. Heterochrony in brontothere horn evolution: Allometric interpretations and the effect of life history scaling. *Paleobiology* 22:481–495.

Barel, C. D. N. 1993. Concepts of an architectonic approach to transformational morphology. *Acta Biotheor.* 41:345–381.

Bargmann, C. I., and H. R. Horvitz. 1991. Control of larval development by chemosensory neurons in *Caenorhabditis elegans*. *Science* 251:1243–1246.

Barinaga, M. 1995. Focusing on the *eyeless* gene. *Science* 267:1766–1767.

Barkal, N., and S. Leibler . 1997. Robustness in simple biochemical networks. *Nature* 387:913–916.

Barker, J. S. F., and R. A. Krebs. 1995. Genetic variation and plasticity of thorax length and wing length in *Drosophila aldrichi* and *D. buzzattii*. *J. Evol. Biol.* 8:689–710.

Barker, J. S. F., and R. H. Thomas. 1987. A quantitative genetic perspective on adaptive evolution. In V. Loeschcke (ed.), *Genetic Constraints on Adaptive Evolution*, pp. 3–23. Springer-Verlag, Berlin.

Barolo, S., and M. Levine. 1997. *hairy* mediates dominant repression in the *Drosophila* embryo. *EMBO J.* 16:2883–2891.

Bartlett, J. 1901. *Familiar Quotations: A Collection of Passages, Phrases, and Proverbs Traced to Their Sources in Ancient and Modern Literature.* 9th ed. Little, Brown, and Company, Boston. On-Line Ed. Columbia University, Academic Information Systems (AcIS), Bartleby Library (http://www.columbia.edu/acis/bartleby/bartlett/).

Barton, N. H. 1992. On the spread of new gene combinations in the third phase of Wright's shifting-balance. *Evolution* 46:551–556.

Barton, N. H., and M. Turelli. 1989. Evolutionary quantitative genetics: How little do we know? *Annu. Rev. Genet.* 23:337–370.

Batterham, P., A. G. Davies, A. Y. Game, and J. A. McKenzie. 1996. Asymmetry: Where evolutionary and developmental genetics meet. *BioEssays* 18:841–845.

Bayer, C. A., L. von Kalm, and J. W. Fristrom. 1997. Relationships between protein isoforms and genetic functions demonstrate functional redundancy at the *Broad-Complex* during *Drosophila* metamorphosis. *Dev. Biol.* 187:267–282.

Bazin, M. J., and P. T. Saunders. 1978. Determination of critical variables in a microbial predator-prey system by catastrophe theory. *Nature* 275:52–54.

Becker, W. A. 1984. *Manual of Quantitative Genetics.* 4th ed. Academic Enterprises, Pullman, WA.

Beckers, J., M. Gérard, and D. Duboule. 1996. Transgenic analysis of a potential Hoxd-11 limb regulatory element present in tetrapods and fish. *Development* 180:543–554.

Beeby, R., and H. Kacser. 1990. Metabolic constraints in evolution. In J. Maynard Smith and G. Vida (eds.), *Organizational Constraints on the Dynamics of Evolution,* pp. 57–75. Manchester University Press, Manchester, UK.

Bell, G. 1997. *Selection: The Mechanism of Evolution.* Chapman and Hall, New York.

Bell, G., and M. J. Lechowicz. 1993. Spatial heterogeneity at small scales and how plants respond to it. In M. M. Caldwell and R. W. Pearcy (eds.), *Exploitation of Environmental Heterogeneity by Plants: Ecophysiological Processes Above and Below Ground.* Academic Press, San Diego, CA.

Bell, G., and A. O. Mooers. 1997. Size and complexity among multicellular organisms. *Biol. J. Linn. Soc.* 60:345–363.

Bell, M. A. 1987. Interacting evolutionary constraints in pelvic reduction of threespine sticklebacks. *Biol. J. Linn. Soc.* 31:347–382.

Bell, M. A. 1988. Stickleback fishes: Bridging the gap between population biology and paleobiology. *Trends Ecol. Evol.* 3:320–325.

Bell, M. A., and G. Orti. 1994. Pelvic reduction in threespine stickleback from Cook Inlet lakes: Geographical distribution and intrapopulation variation. *Copeia* 2:314–325.

Benfey, P. N., and J. W. Schiefelbein. 1994a. Getting to the root of plant development: The genetics of *Arabidopsis* root. *Trends Genet.* 10:84–88.

Benfey, P. N., and J. W. Schiefelbein. 1994b. Insights into root development from *Arabidopsis* root mutants. *Plant Cell Envir.* 17:675–680.

Benkman, C. W. 1993. Adaptation to single resources and the evolution of crossbill (*Loxia*) diversity. *Ecol. Monogr.* 63:305–325.

Benkman, C. W., and R. E. Miller. 1996. Morphological evolution in response to fluctuating selection. *Evolution* 50:2499–2504.

Bennett, A. F., and R. E. Lenski. 1993. Evolutionary adaptation to temperature. II. Thermal niches of experimental lines of *Escherichia coli. Evolution* 47:1–12.

Bennett, A. F., and R. E. Lenski. 1996. Evolutionary adaptation to temperature. V. Adaptive mechanisms and correlated responses in experimental lines of *Escherichia coli. Evolution* 50:493–503.

Bennett, A. F., and R. E. Lenski. 1997. Evolutionary adaptation to temperature. VI. Phenotypic acclimation and its evolution in *Escherichia coli. Evolution* 51:36–44.

Bennett, A. F., R. E. Lenski, and J. E. Mittler. 1992. Evolutionary adaptation to temperature. I. Fitness responses of *Escherichia coli* to changes in its thermal environment. *Evolution* 46:16–30.

Bennington, C. C., and J. B. McGraw. 1995. Natural selection and ecotypic differentiation in *Impatiens pallida. Ecol. Monogr.* 65:302–323.

Bennington, C. C., and J. B. McGraw. 1996. Environment-dependence of quantitative genetic parameters in *Impatiens pallida. Evolution* 50: 1083–1097.

Berenbaum, M. R., A. R. Zangerl, and J. K. Nitao. 1986. Constraints on chemical evolution: Wild parsnip and the parsnip webworm. *Evolution* 40:1215–1228.

Berg, R. L. 1959. A general evolutionary principle underlying the origin of developmental homeostasis. *Am. Nat.* 93:103–105.

Berg, R. L. 1960. The ecological significance of correlation pleiades. *Evolution* 14:171–180.

Bergelson, J. 1994. The effects of genotype and the environment on costs of resistance in lettuce. *Am. Nat.* 143:349–359.

Berleth, T., and G. Jurgens. 1993. The role of the *monopteros* gene in organising the basal body region of the *Arabidopsis* embryo. *Development* 118:575–587.

Bernays, E. A. 1986. Diet-induced head allometry among foliage-chewing insects and its importance for graminivores. *Science* 231:495–497.

Berntson, G. M., and F. A. Bazzaz. 1996. The allometry of root production and loss in seedlings of *Acer rubrum* (Aceraceae) and *Betula papyrifera* (Betulaceae): Implications for root dynamics in elevated CO_2. *Am. J. Bot.* 83:608–616.

Berven, K. A., D. E. Gill, and S. J. Smith-Gill. 1979. Countergradient selection in the green frog, *Rana clamitans. Evolution* 33:609–623.

Bharathan, G., B.-J. Janssen, E. A. Kellogg, and N. Sinha. 1997. Did homeodomain proteins duplicate before the origin of angiosperms, fungi and metazoa? *Proc. Natl. Acad. Sci. USA* 94:13749–13753.

Biere, A. 1995. Genotypic and plastic variation in plant size: Effects on fecundity and allocation patterns in *Lychnis flos-cuculi* along a gradient of natural soil fertility. *J. Ecol.* 83:629–642.

Biggin, M. D., and W. McGinnis. 1997. Regulation of segmentation and segmental identity by *Drosophila* homeoproteins: The role of DNA binding in functional activity and specificity. *Development* 124: 4425–4433.

Bird, A. P. 1995. Gene number, noise reduction and biological complexity. *Trends Genet.* 11:94–100.

Birot, Y., and C. Christophe. 1983. Genetic structures and expected gains from multitrait selection in wild populations of Douglas fir and Sitka spruce. *Silvae Genetica* 32:141–151.

Black-Samuelsson, S., and S. Andersson. 1997. Reaction norm variation between and within populations of two rare plant species, *Vicia pisiformis* and *V. dumetorum* (Fabaceae). *Heredity* 79:268–276.

Blouin, M. S. 1992. Genetic correlations among morphometric traits and rates of growth and differentiation in the green tree frog, *Hyla cinerea*. *Evolution* 46:735–744.

Blows, M. W., and M. B. Sokolowski. 1995. The expression of additive and nonadditive genetic variation under stress. *Genetics* 140:1149–1159.

Boake, C. R. B. 1994. *Quantitative Genetic Studies of Behavioral Evolution*. University of Chicago Press, Chicago.

Bolker, J. A., and R. A. Raff. 1996. Developmental genetics and traditional homology. *BioEssays* 18:489–494.

Bonner, J. T. 1988. *The Evolution of Complexity*. Princeton University Press, Princeton, NJ.

Bonneton, F., P. J. Shaw, C. Fazakerley, M. Shi, and G. A. Dover. 1997. Comparison of *bicoid*-dependent regulation of *hunchback* between *Musca domestica* and *Drosophila melanogaster*. *Mech. Dev.* 66:143–156.

Bookstein, F. L. 1989. "Size and shape": A comment on semantics. *Syst. Zool.* 38:173–180.

Bookstein, F. L. 1991. *Morphometric Tools for Landmark Data: Geometry and Biology*. Cambridge University Press, Cambridge.

Bowler, P. J. 1983. *The Eclipse of Darwinism*. Johns Hopkins University Press, Baltimore.

Bowman, J. L., G. N. Drews, and E. M. Meyerowitz. 1991. Expression of the *Arabidopsis* floral homeotic gene AGAMOUS is restricted to specific cell types late in flower development. *Plant Cell* 3:749–758.

Bowman, J. L., J. Alvarez, D. Weigel, E. M. Meyerowitz, and D. R. Smyth. 1993. Control of flower development in *Arabidopsis thaliana* by APETALA1 and interacting genes. *Development* 119:721–743.

Braam, J., and R. W. Davis. 1990. Rain-, wind-, and touch-induced expression of calmodulin and calmodulin-related genes in *Arabidopsis*. *Cell* 60:357–364.

Bradshaw, A. D. 1965. Evolutionary significance of phenotypic plasticity in plants. *Adv. Genet.* 13:115–155.

Bradshaw, A. D. 1973. Environment and phenotypic plasticity. *Brookhaven Symp. Biol.* 25:74–94.

Bradshaw, A. D. 1991. Genostasis and the limits to evolution. *Phil. Trans. R. Soc. Lond. B* 333:289–305.

Bradshaw, A. D., and K. Hardwick. 1989. Evolution and stress: Genotypic and phenotypic components. *Biol. J. Linn. Soc.* 37:137–155.

Bradshaw, W. E., and C. M. Holzapfel. 1996. Genetic constraints to life-history evolution in the pitcher plant mosquito *Wyeomyia smithii*. *Evolution* 50:1176–1181.

Brakefield, P. B., and C. J. Breuker. 1996. The genetical basis of fluctuating asymmetry for developmentally integrated traits in a butterfly eyespot pattern. *Proc. R. Soc. Lond. B* 263:1557–1563.

Brakefield, P. B., and F. Kesbeke. 1997. Genotype-environment interactions for insect growth in constant and fluctuating temperature regimes. *Proc. R. Soc. Lond. B* 264:717–723.

Brakefield, P. M. 1997. Phenotypic plasticity and fluctuating asymmetry as responses to environmental stress in the butterfly *Bicyclus anynana*. In R. Biljsma and V. Loeschcke (eds.), *Environmental Stress, Adaptation and Evolution*, pp. 65–78. Birkhauser Verlag, Basel.

Brakefield, P. M., and T. B. Larsen. 1984. The evolutionary significance of dry and wet season forms in some tropical butterflies. *Biol. J. Linn. Soc.* 22:1–12.

Brakefield, P. M., and N. Reitsma. 1991. Phenotypic plasticity, seasonal climate and the population biology of *Bicyclus* butterflies (Satyridae) in Malawi. *Ecol. Entomol.* 16:291–303.

Brakefield, P. M., J. Gates, D. Keys, F. Kesbeke, P. J. Wijngaarden, A. Monteiro, V. French, and S. B. Carroll. 1996. Development, plasticity and evolution of butterfly eyespot patterns. *Nature* 384:236–242.

Brassard, J. T., and D. J. Schoen. 1990. Analysis of phenotypic selection among locations in *Impatiens pallida* and *Impatiens capensis*. *Can. J. Bot.* 68:1098–1105.

Brennan, M. D., C.-Y. Wu, and A. J. Berry. 1988. Tissue-specific regulatory differences for the alcohol dehydrogenase genes of Hawaiian *Drosophila* are conserved in *Drosophila melanogaster* transformants. *Proc. Natl. Acad. Sci. USA* 85:6866–6869.

Brenner, S. E., T. Hubbard, A. Murzin, and C. Chothia. 1995. Gene duplications in *H. influenzae*. *Nature* 378:140.

Britten, H. B. 1996. Meta-analysis of the association between multilocus heterozygosity and fitness. *Evolution* 50:2158–2164.

Brock, W. A., and S. M. Potter. 1992. Diagnostic testing for nonlinearity, chaos, and general dependence in time-series data. In M. Casdagli and S. Eubank (eds.), *Nonlinear Modeling and Forecasting*, pp. 137–161. Addison-Wesley, Redwood City, CA.

Brodie III, E. D. 1992. Correlational selection for color pattern and antipredator behavior in the garter snake *Thamnophis ordinoides*. *Evolution* 46:1284–1298.

Brodie III, E. D., A. J. Moore, and F. J. Janzen. 1995. Visualizing and quantifying natural selection. *Trends Ecol. Evol.* 10:313–318.

Brody, T. B. (ed.) 1997. The Interactive Fly. At http://sdb.bio.purdue.edu/fly/aimain/1aahome.htm

Bronmark, C., and J. G. Miner. 1992. Predator-induced phenotypical change in body morphology in crucian carp. *Science* 258:1348–1350.

Brower, A. V. Z. 1996. Parallel race formation and the evolution of mimicry in *Heliconius* butterflies: A phylogenetic hypothesis from mitochondrial DNA sequences. *Evolution* 50:195–221.

Brown, J. S., and N. B. Pavlovic. 1992. Evolution in heterogeneous environments: Effects of migration on habitat specialization. *Evol. Ecol.* 6:360–382.

Brumpton, R. J., H. Boughey, and J. L. Jinks. 1977. Joint selection for both extremes of performance and of sensitivity to a macroenvironmental variable. I. Family selection. *Heredity* 38:219–226.

Bry, L., P. G. Falk, T. Midtvedt, and J. I. Gordon. 1996. A model of host-microbial interactions in an open mammalian ecosystem. *Science* 273:1380–1383.

Bryant, E. H., and L. M. Meffert. 1992. The effect of serial founder-flush cycles on quantitative genetic variation in the housefly. *Heredity* 70:122–129.

Bryant, E. H., and L. M. Meffert. 1995. An analysis of selectional response in relation to a population bottleneck. *Evolution* 49:626–634.

Bryant, E. H., S. A. McCommas, and L. M. Combs. 1986. The effect of an experimental bottleneck upon quantitative genetic variation in the housefly. *Genetics* 114:1191–1211.

Brylski, P., and J. Patton. 1988. Ontogeny of a macroevolutionary phenotype: The external cheek pouches of geomyoid rodents. *Evolution* 42:391–394.

Buchholz, J. T. 1922. Developmental selection in vascular plants. *Bot. Gaz.* 73:249–286.

Buis, R., H. Barthou, P. Castillon, and J. P. Lacombe. 1978. Some general properties of intra- and interorganic correlations in plant morphogenesis. *Phytomorphology* 28:384–395.

Bull, J. J. 1987. Evolution of phenotypic variance. *Evolution* 41:303–315.

Bullini, L. 1994. Origin and evolution of animal hybrid species. *Trends Ecol. Evol.* 9:422–425.

Burdon, R. D. 1977. Genetic correlation as a concept for studying genotype-environment interaction in forest tree breeding. *Silvae Genetica* 26:168–175.

Burger, R. 1986. Constraints for the evolution of functionally coupled characters: A nonlinear analysis of a phenotypic model. *Evolution* 40:182–193.

Burger, R., and M. Lynch. 1995. Evolution and extinction in a changing environment. *Evolution* 49:151–163.

Burns, R. From "Death and Dr. Hornbook." *Bartlett's Familiar Quotations.* Columbia University, Academic Information Systems. http://gserv.cc.columbia.edu/cgi-bin/texis/webinator/search/bartlett

Burstein, Z. 1995. A network model of developmental gene hierarchy. *J. Theor. Biol.* 174:1–11.

Burton, R. S. 1987. Differentiation and integration of the genome in populations of the marine copepod *Tigriopus californicus. Evolution* 41:504–513.

Burton, R. S. 1990a. Hybrid breakdown in developmental time in the copepod *Tigriopus californicus. Evolution* 44:1814–1822.

Burton, R. S. 1990b. Hybrid breakdown in physiological response: A mechanistic approach. *Evolution* 44:1806–1813.

Burton, R. S., and B. N. Lee. 1994. Nuclear and mitochondrial gene genealogies and allozyme polymorphism across a major phylogeographic break in the copepod *Tigriopus californicus. Proc. Natl. Acad. Sci. USA* 91:5197–5201.

Bush, R. M., and K. Paigen. 1992. Evolution of beta-glucuronidase regulation in the genus *Mus. Evolution* 46:1–15.

Callahan, H. S., M. Pigliucci, and C. D. Schlichting. 1997. Developmental phenotypic plasticity: Where ecology and evolution meet molecular biology. *BioEssays* 19:519–525.

Campbell, D. R. 1996. Evolution of floral traits in a hermaphroditic plant: Field measurements of heritabilities and genetic correlations. *Evolution* 50:1442–1453.

Campbell, D. R. 1997. Genetic and environmental variation in life-history traits of a monocarpic perennial: A decade-long field experiment. *Evolution* 51:373–382.

Campbell, D., M. J. Eriksson, G. Oquist, P. Gustafsson, and A. K. Clarke. 1998. The cyanobacterium *Synechococcus* resists UV-B by exchanging photosystem II reaction-center D1 proteins. *Proc. Natl. Acad. Sci. USA* 95:364–369.

Cane, W. P. 1993. The ontogeny of postcranial integration in the common tern, *Sterna hirundo. Evolution* 47:1138–1151.

Cannon, W. B. 1932. *The Wisdom of the Body.* Norton, New York.

Carmichael, J. S., and W. E. Friedman. 1996. Double fertilization in *Gnetum gnemon* (Gnetaceae): Its bearing on the evolution of sexual reproduction within the Gnetales and the anthophyte clade. *Am. J. Bot.* 83:767–780.

Carriere, Y. 1994. Evolution of phenotypic variance: Non-Mendelian parental influences on phenotypic and genotypic components of life-history traits in a generalist herbivore. *Heredity* 72:420–430.

Carriere, Y., S. Masaki, and D. A. Roff. 1997. The coadaptation of female morphology and offspring size: A comparative analysis in crickets. *Oecologia* 110:197–204.

Carroll, S. B. 1994. Developmental regulatory mechanisms in the evolution of insect diversity. *Development Suppl.*:217–223.

Carroll, S. B. 1995. Homeotic genes and the evolution of arthropods and chordates. *Nature* 376:479–485.

Carroll, S. B., J. Gates, D. N. Keys, S. W. Paddock, G. E. F. Panganiban, J. E. Selegue, and J. A. Williams. 1994. Pattern formation and eyespot determination in butterfly wings. *Science* 265:109–114.

Carroll, S. B., S. D. Weatherbee, and J. A. Langeland. 1995. Homeotic genes and the regulation and evolution of insect wing number. *Nature* 375:58–61.

Carvalho, G. R., P. W. Shaw, L. Hauser, B. H. Seghers, and A. E. Magurran. 1996. Artificial introductions, evolutionary change and population differentiation in Trinidadian guppies (*Poecilia reticulata*: Poeciliidae). *Biol. J. Linn. Soc.* 57:219–234.

Casal, J. J., and H. Smith. 1989. The function, action and adaptive significance of phytochrome in light-grown plants. *Plant Cell Envir.* 12:855–862.

Castelli-Gair, J., and M. Akam. 1995. How the Hox gene *Ultrabithorax* specifies two different segments: The significance of spatial and temporal regulation within metameres. *Development* 121:2973–2982.

Chalmers, A. F. 1976. *What Is This Thing Called Science?: An Assessment of the Nature and Status of Science and its Methods*. University of Queensland Press, St. Lucia, Queensland.

Chandler, J., and C. Dean. 1994. Factors influencing the vernalization response and flowering time of late flowering mutants of *Arabidopsis thaliana* (L.) Heynh. *J. Exp. Bot.* 45:1279–1288.

Chanoine, C., and C. L. Gallien. 1989. Myosin isoenzymes and their subunits in urodelan amphibian fast skeletal muscle. Coexistence of larval and adult heavy chains in neotenic individuals. *Eur. J. Biochem.* 181:125–128.

Chapin III, F. S. 1991. Integrated responses of plants to stress. *BioScience* 41:29–36.

Charlesworth, B. 1990a. The evolutionary genetics of adaptation. In M. H. Nitecki (ed.), *Evolutionary Innovations*, pp. 47–70. University of Chicago Press, Chicago.

Charlesworth, B. 1990b. Optimization models, quantitative genetics, and mutation. *Evolution* 44:520–538.

Charlesworth, B., R. Lande, and M. Slatkin. 1982. A neo-Darwinian commentary on macroevolution. *Evolution* 36:474–498.

Chazdon, R. L., and R. W. Pearcy. 1991. The importance of sunflecks for forest understory plants: Photosynthetic machinery appears adapted to brief, unpredictable periods of radiation. *BioScience* 41:760–766.

Cheetham, A. H., J. B. C. Jackson, and L.-A. C. Hayek. 1994. Quantitative genetics of bryozoan phenotypic evolution. II. Analysis of selection and random change in fossil species using reconstructed genetic parameters. *Evolution* 48:360–375.

Chen, L., A. L. De Vries, and C.-H. C. Cheng. 1997a. Convergent evolution of antifreeze glycoproteins in Antarctic notothenioid fish and Arctic cod. *Proc. Natl. Acad. Sci. USA* 94:3817–3822.

Chen, L., A. L. De Vries, and C.-H. C. Cheng. 1997b. Evolution of antifreeze glycoprotein gene from a trypsinogen gene in Antarctic notothenioid fish. *Proc. Natl. Acad. Sci. USA* 94:3811–3816.

Cheplick, G. P. 1995. Genotypic variation and plasticity of clonal growth in relation to nutrient availability in *Amphibromus scabrivalvis*. *J. Ecol.* 83:459–468.

Cheverud, J. M. 1982a. Relationships among ontogenetic, static, and evolutionary allometry. *Am. J. Phys. Anthro.* 59:139–149.

Cheverud, J. M. 1982b. Phenotypic, genetic, and environmental morphological integration in the cranium. *Evolution* 36:499–516.

Cheverud, J. M. 1984. Quantitative genetics and developmental constraints on evolution by selection. *J. Theor. Biol.* 110:155–171.

Cheverud, J. M. 1988. The evolution of genetic correlation and developmental constraints. In G. de Jong (ed.), *Population Genetics and Evolution*, pp. 94–101. Springer, Berlin.

Cheverud, J. M. 1995. Morphological integration in the saddle-back tamarin (*Saguinus fuscicollis*) cranium. *Am. Nat.* 145:63–89.

Cheverud, J. M. 1996. Developmental integration and the evolution of pleiotropy. *Am. Zool.* 36:44–50.

Cheverud, J. M., and E. J. Routman. 1995. Epistasis and its contribution to genetic variance components. *Genetics* 139:1455–1461.

Cheverud, J. M., and E. J. Routman. 1996. Epistasis as a source of increased additive genetic variance at population bottlenecks. *Evolution* 50:1042–1051.

Cheverud, J. M., J. J. Rutledge, and W. R. Atchley. 1983. Quantitative genetics of development: Genetic correlations among age-specific trait values, and the evolution of ontogeny. *Evolution* 37:895–905.

Cheverud, J. M., G. P. Wagner, and M. M. Dow. 1989. Methods for the comparative analysis of variation patterns. *Syst. Zool.* 38:201–213.

Cheverud, J. M., E. J. Routman, and D. J. Irschick. 1997. Pleiotropic effects of individual gene loci on mandibular morphology. *Evolution* 51:2006–2016.

Chippindale, A. K., A. M. Leroi, S. B. Kim, and M. R. Rose. 1993. Phenotypic plasticity and selection in *Drosophila* life-history evolution. I. Nutrition and the cost of reproduction. *J. Evol. Biol.* 6:171–193.

Chippindale, A. K., D. T. Hoang, P. M. Service, and M. R. Rose. 1994. The evolution of development in *Drosophila melanogaster* selected for postponed senescence. *Evolution* 48:1880–1899.

Chippindale, A. K., T. J. F. Chu, and M. R. Rose. 1996. Complex trade-offs and the evolution of starvation resistance in *Drosophila melanogaster*. *Evolution* 50:753–766.

Chippindale, A. K., A. M. Leroi, H. Saing, D. J. Borash, and M. R. Rose. 1997. Phenotypic plasticity and selection in *Drosophila* life history evolution. 2. Diet, mates and the cost of reproduction. *J. Evol. Biol.* 10:269–293.

Chory, J. 1993. Out of darkness: Mutants reveal pathways controlling light-regulated development in plants. *Trends Genet.* 9:167–172.

Chothia, C. 1994. Protein families in the metazoan genome. *Development Suppl.*:27–33.

Christiansen, F. B., and O. Frydenberg. 1973. Selection component analysis of natural polymorphisms using population samples including mother-offspring combinations. *Theor. Popul. Biol.* 4:425–445.

Churchill, S. E. 1996. Particulate versus integrated evolution of the upper body in late Pleistocene humans: A test of two models. *Am. J. Phys. Anthropol.* 100:559–583.

Clark, A. G. 1987a. Genetic correlations: The quantitative genetics of evolutionary constraints. In V. Loeschcke (ed.), *Genetic Constraints on Adaptive Evolution*, pp. 25–45. Springer-Verlag, Berlin.

Clark, A. G. 1987b. Senescence and the genetic-correlation hang-up. *Am. Nat.* 129:932–940.

Clark, A. G. 1990. Genetic components of variation in energy storage in *Drosophila melanogaster*. *Evolution* 44:637–650.

Clark, A. G. 1994. Invasion and maintenance of a gene duplication. *Proc. Natl. Acad. Sci. USA* 91:2950–2954.

Clark, A. G., and L. Wang. 1994. Comparative evolutionary analysis of metabolism in nine *Drosophila* species. *Evolution* 48:1230–1243.

Clark, A. G., and L. Wang. 1997. Epistasis in measured genotypes: *Drosophila* P-element insertions. *Genetics* 147:157–163.

Clark, C. W. 1993. Dynamic models of behavior: An extension of life history theory. *Trends Ecol. Evol.* 8:205–208.

Clarke, G. M. 1993. The genetic basis of developmental stability. I. Relationship between stability, heterozygosity and genomic coadaptation. *Genetica* 89:15–23.

Clarke, G. M. 1995. Relationships between developmental stability and fitness: Application for conservation biology. *Cons. Biol.* 9:18–24.

Clarke, G. M. 1997. The genetic and molecular basis of developmental stability: The *Lucilia* story. *Trends Ecol. Evol.* 12:89–91.

Clarke, G. M., and J. A. McKenzie. 1987. Developmental stability of insecticide resistant phenotypes in blowfly: A result of canalizing natural selection. *Nature* 325:345–346.

Clausen, J. 1951. *Stages in the Evolution of Plant Species.* Cornell University Press, Ithaca, NY.

Clausen, J., and W. M. Hiesey. 1958. *Experimental Studies on the Nature of Plant Species. IV. Genetic Structure of Ecological Races.* Carnegie Institution of Washington, Washington, DC.

Clausen, J., and W. M. Hiesey. 1960. The balance between coherence and variation in evolution. *Proc. Natl. Acad. Sci. USA*, 46:494–506.

Clausen, J., D. Keck, and W. M. Hiesey. 1940. *Experimental Studies on the Nature of Plant Species. I. Effect of Varied Environment on Western North American Plants.* Carnegie Institution of Washington, Publication 520. Washington, DC.

Clausen, J., D. Keck, and W. M. Hiesey. 1945. *Experimental Studies on the Nature of Plant Species. II. Plant Evolution through Amphiploidy and Autoploidy, With Examples from the Madiinae.* Carnegie Institution of Washington, Publication 564. Washington, DC.

Clausen, J., D. Keck, and W. M. Hiesey. 1947. Heredity of geographically and ecologically isolated races. *Am. Nat.* 81:114–133.

Clausen, J., D. Keck, and W. M. Hiesey. 1948. *Experimental Studies on the Nature of Plant Species. III. Environmental Responses of Climatic Races of Achillea.* Carnegie Institution of Washington, Publication 581. Washington, DC.

Clegg, M. T. 1997. Plant genetic diversity and the struggle to measure selection. *J. Hered.* 88:1–7.

Clegg, M. T., and R. W. Allard. 1973. Viability versus fecundity selection in the Slender Wild Oat, *Avena barbata* L. *Science* 181:667–668.

Clegg, M. T., A. T. Kahler, and R. W. Allard. 1978. Estimation of life cycle components of selection in an experimental plant population. *Genetics* 89:765–792.

Clegg, M. T., M. P. Cummings, and M. L. Durbin. 1997. The evolution of plant nuclear genes. *Proc. Natl. Acad. Sci. USA* 94:7791–7798.

Cock, A. G. 1966. Genetical aspects of metrical growth and form in animals. *Q. Rev. Biol.* 41:131–190.

Coen, E. S., and E. M. Meyerowitz. 1991. The war of the whorls: Genetic interactions controlling flower development. *Nature* 353:31–37.

Cohan, F. M., A. A. Hoffman, and T. W. Gayley. 1989. A test of the role of epistasis in divergence under uniform selection. *Evolution* 43:766–774.

Cohan, F. M., E. C. King, and P. Zawadski. 1994. Amelioration of the deleterious pleiotropic effects of an adaptive mutation in *Bacillus subtilis*. *Evolution* 48:81–95.

Coleman, J. S., and K. D. M. McConnaughay. 1995. A non-functional interpretation of a classical optimal-partitioning example. *Func. Ecol.* 9:951–954.

Coleman, J. S., K. D. M. McConnaughay, and D. D. Ackerly. 1994. Interpreting phenotypic variation in plants. *Trends Ecol. Evol.* 9:187–190.

Coleman, W. 1980. Morphology in the evolutionary synthesis. In E. Mayr and W. B. Provine (eds.), *The Evolutionary Synthesis*, pp. 174–179. Harvard University Press, Cambridge, MA.

Collazo, A. 1994. Molecular heterochrony in the pattern of fibronectin expression during gastrulation in amphibians. *Evolution* 48:2037–2045.

Comes, H. P. 1998. Major gene effects during weed evolution: Phenotypic characters cosegregate with alleles at the ray floret locus in *Senecio vulgaris* L. (Asteraceae). *J. Hered.* 89:54–61.

Comstock, R. E. 1996. (ed.) *Quantitative Genetics with Special Reference to Plant and Animal Breeding.* Iowa State University Press, Ames, IA.

Conner, J. K., and A. Sterling. 1995. Testing hypotheses of functional relationships: A comparative survey of correlation patterns among floral traits in five insect-pollinated plants. *Am. J. Bot.* 82:1399–1406.

Conner, J. K., S. Rush, and P. Jennetten. 1996. Measurements of natural selection on floral traits in wild radish (*Raphanus raphanistrum*). I. Selection through lifetime female fitness. *Evolution* 50:1127–1136.

Conover, D. O., and E. T. Schultz. 1995. Phenotypic similarity and the evolutionary significance of countergradient variation. *Trends Ecol. Evol.* 10:248–252.

Conway, L. J., and R. S. Poethig. 1993. Heterochrony in plant development. *Dev. Biol.* 4:65–72.

Cook, C. D. K. 1968. Phenotypic plasticity with particular reference to three amphibious plant species. In V. Heywood (ed.), *Modern Methods in Plant Taxonomy*, pp. 97–111. Academic Press, London.

Cook, S. A., and M. P. Johnson. 1968. Adaptation to heterogeneous environments. I. Variation in heterophylly in *Ranunculus flammula* L. *Evolution* 22:496–516.

Cooke, J., M. A. Nowak, M. C. Boerlijst, and J. Maynard Smith. 1997. Evolutionary origins and maintenance of redundant gene expression during metazoan development. *Trends Genet.* 13:360–364.

Cooper, W. S., and R. H. Kaplan. 1981. Adaptive "coin-flipping": A decision-theoretic examination of natural selection for random individual variation. *J. Theor. Biol.* 93:135–151.

Counts, R. L. 1993. Phenotypic plasticity and genetic variability in annual *Zizania* spp. along a latitudinal gradient. *Can. J. Bot.* 71:145–154.

Coupland, G. 1995. Genetic and environmental control of flowering time in *Arabidopsis*. *Trends Genet.* 11:393–397.

Cowley, D. E., and W. R. Atchley. 1990. Development and quantitative genetics of correlation structure among body parts of *Drosophila melanogaster*. *Am. Nat.* 135:242–268.

Cowley, D. E., and W. R. Atchley. 1992. Quantitative genetic models for development, epigenetic selection, and phenotypic evolution. *Evolution* 46:495–518.

Coyne, J. A., and B. Charlesworth. 1997. On punctuated equilibria. *Science* 276:338.

Coyne, J., N. H. Barton, and M. Turelli. 1997. A critique of Sewall Wright's shifting balance theory of evolution. *Evolution* 51:643–671.

Crespi, B. J., and F. L. Bookstein. 1989. A path-analytic model for the measurement of selection on morphology. *Evolution* 43:18–28.

Crone, W., and E. M. Lord. 1994. Floral organ initiation and development in a wild-type *Arabidopsis thaliana* (Brassicaceae) and in the organ identity mutants *apetala2-1* and *agamous-1*. *Can. J. Bot.* 72:384–401.

Crow, J. F., W. R. Engels, and C. Denniston. 1990. Phase three of Wright's shifting-balance theory. *Evolution* 44:233–247.

Cruzan, M. B., and M. L. Arnold. 1993. Ecological and genetic associations in an *Iris* hybrid zone. *Evolution* 47:1432–1445.

Cullis, C. A. 1986. Unstable genes in plants. In D. H. Jennings and A. J. Trewavas (eds.), *Plasticity in Plants*, pp. 77–84. Pindar, Scarborough, UK.

Culver, D. C., R. W. Jernigan, and J. O'Connell. 1994. The geometry of natural selection in cave and spring populations of the amphipod *Gammarus minus* Say (Crustacea: Amphipoda). *Biol. J. Linn. Soc.* 52:49–67.

Danjon, F. 1995. Observed selection effects on height growth, diameter and stem form in maritime pine. *Silvae Genetica* 44:1–11.

Davidson, E. H., K. J. Peterson, and R. A. Cameron. 1995. Origin of bilaterian body plans: Evolution of developmental regulatory mechanisms. *Science* 270:1319–1325.

Davies, M. S., and R. W. Snaydon. 1972. Physiological differences among populations of *Anthoxanthum odoratum* L. collected from the Park Grass experiment, Rothamsted. III. Response to phosphorus. *J. Ecol.* 60:699–707.

Davies, M. S., and R. W. Snaydon. 1976. Rapid population differentiation in a mosaic environment. III. Measures of selection pressures. *Heredity* 36:59–66.

Davis, A. P., and M. R. Capecchi. 1996. A mutational analysis of the 5′ *HoxD* genes: Dissection of genetic interactions during limb development in the mouse. *Development* 122:1175–1185.

Day, T., J. Pritchard, and D. Schluter. 1994. Ecology and genetics of phenotypic plasticity: A comparison of two sticklebacks. *Evolution* 48:1723–1734.

de Beer, G. 1930. *Embryology and Evolution*. Clarendon Press, Oxford.

de Beer, G. 1971. *Homology, an Unsolved Problem*. Oxford University Press, London.

de Duve, C. 1987. Selection by differential molecular survival: A possible mechanism of early chemical evolution. *Proc. Natl. Acad. Sci. USA* 84:8253–8256.

de Jong, G. 1990. Quantitative genetics of reaction norms. *J. Evol. Biol.* 3:447–468.

de Jong, G. 1995. Phenotypic plasticity as a product of selection in a variable environment. *Am. Nat.* 145:493–512.

de Jong, G., and A. J. van Noordwijk. 1992. Acquisition and allocation of resources: Genetic (co)variances, selection, and life histories. *Am. Nat.* 139:749–770.

de Laguerie, P., I. Olivieri, and P. H. Gouyon. 1993. Environmental effects on fitness-sets shape and evolutionarily stable strategies. *J. Theor. Biol.* 163:113–125.

Delph, L. F., and D. G. Lloyd. 1991. Environmental and genetic control of gender in the dimorphic shrub *Hebe subalpina*. *Evolution* 45:1957–1964.

Demeester, L. 1993. Genotype, fish-mediated chemicals, and phototactic behavior in *Daphnia magna*. *Ecology* 74:1467–1474.

Depew, D. J., and B. H. Weber. 1995. *Darwinism Evolving: Systems Dynamics and the Genealogy of Natural Selection*. MIT Press, Cambridge, MA.

DeRocher, A. E., K. W. Helm, L. M. Lauzon, and E. Vierling. 1991. Expression of a conserved family of cytoplasmic low molecular weight heat shock proteins during heat stress and recovery. *Plant Physiol.* 96:1038–1047.

Deschamp, P. A., and T. J. Cooke. 1983. Leaf dimorphism in aquatic angiosperms: Significance of turgor pressure and cell expansion. *Science* 219:505–507.

DeWitt, T. J., A. Sih, and D. S. Wilson. 1998. Costs and limits to benefits as constraints on the evolution of phenotypic plasticity. *Trends Ecol. Evol.* 13:77–81.

Dickinson, W. J. 1995. Molecules and morphology: Where's the homology? *Trends Genet.* 11:119–121.

Dickinson, W. J., Y. Yang, K. Schuske, and M. Akam. 1993. Conservation of molecular prepatterns during the evolution of cuticle morphology in *Drosophila* larvae. *Evolution* 47:1396–1406.

Diggle, P. K. 1991. Labile sex expression in andromonoecious *Solanum hirtum*: Floral morphogenesis and sex determination. *Am. J. Bot.* 78:377–393.

Diggle, P. K. 1993. Developmental plasticity, genetic variation, and the evolution of andromonoecy in *Solanum hirtum* (Solanaceae). *Am. J. Bot.* 80:967–973.

Diggle, P. K. 1994. The expression of andromonoecy in *Solanum hirtum* (Solanaceae): Phenotypic plasticity and ontogenetic contingency. *Am. J. Bot.* 81:1354–1365.

Diggle, P. K. 1995. Architectural effects and the interpretation of patterns of fruit and seed development. *Annu. Rev. Ecol. Syst.* 26:531–552.

Dillon, W. R., and M. Goldstein. 1984. *Multivariate Analysis: Methods and Applications*. John Wiley & Sons, New York.

DiMichele, W. A., J. I. Davis, and R. G. Olmstead. 1989. Origins of heterospory and the seed habit: The role of heterochrony. *Taxon* 38:1–11.

Dobzhansky, T. 1937. *Genetics and the Origin of Species*. Columbia University Press, New York.

Dobzhansky, T. 1951. *Genetics and the Origin of Species*. 3rd ed. Columbia University Press, New York.

Dobzhansky, T., and H. Levene. 1955. Genetics of natural populations. XXIV. Developmental homeostasis in natural populations of *Drosophila pseudoobscura*. *Genetics* 40:797–808.

Dobzhansky, T., and B. Spassky. 1944. Genetics of natural populations. XI. Manifestation of genetic variants in *Drosophila pseudoobscura* in different environments. *Genetics* 29:270–290.

Dobzhansky, T., and B. Wallace. 1953. The genetics of homeostasis in *Drosophila*. *Proc. Natl. Acad. Sci. USA* 439:162–171.

Doebley, J., A. Stec, and C. Gustus. 1995. *teosinte branched1* and the origin of maize: Evidence for epistasis and the evolution of dominance. *Genetics* 141:333–346.

Doebley, J., A. Stec, and L. Hubbard. 1997. The evolution of apical dominance in maize. *Nature* 386:485–488.

Dolferus, R., and M. Jacobs. 1984. Polymorphism of alcohol dehydrogenase in *Arabidopsis thaliana* (L.) Heynh.: Genetical and biochemical characterization. *Biochem. Genet.* 22:817–838.

Dolferus, R., G. de Bruxelles, E. S. Dennis, and W. J. Peacock. 1994a. Regulation of the *Arabidopsis Adh* gene by anaerobic and other environmental stresses. *Ann. Bot.* 74:301–308.

Dolferus, R., M. Jacobs, W. J. Peacock, and E. S. Dennis. 1994b. Differential interactions of promoter elements in stress responses of the *Arabidopsis Adh* gene. *Plant Physiol.* 105:1075–1087.

Dommergues, J.-L., B. Laurin, and C. Meister. 1996. Evolution of ammonoid morphospace during the Early Jurassic radiation. *Paleobiology* 22:200–240.

Doughty, P. 1995. Testing the ecological correlates of phenotypically plastic traits within a phylogenetic framework. *Acta Oecol.* 16:519–524.

Doyle, J. J. 1994. Evolution of a plant homeotic multigene family: Toward connecting molecular systematics and molecular *Dev. Genet. Syst. Biol.* 43:307–328.

Dragavstev, V. A., and A. F. Aver'yanova. 1984a. Mechanism of genotype-environment interaction and homeostasis of quantitative characters in plants. *Soviet Genet.* 19:1420–1424.

Dragavstev, V. A., and A. F. Aver'yanova. 1984b. Redetermination of the genetic formula of quantitative characters of wheat in different environmental conditions. *Soviet Genet.* 19:1424–1430.

Druger, M. 1967. Selection and the effects of temperature on scutellar bristle number in *Drosophila*. *Genetics* 56:39–47.

Duboule, D. 1994. Temporal colinearity and the phylotypic progression: A basis for the stability of a vertebrate Bauplan and the evolution of morphologies through heterochrony. *Development Suppl.*:135–142.

Duboule, D., and A. S. Wilkins. 1998. The evolution of 'bricolage.' *Trends Genet.* 14:54–59.

Dudley, S. A., and J. Schmitt. 1996. Testing the adaptive plasticity hypothesis: Density-dependent selection on manipulated stem length in *Impatiens capensis*. *Am. Nat.* 147:445–465.

Dun, R. B., and A. S. Fraser. 1959. Selection for an invariant character, vibrissa number, in the house mouse. *Aust. J. Biol. Sci.* 12:506–523.

Ebert, D., L. Yampolsky, and S. C. Stearns. 1993. Genetics of life history in *Daphnia magna*. I. Heritabilities at two food levels. *Heredity* 70:335–343.

Egan, S. E., and R. A. Weinberg. 1993. The pathway to signal achievement. *Nature* 365:781–783.

Eizinger, A., and R. J. Sommer. 1997. The homeotic gene *lin-39* and the evolution of nematode epidermal cell fates. *Science* 278:452–455.

Ekstig, B. 1994. Condensation of developmental stages in evolution. *BioScience* 44:158–164.

Eldredge, N., and S. J. Gould. 1972. Punctuated equilibria: An alternative to phyletic gradualism. In T. J. M. Schopf (ed.), *Models in Paleobiology*, pp. 82–115. Freeman, Cooper and Co., San Francisco.

Eldredge, N., and S. J. Gould. 1997. On punctuated equilibria. *Science* 276:337.

Ellner, S., and P. Turchin. 1995. Chaos in a noisy world: New methods and evidence from time-series analysis. *Am. Nat.* 145:343–375.

Ellstrand, N. C. 1983. Floral formula inconstancy within and among plants and populations of *Ipomopsis aggregata* (Pursh) V. Grant (Polemoniaceae). *Bot. Gaz.* 144:119–123.

Ellstrand, N. C., and R. B. Mitchell. 1988. Spatial and temporal patterns of floral inconstancy within and among plants and populations of *Ipomopsis aggregata* (Polemoniaceae). *Bot. Gaz.* 149:209–212.

Emery, R. J. N., D. M. Reid, and C. C. Chinnappa. 1994. Phenotypic plasticity of stem elongation in two ecotypes of *Stellaria longipes*: the role of ethylene and response to wind. *Plant Cell Envir.* 17:691–700.

Emlen, D. J. 1996. Artificial selection on horn length-body size allometry in the horned beetle *Onthophagus acuminatus* (Coleoptera: Scarabaeidae). *Evolution* 50:1219–1230.

Endler, J. A. 1986. *Natural Selection in the Wild*. Princeton University Press, Princeton, NJ.

Endress, P. K. 1992. Evolution and floral diversity: The phylogenetic surroundings of *Arabidopsis* and *Antirrhinum*. *Int. J. Plant Sci.* 153:S106-S122.

Evans, A. S., and M. Marshall. 1996. Developmental instability in *Brassica campestris* (Cruciferae): Fluctuating asymmetry of foliar and floral traits. *J. Evol. Biol.* 9:717–736.

Fairbairn, D. J., and D. E. Yadlowski. 1997. Coevolution of traits determining migratory tendency: Correlated response of a critical enzyme, juvenile hormone esterase, to selection on wing morphology. *J. Evol. Biol.* 10:495–513.

Falconer, D. S. 1952. The problem of environment and selection. *Am. Nat.* 86:293–298.

Falconer, D. S. 1957. Selection for phenotypic intermediates in *Drosophila*. *J. Genet.* 55:551–561.

Falconer, D. S. 1989. *Introduction to Quantitative Genetics*. 3rd ed. Longman Scientific and Technical, Essex, UK.

Falconer, D. S. 1990. Selection in different environments: Effects on environmental sensitivity (reaction norm) and on mean performance. *Genet. Res.* 56:57–70.

Falconer, D. S., and T. F. C. Mackay. 1996. *Introduction to Quantitative Genetics*. 4th ed. Longman Scientific and Technical, Essex, UK.

Fang, X.-M., C.-Y. Wu, and M. D. Brennan. 1991. Complexity in evolved regulatory variation for alcohol dehydrogenase genes in Hawaiian *Drosophila. J. Mol. Evol.* 32:220–226.

Farris, M. A., and M. J. Lechowicz. 1990. Functional interactions among traits that determine reproductive success in a native annual plant. *Ecology* 71: 548–557.

Feder, M. E., and W. B. Watt. 1992. Functional biology of adaptation. In R. J. Berry, T. J. Crawford, G. M. Hewitt (ed.), *Genes in Ecology*, pp. 365–392. Blackwell Scientific Publications, Oxford, UK.

Feder, M. E., N. V. Cartano, L. Milos, R. A. Krebs, and S. L. Lindquist. 1996. Effect of engineering Hsp70 copy number on Hsp70 expression and tolerance of ecologically relevant heat shock in larvae and pupae of *Drosophila melanogaster. J. Exp. Biol.* 199:1837–1844.

Felsenstein, J. 1976. The theoretical population genetics of variable selection and migration. *Annu. Rev. Genet.* 10:253–280.

Felsenstein, J. 1985. Phylogenies and the comparative method. *Am. Nat.* 125:1–15.

Felsenstein, J. 1988. Phylogenies and quantitative characters. *Annu. Rev. Ecol. Syst.* 19:445–471.

Fenster, C. B., L. F. Galloway, and L. Chao. 1997. Epistasis and its consequences for the evolution of natural populations. *Trends Ecol. Evol.* 12:282–286.

Ferriere, R., and G. A. Fox. 1995. Chaos and evolution. *Trends Ecol. Evol.* 10:480–484.

Fink, W. L., and M. L. Zelditch. 1996. Historical patterns of developmental integration in piranhas. *Am. Zool.* 36:61–69.

Finnegan, J., W. J. Peacock, and E. S. Dennis. 1996. Reduced DNA methylation in *Arabidopsis thaliana* results in abnormal plant development. *Proc. Natl. Acad. Sci. USA* 93:8449–8454.

Fiorella, C. V., and R. Z. German. 1997. Heterochrony within species: Craniofacial growth in giant, standard, and dwarf rabbits. *Evolution* 51:250–261.

Fisher, R. A. 1918. The correlation between relatives on the supposition of Mendelian inheritance. *Transactions of the Royal Society of Edinburgh* 52:399–433.

Fisher, R. A. 1930. *The Genetical Theory of Natural Selection*. Clarendon Press, Oxford, England.

Foote, M. 1997. The evolution of morphological diversity. *Annu. Rev. Ecol. Syst.* 28:129–152.

Ford, E. B. 1931. *Mendelism and Evolution*. Methuen, London.

Frank, S. A., and M. Slatkin. 1992. Fisher's Fundamental Theorem of natural selection. *Trends Ecol. Evol.* 7:92–95.

Frazzetta, T. H. 1975. *Complex Adaptations in Evolving Populations*. Sinauer Associates, Sunderland, MA.

French, V. 1997. Pattern formation in colour on butterfly wings. *Curr. Opin. Genet. Dev.* 7:524–529.

Friedman, W. E. 1992. Evidence of a pre-Angiosperm origin of endosperm: Implications for the evolution of flowering plants. *Science* 255:336–339.

Friedman, W. E. 1993. The evolutionary history of the seed plant male gametophyte. *Trends Ecol. Evol.* 8:15–20.

Friedman, W. E. 1998. The evolution of double fertilization and endosperm: An "historical" perspective. *Sexual Plant Repro.* 11:6–16.

Friedman, W. E. In press. Heterochrony and developmental innovation: Evolution of female gametophyte ontogeny in *Gnetum*, a highly apomorphic seed plant. *Evolution*.

Friedman, W. E., and J. S. Carmichael. 1996. Evolution of fertilization patterns in the Gnetales: Implications for understanding reproductive diversification among anthophytes. *Int. J. Plant. Sci. Suppl.* 157:77–94.

Fry, J. D. 1996. The evolution of host specialization: Are trade-offs overrated? *Am. Nat. Suppl.* 148:S84-S107.

Fry, J. D., S. L. Heinsohn, and T. F. C. Mackay. 1996. The contribution of new mutations to genotype-environment interaction for fitness in *Drosophila melanogaster. Evolution* 50:2316–2327.

Furuya, M. 1993. Phytochromes: Their molecular species, gene families, and functions. *Annu. Rev. Plant Physiol. Mol. Biol.* 44:617–745.

Futuyma, D. J. 1979. *Evolutionary Biology*. Sinauer Associates, Sunderland, MA.

Futuyma, D. J. 1986. *Evolutionary Biology*. 2nd ed. Sinauer Associates, Sunderland, MA.

Futuyma, D. J. 1998. *Evolutionary Biology*. 3rd ed. Sinauer Associates, Sunderland, MA.

Futuyma, D. J., and G. Moreno. 1988. The evolution of ecological specialization. *Annu. Rev. Ecol. Syst.* 19: 207–233.

Futuyma, D. J., M. C. Keese, and S. J. Scheffer. 1993. Genetic constraints and the phylogeny of insect-plant associations: Responses of *Ophraella communa* (Coleoptera: Chrysomelidae) to host plants of its congeners. *Evolution* 47:888–905.

Futuyma, D. J., M. C. Keese, and D. J. Funk. 1995. Genetic constraints on macroevolution: The evolution of host affiliation in the leaf beetle genus *Ophraella. Evolution* 49:797–809.

Gabriel, W., and M. Lynch. 1992. The selective advantage of reaction norms for environmental tolerance. *J. Evol. Biol.* 5:41–59.

Galis, F. 1996. The application of functional morphology to evolutionary studies. *Trends Ecol. Evol.* 11:124–129.

Galis, F., and E. G. Drucker. 1996. Pharyngeal biting mechanics in centrarchid and cichlid fishes: Insights into a key evolutionary innovation. *J. Evol. Biol.* 9:641–670.

Galis, F., and J. A. J. Metz. 1998. Why are there so many cichlid species? *Trends Ecol. Evol.* 13:1–2.

Garland, T., A. W. Dickerman, C. M. Janis, and J. A. Jones. 1993. Phylogenetic analysis of covariance by computer simulation. *Syst. Biol.* 42:265–292.

Garrido, B., R. Nodari, D. G. Debouck, and P. Gepts. 1991. Uni-2—A dominant mutation affecting leaf development in *Phaseolus vulgaris. J. Hered.* 82:181–183.

Gatesy, S. M., and K. P. Dial. 1996. Locomotor modules and the evolution of avian flight. *Evolution* 50:331–340.

Gavrilets, S. 1996. On phase three of the shifting-balance theory. *Evolution* 50:1034–1041.

Gavrilets, S. 1997. Evolution and speciation on holey adaptive landscapes. *Trends Ecol. Evol.* 12:307–312.

Gavrilets, S., and S. M. Scheiner. 1993a. The genetics of phenotypic plasticity. V. Evolution of reaction norm shape. *J. Evol. Biol.* 6:31–48.

Gavrilets, S., and S. M. Scheiner. 1993b. The genetics of phenotypic plasticity. VI. Theoretical predictions for directional selection. *J. Evol. Biol.* 6:49–68.

Gebhardt, M. D., and S. C. Stearns. 1988. Reaction norms for developmental time and weight at eclosion in *Drosophila mercatorum. J. Evol. Biol.* 1:335–354.

Gedroc, J. J., K. D. M. McConnaughay, and J. S. Coleman. 1996. Plasticity in root/shoot partitioning: Optimal, ontogenetic, or both? *Func. Ecol.* 10:44–50.

Gerhart, J., and M. Kirschner. 1997. *Cells, Embryos, and Evolution.* Blackwell Science, Malden, MA.

Getty, T. 1996. The maintenance of phenotypic plasticity as a signal detection problem. *Am. Nat.* 148:378–385.

Ghysen, A., and C. Dambly-Chaudiere. 1988. From DNA to form: The achaete-scute complex. *Genes Dev.* 2:495–501.

Gibbs, H. L., and P. R. Grant. 1987. Oscillating selection on Darwin's finches. *Nature* 327:511–513.

Gibson, G. 1996. Epistasis and pleiotropy as natural properties of transcriptional regulation. *Theor. Popul. Biol.* 49:58–89.

Gibson, G., and D. S. Hogness. 1996. Effect of polymorphism in the *Drosophila* regulatory gene *Ultrabithorax* on homeotic stability. *Science* 271:200–203.

Giesel, J. T., P. A. Murphy, and M. N. Manlove. 1982. The influence of temperature on genetic interrelationships of life history traits in a population of *Drosophila melanogaster*: What tangled data sets we weave. *Am. Nat.* 119:464–479.

Gilbert, S. F. 1994. Dobzhansky, Waddington, and Schmalhausen: Embryology and the modern synthesis. In M. B. Adams (ed.), *The Evolution of Theodosius Dobzhansky*, pp. 143–154. Princeton University Press, Princeton, NJ.

Gilchrist, G. W. 1995. Specialists and generalists in changing environments. I. Fitness landscapes of thermal sensitivity. *Am. Nat.* 146:252–270.

Gillespie, J. H., and M. Turelli. 1989. Genotype-environment interactions and the maintenance of polygenic variation. *Genetics* 121:129–138.

Gittleman, J. L., and D. M. Decker. 1994. The phylogeny of behaviour. In P. J. B. Slater and T. R. Halliday (eds.), *Behaviour and Evolution*, pp. 80–105. Cambridge University Press, Cambridge.

Gittleman, J. L., and M. Kot. 1990. Adaptation: Statistics and a null model for estimating phylogenetic effects. *Syst. Zool.* 39:227–241.

Gittleman, J. L., and H.-K. Luh. 1992. On comparing comparative methods. *Annu. Rev. Ecol. Syst.* 23:383–404.

Godfray, H. C. J. 1995. Field experiments with genetically manipulated insect viruses: Ecological issues. *Trends Ecol. Evol.* 10:465–469.

Goffeau, A., B. G. Barrell, H. Bussey, R. W. Davis, B. Dujon, H. Feldmann, F. Galibert, J. D. Hoheisel, C. Jacq, M. Johnston, E. J. Louis, H. W. Mewes, Y. Murakami, P. Philippsen, H. Tettelin, and S. G. Oliver. 1996. Life with 6000 genes. *Science* 274:562–567.

Goldschmidt, R. B. 1940. *The Material Basis of Evolution.* Yale University Press, New Haven, CT.

Goldstein, D. B., and K. E. Holsinger. 1992. Maintenance of polygenic variation in spatially structured populations: Roles for local mating and genetic redundancy. *Evolution* 46:412–429.

Goliber, T. E., and L. J. Feldman. 1990. Developmental analysis of leaf plasticity in the heterophyllous aquatic plant *Hippuris vulgaris. Am. J. Bot.* 77:399–412.

Gomulkiewicz, R., and M. Kirkpatrick. 1992. Quantitative genetics and the evolution of reaction norms. *Evolution* 46:390–411.

Goodnight, C. J. 1987. On the effect of founder events on epistatic genetic variance. *Evolution* 41:80–91.

Goodnight, C. J. 1988. Epistasis and the effect of founder events on the additive genetic variance. *Evolution* 42:399–403.

Goodnight, C. J. 1995. Epistasis and the increase in additive genetic variance: Implications for phase I of Wright's shifting balance process. *Evolution* 49:502–511.

Goodrich, J., P. Puangsomlee, M. Martin, D. Long, E. M. Meyerowitz, and G. Coupland. 1997. A polycomb-group gene regulates homeotic gene expression in *Arabidopsis. Nature* 386:44–51.

Goodwin, B. C. 1989. Evolution and the generative order. In B. Goodwin and P. Saunders (eds.), *Theoretical Biology: Epigenetic and Evolutionary Order from Complex Systems*, pp. 89–100. Edinburgh University Press, Edinburgh.

Goodwin, B. C. 1994. *How the Leopard Changed Its Spots: The Evolution of Complexity.* Charles Scribner's Sons, New York.

Gordon, D. M. 1991. Variation and change in behavioral ecology. *Ecology* 72:1196–1203.

Gotthard, K. 1998. Life history plasticity in the satyrine butterfly *Lasiommata petropolitana*: Investigating an adaptive reaction norm. *J. Evol. Biol.* 11:21–39.

Gottlieb, G. 1992. *Individual Development and Evolution: The Genesis of Novel Behavior.* Oxford University Press, New York.

Gottlieb, L. D. 1977. Phenotypic variation in *Stephanomeria exigua* ssp *coronaria* (Compositae) and its recent derivative species "Malheurensis." *Am. J. Bot.* 64:873–880.

Gottlieb, L. D. 1984. Genetics and morphological evolution in plants. *Am. Nat.* 123:681–709.

Gould, S. J. 1966. Allometry and size in ontogeny and phylogeny. *Biol. Rev.*, 41:587–640.

Gould, S. J. 1974. The evolutionary significance of "bizarre" structures: Antler size and skull size in the "Irish Elk," *Megaloceros giganteus. Evolution* 28:191–220.

Gould, S. J. 1977. *Ontogeny and Phylogeny.* Harvard University Press, Cambridge, MA.

Gould, S. J. 1979. Mickey Mouse meets Konrad Lorenz. *Nat. Hist.* 88:30–36.

Gould, S. J. 1980a. Is a new and general theory of evolution emerging? *Paleobiology* 3:115–151.

Gould, S. J. 1980b. The evolutionary biology of constraint. *Daedalus* 109:39–52.

Gould, S. J. 1982. Darwinism and the expansion of evolutionary theory. *Science* 216:380–387.

Gould, S. J. 1984. Covariance sets and ordered geographic variation in *Cerion* from Aruba, Bonaire and Curaçao: A way of studying nonadaptation. *Syst. Zool.* 33:217–237.

Gould, S. J. 1989. A developmental constraint in *Cerion*, with comments on the definition and interpretation of constraint in evolution. *Evolution* 43:516–539.

Gould, S. J., and N. Eldredge. 1993. Punctuated equilibrium come of age. *Nature* 366:223–227.

Gould, S. J., and R. C. Lewontin. 1979. The spandrels of San Marco and the Panglossian paradigm: A critique of the adaptationist programme. *Proc. R. Soc. Lond. B* 205:581–598.

Gould, S. J., and E. S. Vrba. 1982. Exaptation—A missing term in the science of form. *Paleobiology* 8:4–15.

Graba, Y., D. Aragnol, and J. Pradel. 1997. *Drosophila* Hox complex downstream targets and the function of homeotic genes. *BioEssays* 19:379–388.

Grace, J. B. 1985. Juvenile versus adult competitive abilities in plants: Size-dependence in cattails (*Typha*). *Ecology* 66:1630–1638.

Gralla, J. D. 1991. Transcriptional control—Lessons from an *E. coli* promoter data base. *Cell* 66:415–418.

Grant, V. 1977. *Organismic Evolution.* Freeman, San Francisco.

Grant, V. 1979. Character coherence in natural hybrid populations in plants. *Bot. Gaz.* 140:443–448.

Grant, V. 1981. *Plant Speciation.* 2nd ed. Columbia University Press, New York.

Grbic, M., and M. R. Strand. 1998. Shifts in the life history of parasitic wasps correlate with pronounced alterations in early development. *Proc. Natl. Acad. Sci. USA* 95:1097–1101.

Greene, E. 1989. A diet-induced developmental polymorphism in a caterpillar. *Science* 243:643–646.

Grenier, J. K., T. L. Garber, R. Warren, P. M. Whitington, and S. Carroll. 1997. Evolution of the entire arthropod Hox gene set predated the origin and radiation of the onychophoran/arthropod clade. *Curr. Biol.* 7:547–553.

Groeters, F. R., B. E. Tabashnik, N. Finson, and M. W. Johnson. 1994. Fitness costs of resistance to *Bacillus thuringiensis* in the diamondback moth (*Plutella xylostella*). *Evolution* 48:197–201.

Gromko, M. H. 1995. Unpredictability of correlated response to selection: Pleiotropy and sampling interact. *Evolution* 49:685–693.

Grossman, A. D. 1995. Genetic networks controlling the initiation of sporulation and the development of genetic competence in *Bacillus subtilis. Annu. Rev. Genet.* 29:477–508.

Guerra, D., M. C. Pezzoli, G. Giorgi, F. Garoia, and S. Cavicchi. 1997. Developmental constraints in the *Drosophila* wing. *Heredity* 79:564–571.

Guerrant, E. O. 1982. Neotenic evolution of *Delphinium nudicaule* (Ranunculaceae): A hummingbird-pollinated larkspur. *Evolution* 36:699–712.

Guerrant, E. O. 1988. Heterochrony in plants: The intersection of evolution, ecology and ontogeny. In M. L. McKinney (ed.), *Heterochrony in Evolution: A Multidisciplinary Approach*, pp. 111–133. Plenum Press, New York.

Haeckel, E. 1866. *Generelle Morphologie der Organismen: Allgemeine Grundzuge der organischen Formen-Wissenschaft, mechanisch begrundet durch die von Charles Darwin reformirte Descendenz-Theorie.* Georg Reimer, Berlin.

Hairston, N. G., Jr., and T. A. Dillon. 1990. Fluctuating selection and response in a population of freshwater copepods. *Evolution* 44:1796–1805.

Hake, S., and N. Sinha. 1991. Genetic analysis of leaf development. *Oxford Surv. Plant Mol. Cell Biol.* 7:187–222.

Haldane, J. B. S. 1932a. *The Causes of Evolution.* Longman, Green and Co., London.

Haldane, J. B. S. 1932b. The time of action of genes, and its bearing on some evolutionary problems. *Am. Nat.* 66:5–24.

Haldane, J. B. S., and S. D. Jayakar. 1963. Polymorphism due to selection of varying direction. *J. Genet.* 58:237–242.

Halder, G., P. Callaerts, and W. J. Gehring. 1995. Induction of ectopic eyes by targeted expression of the *eyeless* gene in *Drosophila. Science* 267:1788–1792.

Hall, B. K. 1992a. *Evolutionary Developmental Biology.* Chapman and Hall. London.

Hall, B. K. 1992b. Waddington's legacy in development and evolution. *Am. Zool.* 32:113–122.

Hall, B. K. (ed.) 1994. *Homology: The Hierarchical Basis of Comparative Biology.* Academic Press, San Diego.

Hall, B. K. 1995. Homology and embryonic development. *Evol. Biol.* 28:1–37.

Hall, B. K. 1996. *Baupläne*, phylotypic stages, and constraint. *Evol. Biol.* 29:215–261.

Hamrick, J. L., and M. J. Godt. 1996. Gene flow and distribution of genetic variation in plant populations. *Phil. Trans. R. Soc. Lond. B* 351:1291–1298.

Hanken, J., D. H. Jennings, and L. Olsson. 1997. Mechanistic basis of life-history evolution in anuran amphibians: Direct development. *Am. Zool.* 37:160–171.

Hard, J. J., W. E. Bradshaw, and C. A. Holzapfel. 1993. Genetic coordination of demography and phenology in the pitcher-plant mosquito, *Wyeomyia smithii. J. Evol. Biol.* 6:707–723.

Harland, S. C. 1936. The genetical conception of the species. *Biol. Rev.* 11:83–112.

Harris, R. N., R. D. Semlitsch, H. M. Wilbur, and J. E. Fauth. 1990. Local variation in the genetic basis of paedomorphosis in the salamander *Ambystoma talpoideum*. *Evolution* 44:1588–1603.

Harris, R. N., J. V. Fondacaro, and L. A. Kasbohm. 1995. Local variation in the genetic basis of morphological traits in the salamander *Ambystoma talpoideum*. *Evolution* 49:565–569.

Hartl, D. L., and A. G. Clark. 1989. *Principles of Population Genetics.* 2nd ed. Sinauer Associates, Sunderland, MA.

Hartl, D. L., and D. E. Dykhuizen. 1981. Potential for selection among nearly neutral allozymes of 6-phosphogluconate dehydrogenase in *Escherichia coli. Proc. Natl. Acad. Sci. USA*, 78:6344–6348.

Harvell, C. D. 1990. The ecology and evolution of inducible defenses. *Q. Rev. Biol.* 65:323–340.

Harvell, C. D. 1994. The evolution of polymorphism in colonial invertebrates and social insects. *Q. Rev. Biol.* 69:155–185.

Harvey, P. H., and M. D. Pagel. 1991. *The Comparative Method in Evolutionary Biology.* Oxford University Press, Oxford.

Harvey, P. H., and A. Purvis. 1991. Comparative methods for explaining adaptations. *Nature* 351:619–624.

Hazel, W. N., R. Smock, and M. D. Johnson. 1990. A polygenic model for the evolution and maintenance of conditional strategies. *Proc. R. Soc. Lond. B* 242:181–187.

Hedrick, P. W. 1986. Genetic polymorphism in heterogeneous environments: A decade later. *Annu. Rev. Ecol. Syst.* 17:535–566.

Hedrick, P. W., M. E. Ginevan, and E. P. Ewing. 1976. Genetic polymorphism in heterogeneous environments. *Annu. Rev. Ecol. Syst.* 7:1–32.

Heikkila, J. J. 1993a. Heat shock gene expression and development. I. An overview of fungal, plant, and poikilothermic animal developmental systems. *Dev. Genet.* 14:1–5.

Heikkila, J. J. 1993b. Heat shock gene expression and development. II. An overview of mammalian and avian developmental systems. *Dev. Genet.* 14:87–91.

Hempel, F. D., and L. J. Feldman. 1994. Bi-directional inflorescence development in *Arabidopsis thaliana*: Acropetal initiation of flowers and basipetal initiation of paraclades. *Planta* 192:276–286.

Henikoff, S., E. A. Greene, S. Pietrokovski, P. Bork, T. K. Attwood, and L. Hood. 1997. Gene families: The taxonomy of protein paralogs and chimeras. *Science* 278:609–614.

Hensley, F. R. 1993. Ontogenetic loss of phenotypic plasticity of age at metamorphosis in tadpoles. *Ecology* 74:2405–2412.

Hiesey, W. M., M. A. Nobs, and O. Björkman. 1971. *Experimental Studies on the Nature of Species. V. BioSystematics, Genetics, and Physiological Ecology of the Erythranthe Section of Mimulus.* Carnegie Institution of Washington, Washington, DC.

Higgins, L. E., and M. A. Rankin. 1996. Different pathways in arthropod postembryonic development. *Evolution* 50:573–582.

Hinton, G. E., and S. J. Nowlan. 1987. How learning can guide evolution. *Complex Systems* 1:495–502.

Ho, M.-W., and P. T. Saunders. 1979. Beyond neo-Darwinism: An epigenetic approach to evolution. *J. Theor. Biol.* 78:573–591.

Hobbs, S. L. A., and J. H. Burnett. 1982. The genetic control of morphological and yield characters in *Vicia faba* L. *Theor. Appl. Genet.* 62:9–15.

Hoekstra, R. F. 1988. Theory of phenotypic evolution: Genetic or non-genetic models ? In G. de Jong (ed.), *Population Genetics and Evolution*, pp. 33–41. Springer-Verlag, Berlin.

Hofer, J., L. Turner, R. Hellens, M. Ambrose, P. Matthews, A. Michael, and N. Ellis. 1997. *UNIFO-LIATA* regulates leaf and flower morphogenesis in pea. *Curr. Biol.* 7:581–587.

Hoffmann, A. A., and P. A. Parsons. 1991. *Evolutionary Genetics and Environmental Stress.* Oxford University Press, NY.

Hoffman, A. A., C. M. Sgro, and S. H. Lawler. 1995. Ecological population genetics: The interface between genes and the environment. *Annu. Rev. Genet.* 29:349–370.

Holland, J. H. 1986. A mathematical framework for studying learning in classifier systems. *Physica* 22D:307–317.

Holland, J. H. 1988. The global economy as an adaptive process. In P. W. Anderson, K. J. Arrow and D. Pines (eds.), *The Economy as an Evolving Complex System*, pp. 117–124. Addison-Wesley, Redwood City, CA.

Holland, P. W. H., and J. Garcia-Fernandez. 1996. *Hox* genes and chordate evolution. *Dev. Biol.* 173:382–395.

Holland, P. W. H., J. Garcia-Fernandez, N. A. Williams, and A. Sidow. 1994. Gene duplications and the origins of vertebrate development. *Development Suppl.*:125–133.

Holloway, G. J., and P. M. Brakefield. 1995. Artificial selection of reaction norms of wing pattern elements in *Bicyclus anynana*. *Heredity* 74:91–99.

Holloway, G. J., S. R. Povey, and R. M. Sibly. 1990. The effect of new environment on adapted genetic architecture. *Heredity* 64:323–330.

Holloway, G. J., P. M. Brakefield, and S. Kofman. 1993. The genetics of wing pattern elements in the polyphenic butterfly, *Bicyclus anynana*. *Heredity* 70:179–186.

Houle, D. 1991. Genetic covariance of fitness correlates: What genetic correlations are made of and why it matters. *Evolution* 45:630–648.

Houle, D. 1992. Comparing evolvability and variability of quantitative traits. *Genetics* 130:195–204.

Houle, D., D. K. Hoffmaster, S. Assimacopoulos, and B. Charlesworth. 1992. The genomic mutation rate for fitness in *Drosophila*. *Nature* 359:58–60.

Houle, D., K. A. Hughes, D. K. Hoffmaster, J. Ihara, S. Assimacopoulos, D. Canada, and B. Charlesworth. 1994. The effects of spontaneous mutation on quantitative traits. I. Variances and covariances of life history traits. *Genetics* 138:773–785.

Houssard, C., and J. Escarre. 1995. Variation and covariation among life-history traits in *Rumex ace-*

tosella from a successional old-field gradient. *Oecologia* 102:70–80.

Houston, A. I., and J. M. McNamara. 1992. Phenotypic plasticity as a state-dependent life-history decision. *Evol. Ecol.* 6:243–253.

Huether, C. A., Jr. 1968. Exposure of natural genetic variability underlying the pentamerous corolla constancy in *Linanthus androsaceus* ssp. *androsaceus*. *Genetics* 60:123–146.

Huether, C. A., Jr. 1969. Constancy of the pentamerous corolla phenotype in natural populations of *Linanthus*. *Evolution* 23:572–588.

Hufford, L. 1996. Ontogenetic evolution, clade diversification, and homoplasy. In M. J. Sanderson and L. Hufford (eds.), *Homoplasy: The Recurrence of Similarity in Evolution*, pp. 271–301. Academic Press, San Diego.

Hughes, A. L. 1994. The evolution of functionally novel proteins after gene duplication. *Proc. R. Soc. Lond. B* 256:119–124.

Hughes, A. L., and M. Yeager. 1997. Molecular evolution of the vertebrate immune system. *BioEssays* 19:777–786.

Hulskamp, M., W. Lukowitz, A. Beermann, G. Glaser, and D. Tautz. 1994. Differential regulation of target genes by different alleles of the segmentation gene *hunchback* in *Drosophila*. *Genetics* 138:125–134.

Humphries, J. M., F. L. Bookstein, B. Chernoff, G. R. Smith, R. L. Elder, and S. G. Poss. 1981. Multivariate discrimination by shape in relation to size. *Syst. Zool.* 30:291–308.

Hunding, A., and R. Engelhardt. 1995. Early biological morphogenesis and nonlinear dynamics. *J. Theor. Biol.* 173:401–413.

Hunkapillar, T., and L. Hood. 1986. The growing immunoglobin supergene family. *Nature* 323:15–16.

Hunter, C. P., and C. Kenyon. 1995. Specification of anteroposterior cell fates in *Caenorhabditis elegans* by *Drosophila Hox* proteins. *Nature* 377:229–232.

Hutchings, J. A. 1993. Adaptive life histories effected by age-specific survival and growth rate. *Ecology* 74:673–684.

Huxley, J. S. 1924. Constant differential growth-ratios and their significance. *Nature* 114:895–896.

Huxley, J. S. 1932. *Problems of Relative Growth*. MacVeagh, London.

Huxley, J. S. 1942. *Evolution, The Modern Synthesis*. Allen and Unwin, London.

Huynen, M., P. F. Stadler, and W. Fontana. 1996. Smoothness within ruggedness: The role of neutrality in adaptation. *Proc. Natl. Acad. Sci. USA* 93:397–401.

Imasheva, A. G., V. Loeschcke, L. A. Zhivotovsky, and O. E. Lazenby. 1997. Effects of extreme temperatures on phenotypic variation and developmental stability in *Drosophila melanogaster* and *Drosophila buzzatii*. *Biol. J. Linn. Soc.* 61:117–126.

Irish, V. F., and I. M. Sussex. 1990. Function of the *apetala-1* gene during *Arabidopsis* floral development. *Plant Cell* 2:741–753.

Iwabe, N., K.-I. Kuma, and T. Miyata. 1996. Evolution of gene families and relationship with organismal evolution: Rapid divergence of tissue-specific genes in the early evolution of chordates. *Mol. Biol. Evol.* 13:483–493.

Jablonka, E., and E. Szathmary. 1995. The evolution of information storage and heredity. *Trends Ecol. Evol.* 10:206–211.

Jablonka, E., B. Oborny, I. Molnár, E. Kisdi, J. Hofbauer, and T. Czárán. 1995. The adaptive advantage of phenotypic memory in changing environments. *Phil. Trans. R. Soc. Lond. B.* 350:133–141.

Jack, T., L. L. Brockman, and E. M. Meyerowitz. 1992. The homeotic gene *APETALA3* of *Arabidopsis thaliana* encodes a MADS box and is expressed in petals and stamens. *Cell* 68:683–697.

Jacob, F. 1977. Evolution and tinkering. *Science* 196:1161–1166.

Jacob, F., and J. Monod. 1961. On the regulation of gene activity. *Cold Spring Harb. Symp. Quant. Biol.* 26:193–210.

Jaenike, J., and D. R. Papaj. 1992. Behavioral plasticity and patterns of host use by insects. In B. D. Roitberg and M. B. Isman (eds.), *Insect Chemical Ecology: An Evolutionary Approach*, pp. 245–264. Routledge, Chapman and Hall, New York.

Jain, S. K. 1978. Inheritance of phenotypic plasticity in soft chess, *Bromus mollis* L. (Gramineae). *Experientia* 34:835–836.

Jain, S. K. 1979. Adaptive strategies: Polymorphism, plasticity, and homeostasis. In O. T. Solbrig, S. Jain, G. B. Johnson and P. H. Raven (eds.), *Topics in Plant Population Biology*. Columbia University Press, New York.

James, F. C., and C. E. McCulloch. 1990. Multivariate analysis in ecology and systematics: Panacea or Pandora's box? *Annu. Rev. Ecol. Syst.* 21:129–166.

Janson, C. H. 1992. Measuring evolutionary constraints: A Markov model for phylogenetic transitions among seed dispersal syndromes. *Evolution* 46:136–158.

Jenks, M. A., A. M. Rashotte, H. A. Tuttle, and K. A. Feldmann. 1996. Mutants in *Arabidopsis thaliana* altered in epicuticular wax and leaf morphology. *Plant Physiol.* 110:377–385.

Jernigan, R. W., D. C. Culver, and D. W. Fong. 1994. The dual role of selection and evolutionary history as reflected in genetic correlations. *Evolution* 48:587–596.

Jinks, J. L., and V. Connolly. 1973. Selection for specific and general response to environmental differences. *Heredity* 30:33–40.

Jinks, J. L., and H. S. Pooni. 1982. Determination of environmental sensitivity of selection lines of *Nicotiana rustica* by the selection environment. *Heredity* 49:291–294.

Jockusch, E. L., and L. M. Nagy. 1997. Insect evolution: How did insect wings originate? *Curr. Biol.* 7:R358-R361.

Johannsen, W. 1911. The genotype conception of heredity. *Am. Nat.* 45:129–159.

Johnson, A. S., and M. A. R. Koehl. 1994. Maintenance of dynamic strain similarity and environmental stress factor in different flow habitats: Thallus allometry and material properties of a giant kelp. *J. Exp. Biol.* 195:381–410.

Johnson, P. A., R. E. Lenski, and F. C. Hoppensteadt. 1995. Analysis of variation in mean fitness between replicate populations: Random and non-random factors in evolution. *Proc. R. Soc. Lond. B* 259:131–136.

Jolicoeur, P. 1989. A simplified model for bivariate complex allometry. *J. Theor. Biol.* 140:41–49.

Jones, C. S. 1992. Comparative ontogeny of a wild cucurbit and its derived cultivar. *Evolution* 46:1827–1847.

Jones, C. S. 1993. Heterochrony and heteroblastic leaf development in two subspecies of *Cucurbita argyrosperma* (Cucurbitaceae). *Am. J. Bot.* 80:778–795.

Jones, C. S., and R. Price. 1996. Diversity and evolution of seedling *Baupläne* in *Pelargonium* (Geraniaceae). *Aliso* 14:281–295.

Jordan, N. 1991. Multivariate analysis of selection in experimental populations derived from hybridization of two ecotypes of the annual plant *Diodia teres* W. (Rubiaceae). *Evolution* 45:1760–1772.

Jordan, N. 1992. Path analysis of local adaptation in two ecotypes of the annual plant *Diodia teres* Walt. (Rubiaceae). *Am. Nat.* 140:149–165.

Jordano, P. 1995. Angiosperm fleshy fruits and seed dispersers: A comparative analysis of adaptation and constraints in plant-animal interactions. *Am. Nat.* 145:163–191.

Joshi, A., and L. D. Mueller. 1993. Directional and stabilizing density-dependent natural selection for pupation height in *Drosophila melanogaster*. *Evolution* 47:176–184.

Joshi, A., and J. N. Thompson. 1995. Alternative routes to evolution of competitive ability in two competing species of *Drosophila*. *Evolution* 49:616–625.

Kacser, H., and R. Beeby. 1984. Evolution of catalytic proteins or on the origin of enzyme species by means of natural selection. *J. Mol. Evol.* 20:38–51.

Kacser, H., and J. A. Burns. 1981. The molecular basis of dominance. *Genetics* 97:639–666.

Kadmon, R. 1995. Plant competition along soil moisture gradients: A field experiment with the desert annual *Stipa capensis*. *J. Ecol.* 83:253–262.

Kalisz, S. 1986. Variable selection on the timing of germination in *Collinsia verna* (Scrophulariaceae). *Evolution* 40:479–491.

Kanki, J. P., and R. K. Ho. 1997. The development of the posterior body in zebrafish. *Development* 124:881–893.

Kaplan, D. R., and W. Hagemann. 1991. The relationship of cell and organism in vascular plants. *BioScience* 41:693–703.

Kappen, C. 1996. Ancient evolutionary origin of homeodomains: Studies on yeasts, plants and animals. In M. Nei and N. Takahata (eds.), *Current Topics on Molecular Evolution*, pp. 199–211. Institute of Molecular Evolutionary Genetics, Pennsylvania State University, University Park, PA.

Karlin, S., and B. Levikson. 1974. Temporal fluctuations in selection intensities: Case of small population size. *Theor. Popul. Biol.* 6:383–412.

Karlsson, B. H., G. R. Sills, and J. Nienhuis. 1993. Effects of photoperiod and vernalization on the number of leaves at flowering in 32 *Arabidopsis thaliana* (Brassicaceae) ecotypes. *Am. J. Bot.* 80:646–648.

Kathiresan, A., K. C. Nagarathna, M. M. Moloney, D. M. Reid, and C. C. Chinnappa. 1998. Differential regulation of 1-aminocyclopropane-1-carboxylate synthase gene family and its role in phenotypic plasticity in *Stellaria longipes*. *Plant Mol. Biol.* 36:265–274.

Kauffman, S. A. 1988. The evolution of economic webs. In P. W. Anderson, K. J. Arrow, and D. Pines (eds.), *The Economy as an Evolving Complex System*, pp. 125–146. Addison-Wesley, Redwood City, CA.

Kauffman, S. A. 1991. Antichaos and adaptation. *Sci. Am.* August:78–84.

Kauffman, S. A. 1993. *The Origins of Order: Self-organization and Selection in Evolution.* Oxford University Press, New York.

Kauffman, S. A., and S. Levin. 1987. Towards a general theory of adaptive walks on rugged landscapes. *J. Theor. Biol.* 128:11–45.

Kauffman, S. A., and R. G. Smith. 1986. Adaptive automata based on Darwinian selection. *Physica* 22D:68–82.

Kawano, S., and T. Hara. 1995. Optimal balance between propagule output, propagule size, and cost of propagule production in plants with special reference to its evolutionary-ecological implications. *Plant Species Biol.* 10:119–125.

Kawecki, T. J., and S. C. Stearns. 1993. The evolution of life histories in spatially heterogeneous environments: Optimal reaction norms revisited. *Evol. Ecol.* 7:155–174.

Kellogg, E. A. 1990. Ontogenetic studies of florets in *Poa* (Gramineae): Allometry and heterochrony. *Evolution* 44:1978–1989.

Kempin, S. A., B. Savidge, and M. F. Yanofsky. 1995. Molecular basis of the *cauliflower* phenotype in *Arabidopsis*. *Science* 267:522–525.

Khan, M. A., J. Antonovics, and A. D. Bradshaw. 1976. Adaptation to heterogeneous environments. III. The inheritance of response to spacing in Flax and Linseed (*Linum usitatissimum*). *Aust. J. Agric. Res.* 27:649–659.

Khavkin, E., and E. H. Coe. 1997. Mapped genomic locations for developmental functions and QTLs reflect concerted groups in maize (*Zea mays* L.). *Theor. Appl. Genet.* 95:343–352.

Kimble, J. 1994. An ancient molecular mechanism for establishing embryonic polarity? *Science* 266:577–578.

Kimura, K. D., H. A. Tissenbaum, Y. Liu, and G. Ruvkun. 1997. *daf-2*, an insulin receptor-like gene that regulates longevity and diapause in *Caenorhabditis elegans*. *Science* 277:942–946.

Kimura, M. 1983. *The Neutral Theory of Molecular Evolution.* Cambridge University Press, Cambridge.

King, R. B. 1993. Color-pattern variation in Lake Erie water snakes: Prediction and measurement of natural selection. *Evolution* 47:1819–1833.

Kingsolver, J. G., and D. C. Wiernasz. 1987. Dissecting correlated characters: Adaptive aspects of phenotypic covariation in melanization pattern of *Pieris* butterflies. *Evolution* 41:491–503.

Kingsolver, J. G., and D. C. Wiernasz. 1991. Development, function, and the quantitative genetics of wing melanin pattern in *Pieris* butterflies. *Evolution* 45:1480–1492.

Kingsolver, J. G., and D. W. Schemske. 1991. Path analyses of selection. *Trends Ecol. Evol.* 6:276–280.

Kirchhammer, C. V., and E. H. Davidson. 1996. Spatial and temporal information processing in the sea urchin embryo: Modular and intramodular organization of the *CyIIIa* gene *cis*-regulatory system. *Development* 122:333–348.

Kirkpatrick, M., and R. Lande. 1989. The evolution of maternal characters. *Evolution* 43:485–503.

Kirkpatrick, M., and D. Lofsvold. 1992. Measuring selection and constraint in the evolution of growth. *Evolution* 46:954–971.

Kirkpatrick, M., D. Lofsvold, and M. Bulmer. 1990. Analysis of the inheritance, selection and evolution of growth trajectories. *Genetics* 124:979–993.

Kitchell, J. A. 1990. The reciprocal interaction of organism and effective environment: Learning more about 'and'. In R. M. Ross and W. D. Allmon (eds.), *Causes of Evolution: A Paleontological Perspective*, pp. 151–169. The University of Chicago Press, Chicago.

Klauck, T. M., M. C. Faux, K. Labudda, L. K. Langenberg, S. Jaken, and J. D. Scott. 1996. Coordination of three signaling enzymes by AKAP79, a mammalian scaffold protein. *Science* 271:1589–1591.

Klingenberg, C. P. 1996. Individual variation of ontogenies: A longitudinal study of growth and timing. *Evolution* 50:2412–2428.

Klingenberg, C. P., and J. R. Spence. 1993. Heterochrony and allometry: Lessons from the water strider genus *Limnoporus*. *Evolution* 47:1834–1853.

Klingenberg, C. P., and M. Zimmermann. 1992. Static, ontogenetic, and evolutionary allometry: A multivariate comparison in nine species of water striders. *Am. Nat.* 140:601–620.

Klingenberg, C. P., B. E. Neuenschwander, and B. D. Flury. 1996. Ontogeny and individual variation: Analysis of patterned covariance matrices with common principal components. *Syst. Biol.* 45:135–150.

Koch, P. B., P. B. Brakefield, and F. Kesbeke. 1996. Ecdysteroids control eyespot size and wing color pattern in the polyphenic butterfly *Bicyclus anynana* (Lepidoptera: Satyridae). *J. Insect Physiol.* 42:223–230.

Kondrashov, A. S. 1997. Evolutionary genetics of life cycles. *Annu. Rev. Ecol. Syst.* 28:391–435.

Kondrashov, A. S., and D. Houle. 1994. Genotype-environment interactions and the estimation of the genomic mutation rate in *Drosophila melanogaster*. *Proc. R. Soc. Lond. B, B* 258:221–227.

Kooi, R. E., P. M. Brakefield, and E. M.-T. Rossie. 1996. Effects of food plant on phenotypic plasticity in the tropical butterfly *Bicyclus anynana*. *Entomol. Exp. Appl.* 80:149–151.

Koornneef, M., C. J. Hanhart, and J. H. v. d. Veen. 1991. A genetic and physiological analysis of late flowering mutants in *Arabidopsis thaliana*. *Mol. Gen. Genet.* 229:57–66.

Kosman, D., and S. Small. 1997. Concentration-dependent patterning by an ectopic expression domain of the *Drosophila* gap gene *knirps*. *Development* 124:1343–1354.

Koufopanou, V., and G. Bell. 1991. Developmental mutants of *Volvox*: Does mutation recreate the patterns of phylogenetic diversity? *Evolution* 45:1806–1822.

Kozlowski, J., and R. G. Wiegert. 1987. Optimal age and size at maturity in annuals and perennials with determinate growth. *Evol. Ecol.* 1:231–244.

Krebs, R. A., and M. E. Feder. 1997. Natural variation in the expression of the heat-shock protein hsp70 in a population of *Drosophila melanogaster* and its correlation with tolerance of ecologically relevant thermal stress. *Evolution* 51:173–179.

Krebs, R. A., and V. Loeschcke. 1994. Costs and benefits of activation of the heat-shock response in *Drosophila melanogaster*. *Func. Ecol.* 8:730–737.

Kuiper, D., and P. J. C. Kuiper. 1988. Phenotypic plasticity in a physiological perspective. *Oecol. Plant.* 9:43–59.

Lacey, E. P. 1996. Parental effects in *Plantago lanceolata* L. I.: A growth chamber experiment to examine pre- and post-zygotic temperature effects. *Evolution* 50:865–878.

Lacey, E. P., L. Real, J. Antonovics, and D. G. Heckel. 1983. Variance models in the study of life histories. *Am. Nat.* 122:114–131.

Lair, K. P., W. E. Bradshaw, and C. M. Holzapfel. 1997. Evolutionary divergence of the genetic architecture underlying photoperiodism in the pitcher-plant mosquito, *Wyeomyia smithii*. *Genetics* 147:1873–1883.

Lakatos, I. 1974. Falsification and the methodology of scientific research programmes. In I. Lakatos and A. Musgrave (eds.), *Criticism and the Growth of Knowledge*. Cambridge University Press, Cambridge.

Lammers, T. G. 1990. Sequential paedomorphosis among the endemic Hawaiian Lobelioideae (Campanulaceae). *Taxon* 39:206–211.

Lande, R. 1979. Quantitative genetic analysis of multivariate evolution, applied to brain:body size allometry. *Evolution* 33:402–416.

Lande, R. 1982. A quantitative genetic theory of life history evolution. *Ecology* 63:607–615.

Lande, R., and S. J. Arnold. 1983. The measurement of selection on correlated characters. *Evolution* 37:1210–1226.

Langton, C. G. 1986. Studying artificial life with cellular automata. *Physica* 22D:120–149.

Langton, C. G. (ed.) 1994. *Artificial Life: An Overview*. MIT Press, Cambridge, MA.

Larsen, P. L., P. S. Albert, and D. L. Riddle. 1995. Genes that regulate both development and longevity in *Caenorhabditis elegans*. *Genetics* 139:1567–83.

Lauder, G. V. R. 1989. Group report: How are feeding systems integrated and how have evolutionary innovations been introduced? In D. B. Wake and G. Roth (eds.), *Complex Organismal Functions:*

Integration and Evolution in Vertebrates, pp. 97–115. John Wiley & Sons, Chichester, UK.

Laux, T., and G. Jurgens. 1994. Establishing the body plan of the *Arabidopsis* embryo. *Acta Bot. Needer.* 43:247–260.

Lawrence, W. E. 1947. Chromosome numbers in *Achillea* in relation to geographic distribution. *Am. J. Bot.* 34:538–545.

Le Corff, J. 1993. Effects of light and nutrient availability on chasmogamy and cleistogamy in an understory tropical herb, *Calathea micans* (Marantaceae). *Am. J. Bot.* 80:1392–1399.

Lechowicz, M. J., and G. Bell. 1991. The ecology and genetics of fitness in forest plants. II. Microspatial heterogeneity of the edaphic environment. *J. Ecol.* 79:687–696.

Lechowicz, M. J., and P. A. Blais. 1988. Assessing the contributions of multiple interacting traits to plant reproductive success: Environmental dependence. *J. Evol. Biol.* 1:255–273.

Lecuit, T., and S. M. Cohen. 1997. Proximal-distal axis formation in the *Drosophila* leg. *Nature* 388:139–145.

Lecuit, T., W. J. Brook, M. Ng, M. Calleja, H. Sun, and S. M. Cohen. 1996. Two distinct mechanisms for long-range patterning by *Decapentaplegic* in the *Drosophila* wing. *Nature* 381:387–393.

Lee, M. S. 1993. The origin of the turtle body plan: Bridging a famous morphological gap. *Science* 261: 1716–1720.

Lee, M. S. Y. 1996. Correlated progression and the origin of turtles. *Nature* 379:812–815.

Lenski, R. E. 1988. Experimental studies of pleiotropy and epistasis in *Escherichia coli*. II. Compensation for maladaptive effects associated with resistance to virus T4. *Evolution* 42:433–440.

Lenski, R. E., and A. F. Bennett. 1993. Evolutionary response of *Escherichia coli* to thermal stress. *Am. Nat.* 142:S47-S64.

Lerner, I. M. 1954. *Genetic Homeostasis*. Dover, New York.

Leroi, A. M., W. R. Chen, and M. R. Rose. 1994a. Long-term laboratory evolution of a genetic life-history trade-off in *Drosophila melanogaster*. 2. Stability of genetic correlations. *Evolution* 48:1258–1268.

Leroi, A. M., A. K. Chippindale, and M. R. Rose. 1994b. Long-term laboratory evolution of a genetic life-history trade-off in *Drosophila melanogaster*. 1. The role of genotype-by-environment interaction. *Evolution* 48:1244–1257.

Leroi, A. M., S. B. Kim, and M. R. Rose. 1994c. The evolution of phenotypic life-history trade-offs: An experimental study using *Drosophila melanogaster*. *Am. Nat.* 144:661–676.

Leroi, A. M., R. E. Lenski, and A. F. Bennett. 1994d. Evolutionary adaptation to temperature. III. Adaptation of *Escherichia coli* to a temporally varying environment. *Evolution* 48:1222–1229.

LeRoy Poff, N., and J. V. Ward. 1995. Herbivory under different flow regimes: A field experiment and test of a model with a benthic stream insect. *Oikos* 71:179–188.

Levene, H. 1953. Genetic equilibrium when more than one ecological niche is available. *Am. Nat.* 87: 331–333.

Levin, D. A. 1970. Developmental instability and evolution in peripheral isolates. *Am. Nat.* 104:343–353.

Levin, D. A. 1983. Polyploidy and novelty in flowering plants. *Am. Nat.* 122:1–25.

Levin, D. A. 1988. Plasticity, canalization and evolutionary stasis in plants. In A. J. Davy, M. J. Hutchings and A. R. Watkinson (eds.), *Plant Population Biology*, pp. 35–45. Blackwell, Oxford.

Levins, R. 1963. Theory of fitness in a heterogeneous environment. II. Developmental flexibility and niche selection. *Am. Nat.* 47:75–90.

Levins, R. 1968. *Evolution in Changing Environments*. Princeton University Press, Princeton, NJ.

Levins, R., and R. C. Lewontin. 1985. *The Dialectical Biologist*. Harvard University Press, Cambridge, MA.

Lewin, B. 1997. *Genes VI*. Oxford University Press, New York.

Lewis, E. B. 1978. A gene complex controlling segmentation in *Drosophila*. *Nature* 276:565–570.

Lewontin, R. C. 1957. The adaptations of populations to varying environments. *Cold Spring Harb. Symp. Quant. Biol.* 22:395–408.

Lewontin, R. C. 1974a. The analysis of variance and the analysis of causes. *Am. J. Hum. Genet.* 26:400–411.

Lewontin, R. C. 1974b. *The Genetic Basis of Evolutionary Change*. Columbia University Press, New York.

Lewontin, R. C., and J. L. Hubby. 1966. A molecular approach to the study of genic heterozygosity in natural populations. II. Amount of variation and degree of heterozygosity in natural populations of *Drosophila pseudoobscura*. *Genetics* 54:595–609.

Li, X., and M. Noll. 1994. Evolution of distinct developmental functions of three *Drosophila* genes by acquisition of different *cis*-regulatory regions. *Nature* 367:83–87.

Liu, Z., and V. Ambros. 1991. Alternative temporal control systems for hypodermal cell differentiation in *Caenorhabditis elegans*. *Nature* 350:162–165.

Lively, C. M. 1986. Canalization versus developmental conversion in a spatially variable environment. *Am. Nat.* 128:561–572.

Lloyd Morgan, C. 1896. *Habit and Instinct*. Arnold, London.

Lloyd, D. G. 1984. Variation strategies of plants in heterogeneous environments. *Biol. J. Linn. Soc.* 21: 357–385.

Lloyd, D. G. 1987. Selection of offspring size at independence and other size-versus-number strategies. *Am. Nat.* 129:800–817.

Lloyd, D. G., and D. L. Venable. 1992. Some properties of natural selection with single and multiple constraints. *Theor. Popul. Biol.* 41:90–110.

Loeschcke, V., and R. A. Krebs. 1996. Selection for heat-shock resistance in larval and in adult *Drosophila buzzatii*: Comparing direct and indirect responses. *Evolution* 50:2354–2359.

Lofsvold, D. 1986. Quantitative genetics of morphological differentiation in Peromyscus. I. Tests of the homogeneity of genetic covariance structure among species and subspecies. *Evolution* 40:559–573.

Logsdon, J. M. J., and W. F. Doolittle. 1997. Origin of antifreeze protein genes: A cool tale in molecular evolution. *Proc. Natl. Acad. Sci. USA* 94:3485–3487.

Lois, R., and B. B. Buchanan. 1994. Severe sensitivity to ultraviolet radiation in an *Arabidopsis* mutant deficient in flavonoid accumulation. II. Mechanisms of UV-resistance in *Arabidopsis*. *Planta* 194:504–509.

Lord, E. M. 1981. Cleistogamy: A tool for the study of floral morphogenesis, function and evolution. *Bot. Rev.* 47:421–449.

Lord, E. M., K. J. Eckard, and W. Crone. 1989. Development of the dimorphic anthers in *Collomia grandiflora*: Evidence for heterochrony in the evolution of the cleistogamous anther. *J. Evol. Biol.* 2:81–93.

Lorenz, E. N. 1963. Deterministic nonperiodic flow. *J. Atmos. Sci.* 20:130–141.

Lortie, C. J., and L. W. Aarssen. 1996. The specialization hypothesis for phenotypic plasticity in plants. *Int. J. Plant. Sci.* 157:484–487.

Lowe, C. J., and G. A. Wray. 1997. Radical alterations in the roles of homeobox genes during echinoderm evolution. *Nature* 389:718–721.

Lucas, J. R., and R. D. Howard. 1995. On alternative reproductive tactics in anurans: Dynamic games with density and frequency dependence. *Am. Nat.* 146:369–397.

Lupski, J. R., J. R. Roth, and G. M. Weinstock. 1996. Chromosomal duplications in bacteria, fruit flies and humans. *Am. J. Hum. Genet.* 58:21–27.

Lynch, M., and W. Gabriel. 1987. Environmental tolerance. *Am. Nat.* 129:283–303.

Lynch, M. V., and J. B. Walsh. 1997. *Fundamentals of Quantitative Genetics.* Sinauer Associates, Sunderland, MA.

Ma, H. 1994. The unfolding drama of flower development: Recent results from genetic and molecular analyses. *Genes Dev.* 8:745–756.

Ma, H. 1998. To be, or not to be, a flower—Control of floral meristem identity. *Trends Genet.* 14:26–32.

Mabee, P. M. 1993. Phylogenetic interpretation of ontogenetic change: Sorting out the actual and artefactual in an empirical case study of centrarchid fishes. *Zoo. J. Linn. Soc.* 107:175–291.

Macdonald, S. E., and C. C. Chinnappa. 1989. Population differentiation for phenotypic plasticity in the *Stellaria longipes* complex. *Am. J. Bot.* 76:1627–1637.

Mackay, T. F. C., R. F. Lyman, and M. S. Jackson. 1992a. Effects of P-element insertions on quantitative traits in *Drosophila melanogaster*. *Genetics* 130:315–332.

Mackay, T. F. C., R. F. Lyman, M. S. Jackson, C. Terzian, and W. G. Hill. 1992b. Polygenic mutation in *Drosophila melanogaster*: Estimates from divergence among inbred strains. *Evolution* 46:300–316.

Maddox, G. D., and J. Antonovics. 1983. Experimental ecological genetics in *Plantago*: A structural equation approach to fitness components in *P. aristata* and *P. patagonica*. *Ecology* 64:1092–1099.

Manak, J. R., and M. P. Scott. 1994. A class act: Conservation of homeodomain protein functions. *Development Suppl.*:61–71.

Mangel, M., and C. W. Clark. 1988. *Dynamic Modeling in Behavioral Ecology.* Princeton University Press, Princeton, NJ.

Mangel, M., and D. Ludwig. 1992. Definition and evaluation of the fitness of behavioral and developmental programs. *Annu. Rev. Ecol. Syst.* 23:507–536.

Manly, F. J. 1986. Randomization and regression methods for testing for associations with geographical, environmental and biological distances between populations. *Res. Pop. Ecol.* 28:201–218.

Mantel, N. A. 1967. The detection of disease clustering and a generalized regression approach. *Cancer Res.* 27:209–220.

Marcus, L. F., M. Corti, A. Loy, G. J. P. Naylor, and D. E. Slice (eds.). 1996. *Advances in Morphometrics.* Plenum Press, New York.

Maresca, B., E. Patriarca, C. Goldenberg, and M. Sacco. 1988. Heat shock and cold adaptation in Antarctic fishes: A molecular approach. *Comp. Biochem. Physiol.* 90B:623–629.

Markow, T. A. 1994. *Developmental Instability: Its Origins and Evolutionary Implications.* Kluwer Academic Publishers, Dordrecht.

Markwell, J., and J. C. Osterman. 1992. Occurrence of temperature-sensitive phenotypic plasticity in chlorophyll-deficient mutants of *Arabidopsis thaliana*. *Plant Physiol.* 98:392–394.

Marler, P., and D. A. Nelson. 1993. Action-based learning: A new form of developmental plasticity in bird song. *Neth. J. Zool.* 43:91–103.

Marshall, D. R., and S. K. Jain. 1968. Phenotypic plasticity of *Avena fatua* and *A. barbata*. *Am. Nat.* 102:457–467.

Martinez-Zapater, J. M., and C. R. Somerville. 1990. Effect of light quality and vernalization on late-flowering mutants of *Arabidopsis thaliana*. *Plant Physiol.* 92:770–776.

Martins, E. P. 1994. Estimating the rate of phenotypic evolution from comparative data. *Am. Nat.* 144:193–209.

Martins, E. P. 1996. Conducting phylogenetic comparative studies when the phylogeny is not known. *Evolution* 50:12–22.

Masterson, J. 1994. Stomatal size in fossil plants: Evidence for polyploidy in majority of Angiosperms. *Science* 264:421–424.

Mastick, G. S., R. McKay, T. Oligino, K. Donovan, and A. Javier López. 1995. Identification of target genes regulated by homeotic proteins in *Drosophila melanogaster* through genetic selection of *Ultrabithorax* protein-binding sites in yeast. *Genetics* 39:349–363.

Mather, K. 1953. Genetical control of stability in development. *Heredity* 7:297–336.

Matsuda, R. 1982. The evolutionary process in talitrid amphipods and salamanders in changing environments, with a discussion of "genetic assimilation" and some other evolutionary concepts. *Can. J. Zool.* 60:733–749.

Mauthe, S., K. Bachmann, K. L. Chambers, and H. J. Price. 1984. Independent responses of two fruit characters to developmental regulation in *Microseris douglasii* (Asteraceae; Lactuceae). *Experientia* 40:1280–1281.

Maves, L., and G. Schubiger. 1998. A molecular basis for transdetermination in *Drosophila* imaginal discs: Interactions between *wingless* and *decapentaplegic* signaling. *Development* 125:115–124.

Mayer, U., R. A. Torres Ruiz, T. Berleth, S. Misera, and G. Jurgens. 1991. Mutations affecting body organization in the *Arabidopsis* embryo. *Nature* 353:402–407.

Mayley, G. 1996. Landscapes, learning costs, and genetic assimilation: Modeling the evolution of motivation. *Evol. Comp.* 4:213–234.

Maynard Smith, J. 1962. Disruptive selection, polymorphism, and sympatric speciation. *Nature* 195:60–62.

Maynard Smith, J. 1979. Game theory and the evolution of behavior. *Proc. R. Soc. Lond. B* 205:41–54.

Maynard Smith, J. 1982. *Evolution and the Theory of Games*. Cambridge University Press, Cambridge.

Maynard Smith, J., and K. C. Sondhi. 1960. The genetics of a pattern. *Genetics* 45:1039–1050.

Maynard Smith, J., R. Burian, S. Kauffman, P. Alberch, J. Campbell, B. Goodwin, R. Lande, D. Raup, and L. Wolpert. 1985. Developmental constraints and evolution. *Q. Rev. Biol.* 60:265–287.

Mayr, E. 1942. *Systematics and the Origin of Species*. Columbia University Press, New York.

Mayr, E. 1960. The emergence of evolutionary novelties. In S. Tax (ed.), *Evolution after Darwin: The Evolution of Life*, pp. 349–380. University of Chicago Press, Chicago.

Mayr, E. 1963. *Animal Species and Evolution*. Harvard University Press, Cambridge, MA.

Mayr, E. 1980. Some thoughts on the history of the evolutionary synthesis. In E. Mayr and W. B. Provine (ed.), *The Evolutionary Synthesis. Perspectives on the Unification of Biology*, pp. 1–48. Harvard University Press, Cambridge, MA.

Mayr, E., and W. B. Provine (eds.). 1980. *The Evolutionary Synthesis: Perspectives on the Unification of Biology*. Harvard University Press, Cambridge, MA.

Mazer, S. J., and C. T. Schick. 1991. Constancy of population parameters for life-history and floral traits in *Raphanus sativus* L. II. Effects of planting density on phenotype and heritability estimates. *Evolution* 45:1888–1907.

McCollum, S. A., and J. Van Buskirk. 1996. Costs and benefits of a predator-induced polyphenism in the gray treefrog *Hyla chrysoscelis*. *Evolution* 50:583–593.

McDade, L. A. 1992. Hybrids and phylogenetic systematics. II. The impact of hybrids on cladistic analysis. *Evolution* 46:1329–1346.

McDonald, J. F. 1983. The molecular basis of adaptation: A critical review of relevant ideas and observations. *Annu. Rev. Ecol. Syst.* 14:77–102.

McEdward, L. R., and D. A. Janies. 1997. Relationships among development, ecology, and morphology in the evolution of Echinoderm larvae and lifecycles. *Biol. J. Linn. Soc.* 60:381–400.

McGhee, G. R., Jr. 1980. Shell form in the biconvex articulate Brachiopoda: A geometric analysis. *Paleobiology* 6:57–76.

McKenzie, J. A. 1993. Measuring fitness and intergenic interactions: The evolution of resistance to diazinon in *Lucilia cuprina*. *Genetica* 90:227–237.

McKenzie, J. A., and P. Batterham. 1994. The genetic, molecular and phenotypic consequences of selection for insecticide resistance. *Trends Ecol. Evol.* 9:166–169.

McKinney, M. L., and K. J. McNamara. 1991. *Heterochrony. The Evolution of Ontogeny*. Plenum Press, New York.

McKitrick, M. C. 1993. Phylogenetic constraint in evolutionary theory: Has it any explanatory power? *Annu. Rev. Ecol. Syst.* 24:307–330.

McKitrick, M. C. 1994. On homology and the ontological relationship of parts. *Syst. Biol.* 43:1–10.

McNamara, J. M. 1994. Timing of entry into diapause: Optimal allocation to "growth" and "reproduction" in a stochastic environment. *J. Theor. Biol.* 168:201–209.

McNamara, J. M., and A. I. Houston. 1996. State-dependent life histories. *Nature* 380:215–221.

McNamara, K. J. (ed.) 1995. *Evolutionary Change and Heterochrony*. John Wiley & Sons, Chichester, UK.

McPeek, M. A. 1995. Morphological evolution mediated by behavior in the damselflies in two communities. *Evolution* 49:749–769.

McShea, D. W. 1996. Metazoan complexity and evolution: Is there a trend? *Evolution* 50:477–492.

Medawar, P. B., and J. S. Medawar. 1983. *Aristotle to Zoos. A Philosophical Dictionary of Biology*. Harvard University Press, Cambridge, MA.

Meinke, D. W., and I. M. Sussex. 1979. Embryo-lethal mutants of *Arabidopsis thaliana*. *Dev. Biol.* 72:50–61.

Meinke, D. W., L. H. Franzmann, T. C. Nickle, and E. C. Young. 1994. *Leafy Cotyledon* mutants of *Arabidopsis*. *Plant Cell* 6:1049–1064.

Merila, J., M. Bjorklund, and L. Gustafsson. 1994. Evolution of morphological differences with moderate genetic correlations among traits as exemplified by two flycatcher species (*Ficedula*; Muscicapidae). *Biol. J. Linn. Soc.* 52:19–30.

Mermall, V., P. L. Post, and M. S. Mooseker. 1998. Unconventional myosins in cell movement, membrane traffic, and signal transduction. *Science* 279:527–533.

Meyer, A. 1987. Phenotypic plasticity and heterochrony in *Cichlasoma managuense* (Pisces, Cichlidae) and their implications for speciation in Cichlid fishes. *Evolution* 41:1357–1369.

Meyer, A. 1993. Phylogenetic relationships and evolutionary processes in East African cichlid fishes. *Trends Ecol. Evol.* 8:279–284.

Meyerowitz, E. M. 1994a. The genetics of flower development. *Sci. Am.* November 1994:56–65.

Meyerowitz, E. M. 1994b. Flower development and evolution: New answers and new questions. *Proc. Natl. Acad. Sci. USA* 91:5735–5737.

Micchelli, C. A., E. J. Rulifson, and S. S. Blair. 1997. The function and regulation of *cut* expression on the wing margin of *Drosophila*: Notch, Wingless and a dominant negative role for Delta and Serrate. *Development* 124:1485–1495.

Michelotti, E. F., S. Sanford, and D. Levens. 1997. Marking of active genes on mitotic chromosomes. *Nature* 388:895–899.

Miller, T. E., A. A. Winn, and D. W. Schemske. 1994. The effects of density and spatial distribution on selection for emergence time in *Prunella vulgaris* (Lamiaceae). *Am. J. Bot.* 81:1–6.

Mitchell, M. 1994. Genetic algorithms and artificial life. In C. G. Langton (ed.), *Artificial Life: An Overview*, pp. 267–289. MIT Press, Cambridge, MA.

Mitchell, R. J. 1992. Testing evolutionary and ecological hypotheses using path analysis and structural equation modelling. *Func. Ecol.* 6:123–129.

Mitchell, R. J. 1993. Adaptive significance of *Ipomopsis aggregata* nectar production: Observation and experiment in the field. *Evolution* 47:25–35.

Mitchell-Olds, T. 1996. Genetic constraints on life-history evolution: Quantitative-trait loci influencing growth and flowering in *Arabidopsis thaliana*. *Evolution* 50:140–145.

Mitchell-Olds, T., and J. Bergelson. 1990. Statistical genetics of an annual plant, *Impatiens capensis*. I. Genetic basis of quantitative variation. *Genetics* 124:407–415.

Mitchell-Olds, T., and D. Bradley. 1996. Genetics of *Brassica rapa*. 3. Costs of disease resistance to three fungal pathogens. *Evolution* 50:1849–1858.

Mitchell-Olds, T., and J. J. Rutledge. 1986. Quantitative genetics in natural plant populations: A review of the theory. *Am. Nat.* 127:379–402.

Mitchell-Olds, T., and R. G. Shaw. 1987. Regression analysis of natural selection: Statistical inference and biological interpretation. *Evolution* 41:1149–1161.

Mitton, J. B. 1997. *Selection in Natural Populations*. Oxford University Press, New York.

Mitton, J. B., and M. C. Grant. 1984. Associations among protein heterozygosity, growth rate, and developmental homeostasis. *Annu. Rev. Ecol. Syst.* 15:479–499.

Mole, S., and A. J. Zera. 1993. Differential allocation of resources underlies the dispersal-reproduction trade-off in the wing-dimorphic cricket, *Gryllus rubens*. *Oecologia* 93:121–127.

Møller, A. P., and M. Eriksson. 1994. Patterns of fluctuating asymmetry in flowers: Implications for sexual selection in plants. *J. Evol. Biol.* 7:97–113.

Møller, A. P., and R. Thornhill. 1997. A meta-analysis of the heritability of developmental stability. *J. Evol. Biol.* 10:1–16.

Monteiro, A., P. M. Brakefield, and V. French. 1997a. Butterfly eyespots: The genetics and development of the color rings. *Evolution* 51:1207–1216.

Monteiro, A., P. M. Brakefield, and V. French. 1997b. The relationship between eyespot shape and wing shape in the butterfly *Bicyclus anynana*: A genetic and morphometrical approach. *J. Evol. Biol.* 10:787–802.

Monteiro, A. F., P. M. Brakefield, and V. French. 1994. The evolutionary genetics and developmental basis of wing pattern variation in the butterfly *Bicyclus anynana*. *Evolution* 48:1147–1157.

Mongold, J. A., A. F. Bennett, and R. E. Lenski. 1996. Evolutionary adaptation to temperature. IV. Adaptation of *Escherichia coli* at a niche boundary. *Evolution* 50:35–43.

Moore, A. J., and C. R. B. Boake. 1994. Optimality and evolutionary genetics: Complementarity procedures for evolutionary analysis in behavioural ecology. *Trends Ecol. Evol.* 9:69–72.

Moore, F. B. G., and S. J. Tonsor. 1994. A simulation of Wright's shifting-balance process: Migration and the three phases. *Evolution* 48:69–80.

Moran, N. A. 1992. The evolutionary maintenance of alternative phenotypes. *Am. Nat.* 139:971–989.

Moran, N. A. 1994. Adaptation and constraint in the complex life cycles of animals. *Annu. Rev. Ecol. Syst.* 25:573–600.

Morishima, H., and H. I. Oka. 1975. Comparison of growth pattern and phenotypic plasticity between wild and cultivated rice strains. *Jap. J. Genet.* 50:53–65.

Motte, P., H. Saedler, and Z. Schwarz-Sommer. 1998. *STYLOSA* and *FISTULATA*: Regulatory components of the homeotic control of *Antirrhinum* floral organogenesis. *Development* 125:71–84.

Moxon, E. R., and C. F. Higgins. 1997. A blueprint for life. *Nature* 389:120–121.

Mozley, D., and B. Thomas. 1995. Developmental and photobiological factors affecting photoperiodic induction in *Arabidopsis thaliana* Heynh. Landsberg *erecta*. *J. Exp. Bot.* 46:173–179.

Mukai, T., S. I. Chigusa, and S.-I. Kusakabe. 1982. The genetic structure of natural populations of *D. melanogaster*. XV. Nature of developmental homeostasis for viability. *Genetics* 101:279–300.

Mulkey, S. S., A. P. Smith, S. J. Wright, J. L. Machado, and R. Dudley. 1992. Contrasting leaf phenotypes control seasonal variation in water loss in a tropical forest shrub. *Proc. Natl. Acad. Sci. USA* 89: 9084–9088.

Müller, G. B. 1990. Developmental mechanisms at the origin of morphological novelty: A side-effect hypothesis. In M. H. Nitecki (ed.), *Evolutionary Innovations*, pp. 99–130. University of Chicago Press, Chicago.

Müller, G. B., and G. P. Wagner. 1996. Homology, Hox genes, and developmental integration. *Am. Zool.* 36:4–13.

Munster, T., J. Pahnke, A. Di Rosa, J. T. Kim, W. Martin, H. Saedler, and G. Theissen. 1997. Floral homeotic genes were recruited from homologous MADS-box genes preexisting in the common ancestor of ferns and seed plants. *Proc. Natl. Acad. Sci. USA* 94:2415–2420.

Myers, P., B. L. Lundrigan, B. W. Gillespie, and M. L. Zelditch. 1996. Phenotypic plasticity in skull and dental morphology in the prairie deer mouse (*Peromyscus maniculatus bairdii*). *J. Morphol.* 229: 229–237.

Nadeau, J. H., and D. Sankoff. 1997. Comparable rates of gene loss and functional divergence after genome duplications early in vertebrate evolution. *Genetics* 147:1259–1266.

Nager, R. G., and A. J. v. Noordwijk. 1995. Proximate and ultimate aspects of phenotypic plasticity in timing of great tit breeding in a heterogeneous environment. *Am. Nat.* 146:454–474.

Nakaya, A., A. Yonezawa, and K. Yamamoto. 1997. Classification of RNA secondary structures using the techniques of cluster analysis. *J. Theor. Biol.* 183:105–117.

Nei, M. 1987. *Molecular Evolutionary Genetics.* Columbia University Press, New York.

Netzer, W. J., and F. U. Hartl. 1997. Recombination of protein domains facilitated by co-translational folding in eukaryotes. *Nature* 388:343–349.

Neumann, C., and S. Cohen. 1997. Morphogens and pattern formation. *BioEssays* 19:721–729.

Newman, R. A. 1988. Adaptive plasticity in development of *Scaphiopus couchii* tadpoles in desert ponds. *Evolution* 42:774–783.

Newman, R. A. 1992. Adaptive plasticity in amphibian metamorphosis. *BioScience* 42:671–678.

Newman, R. A. 1994a. Effects of changing density and food level on metamorphosis of a desert amphibian, *Scaphiopus couchii. Ecology* 75:1085–1096.

Newman, R. A. 1994b. Genetic variation for phenotypic plasticity in the larval life history of spadefoot toads (*Scaphiopus couchii*). *Evolution* 48:1773–1785.

Neyfakh, A. A., and D. L. Hartl. 1993. Genetic control of the rate of embryonic development: Selection for faster development at elevated temperatures. *Evolution* 47:1625–1631.

Nijhout, H. F. 1990. Metaphors and the roles of genes in development. *Bioessays* 12:441–446.

Nijhout, H. F. 1991. *The Development and Evolution of Butterfly Wing Patterns.* Smithsonian Institution Press, Washington, DC.

Nijhout, H. F. 1994a. Developmental perspectives on evolution of butterfly mimicry. *BioScience* 44: 148–156.

Nijhout, H. F. 1994b. Symmetry patterns and compartments in Lepidopteran wings: The evolution of a patterning mechanism. *Development Suppl.*:225–233.

Nijhout, H. F. 1996. Focus on butterfly eyespot development. *Nature* 384:209–210.

Nijhout, H. F., and D. E. Wheeler. 1996. Growth models of complex allometries in holometabolous insects. *Am. Nat.* 148:40–56.

Nijhout, H. F., G. A. Wray, C. Kremen, and C. K. Teragawa. 1986. Ontogeny, phylogeny and evolution of form: An algorithmic approach. *Syst. Zool.* 35:445–457.

Niklas, K. J. 1994. Morphological evolution through complex domains of fitness. *Proc. Natl. Acad. Sci. USA* 91:6772–6779.

Niklas, K. J. 1997. Adaptive walks through fitness landscapes for early vascular land plants. *Am. J. Bot.* 84:16–25.

Nitecki, M. H. (ed.) 1990. *Evolutionary Innovations.* University of Chicago Press, Chicago.

Niu, D. K., and J.-K. Chen. 1997. Evolutionary advantages of cell specialization: Save and protect DNA. *J. Theor. Biol.* 187:39–43.

Nordin, K., P. Heino, and E. T. Palva. 1991. Separate signal pathways regulate the expression of a low-temperature-induced gene in *Arabidopsis thaliana* (L.) Heynh. *Plant Mol. Biol.* 16:1061–1071.

Novoplansky, A., D. Cohen, and T. Sachs. 1994. Responses of an annual plant to temporal changes in light environment: An interplay between plasticity and determination. *Oikos* 69:437–446.

Nowak, M. A., M. C. Boerlijst, J. Cooke, and J. Maynard Smith. 1997. Evolution of genetic redundancy. *Nature* 388:167–171.

Nüsslein-Volhard, C. 1994. Of flies and fishes. *Science* 266:572–574.

Nylin, S. 1992. Seasonal plasticity in life history traits: Growth and development in *Polygonia c-album* (Lepidoptera: Nymphalidae). *Biol. J. Linn. Soc.* 47:301–323.

Nylin, S., K. Gotthard, and C. Wiklund. 1996. Reaction norms for age and size at maturity in *Lasiommata* butterflies: Prediction and tests. *Evolution* 50: 1351–1358.

Ohno, S. 1970. *Evolution by Gene Duplication.* Springer-Verlag, New York.

Ohta, T. 1991. Multigene families and the evolution of complexity. *J. Mol. Evol.* 33:34–41.

Ohta, T. 1993. Pattern of nucleotide substitutions in growth hormone-prolactin gene family: A paradigm for evolution by gene duplication. *Genetics* 134:1271–1276.

Oka, H.-I., and W.-T. Chang. 1964. Evolution of responses to growing conditions in wild and cultivated rice forms. *Bot. Bull. Acad. Sinica* 5:120–138.

Okada, K., and Y. Shimura. 1994. Genetic analyses of signalling in flower development using *Arabidopsis. Plant Mol. Biol.* 26:1357–1377.

Oliver, S. G. 1996. From DNA sequence to biological function. *Nature* 379:597–600.

Olson, E. C., and R. L. Miller. 1958. *Morphological Integration.* University of Chicago Press, Chicago.

Oostermeijer, J. G. B., M. W. van Eijck, N. C. van Leeuwen, and J. C. M. den Nijs. 1995. Analysis of the relationship between allozyme heterozygosity and fitness in the rare *Gentiana pneumonanthe* L. *J. Evol. Biol.* 8:739–759.

Orr, H. A., and J. A. Coyne. 1992. The genetics of adaptation: A reassessment. *Am. Nat.* 140:725–742.

Orzack, S. H. 1985. Population dynamics in variable environments. V. The genetics of homeostasis revisited. *Am. Nat.* 125:550–572.

Osborn, H. F. 1897. Organic selection. *Science* 15: 583–587.

Oster, G., and P. Alberch. 1982. Evolution and bifurcation of developmental programs. *Evolution* 36: 444–459.

Padilla, D. K., and S. C. Adolph. 1996. Plastic inducible morphologies are not always adaptive: The importance of time delays in a stochastic environment. *Evol. Ecol.* 10:105–117.

Panganiban, G., A. Sebring, L. Nagy, and S. Carroll. 1995. The development of crustacean limbs and the evolution of arthropods. *Science* 270:1363–1366.

Panganiban, G., S. M. Irvine, C. Lowe, H. Roehl, L. S. Corley, B. Sherbon, J. K. Grenier, J. F. Fallon, J.

Kimble, M. Walker, G. A. Wray, B. J. Swalla, M. Q. Martindale, and S. B. Carroll. 1997. The origin and evolution of animal appendages. *Proc. Natl. Acad. Sci. USA* 94:5162–5166.

Parker, G. A., and J. Maynard Smith. 1990. Optimality theory in evolutionary biology. *Nature* 348:27–33.

Parker, G. A., and W. L. Simmons. 1994. Evolution of phenotypic optima and copula duration in dungflies. *Nature* 370:53–56.

Parsons, P. A. 1987. Evolutionary rates under environmental stress. *Evol. Theory* 21:311–347.

Parsons, P. A. 1993. Developmental variability and the limits of adaptation: Interactions with stress. *Genetica* 89:245–253.

Partridge, L., and K. Fowler. 1992. Direct and correlated responses to selection on age at reproduction in *Drosophila melanogaster*. *Evolution* 46:76–91.

Partridge, L., and P. H. Harvey. 1988. The ecological context of life history evolution. *Science* 241: 1449–1455.

Partridge, L., B. Barrie, N. H. Barton, K. Fowler, and V. French. 1995. Rapid laboratory evolution of adult life-history traits in *Drosophila melanogaster* in response to temperature. *Evolution* 49:538–544.

Patel, N. H. 1994. The evolution of arthropod segmentation: Insights from comparisons of gene expression patterns. *Development Suppl.*:201–207.

Paterson, A. H., Y. Lin, Z. Li, K. F. Scertz, J. F. Doebley, S. R. M. Pinson, S. Liu, J. W. Stensel, and J. E. Irvine. 1995. Convergent domestication of cereal crops by independent mutations at corresponding genetic loci. *Science* 269:1714–1718.

Patterson, C. 1982. Morphological characters and homology. In K. A. Joysey and A. E. Friday (eds.), *Problems of Phylogenetic Reconstructions*, pp. 21–74. Academic Press, London.

Patton, D. A., L. H. Franzmann, and D. W. Meinke. 1991. Mapping genes essential for embryo development in *Arabidopsis thaliana*. *Mol. Gen. Genet.* 227:337–347.

Paulsen, S. M. 1994. Quantitative genetics of butterfly wing color patterns. *Dev. Genet.* 15:79–91.

Pederson, D. G. 1968. Environmental stress, heterozygote advantage and genotype-environment interaction in *Arabidopsis*. *Heredity* 23:127–138.

Perkins, J. M., and J. L. Jinks. 1973. The assessment and specificity of environmental and genotype-environmental components of variability. *Heredity* 30:111–126.

Perrin, N., and J. Travis. 1992. On the use of constraints in evolutionary biology and some allergic reactions to them. *Func. Ecol.* 6:361–363.

Perry, D. A. 1995. Self-organizing systems across scales. *Trends Ecol. Evol.* 10:241–244.

Pettersson, L. B., and C. Bronmark. 1997. Density-dependent costs of an inducible morphological defense in crucian carp. *Ecology* 78:1805–1812.

Pfennig, D. W. 1992a. Polyphenism in spadefoot toad tadpoles as a locally adjusted evolutionary stable strategy. *Evolution* 46:1408–1420.

Pfennig, D. W. 1992b. Proximate and functional causes of polyphenism in an anuran tadpole. *Func. Ecol.* 6:167–174.

Phillips, P. C. 1993. Peak shifts and polymorphism during phase three of Wright's shifting balance process. *Evolution* 47:1733–1743.

Pickett, F. B., and D. R. Meeks-Wagner. 1995. Seeing double: Appreciating genetic redundancy. *Plant Cell* 7:1347–1356.

Piersma, T., and A. Lindstrom. 1997. Rapid reversible changes in organ size as a component of adaptive behaviour. *Trends Ecol. Evol.* 12:134–138.

Pigliucci, M. 1996a. How organisms respond to environmental change: From phenotypes to molecules (and vice versa). *Trends Ecol. Evol.* 11:168–173.

Pigliucci, M. 1996b. Modelling phenotypic plasticity. II. Do genetic correlations matter? *Heredity* 77:453–460.

Pigliucci, M. In press. Plasticity genes: What are they, and why should we care? In H. Greppin (ed.), *Responses of Organisms to the Environment*. University of Geneva, Geneva.

Pigliucci, M., and C. D. Schlichting. 1995. Ontogenetic reaction norms in *Lobelia siphilitica* (Lobeliaceae): Response to shading. *Ecology* 76:2134–2144.

Pigliucci, M., and C. D. Schlichting. 1996. Reaction norms of *Arabidopsis*. IV. Relationships between plasticity and fitness. *Heredity* 76:427–436.

Pigliucci, M., and C. D. Schlichting. 1997. On the limits of quantitative genetics for the study of phenotypic evolution. *Acta Biotheor.* 45:143–160.

Pigliucci, M., C. Paoletti, S. Fineschi, and L. Malvolti. 1991. Phenotypic integration in chestnut (*Castanea sativa* Mill)—Leaves versus fruits. *Bot. Gaz.* 152:514–521.

Pigliucci, M., C. D. Schlichting, and J. Whitton. 1995. Reaction norms of *Arabidopsis*. II. Response to stress and unordered environmental variation. *Func. Ecol.* 9:537–547.

Pigliucci, M., J. Whitton, and C. D. Schlichting. 1995. Reaction norms of *Arabidopsis*. I. Plasticity of characters and correlations across water, nutrient and light gradients. *J. Evol. Biol.* 8:421–438.

Pigliucci, M., C. D. Schlichting, C. S. Jones, and K. Schwenk. 1996. Developmental reaction norms: The interactions among allometry, ontogeny and plasticity. *Plant Species Biol.* 11:69–85.

Pigliucci, M., P. diIorio, and C. D. Schlichting. 1997. Phenotypic plasticity of growth trajectories in two species of *Lobelia* in response to nutrient availability. *J. Ecol.* 85:265–276.

Poethig, R. S. 1988. Heterochronic mutations affecting shoot development in maize. *Genetics* 119:959–973.

Pollak, E., O. Kempthorne, and T. B. Bailey. 1977. Proceedings of the International Conference on Quantitative Genetics. Iowa State University Press, Ames, IA.

Prandl, R., E. Kloske, and F. Schoffl. 1995. Developmental regulation and tissue-specific differences of heat shock gene expression in transgenic tobacco and *Arabidopsis* plants. *Plant Mol. Biol.* 28:73–82.

Price, C. S. C. 1995. Structurally unsound. *Evolution* 49:1298–1302.

Price, T., and T. Langen. 1992. Evolution of correlated characters. *Trends Ecol. Evol.* 7:307–310.

Price, T., M. Turelli, and M. Slatkin. 1993. Peak shifts produced by correlated response to selection. *Evolution* 47:280–290.

Primack, R. B. 1978. Regulation of seed yield in *Plantago*. *J. Ecol.* 66:835–847.

Primack, R. B., and J. Antonovics. 1981. Experimental ecological genetics in *Plantago*. V. Components of seed yield in the ribwort plantain *Plantago lanceolata* L. *Evolution* 35:1069–1079.

Prout, T., and O. Savolainen. 1996. Genotype-by-environment interaction is not sufficient to maintain variation: Levene and the leafhopper. *Am. Nat.* 148:930–936.

Provine, W. B. 1971. *The Origins of Theoretical Population Genetics.* University of Chicago Press, Chicago.

Purugganan, M. D. 1997. The MADS-box floral homeotic gene lineages predate the origin of seed plants: Phylogenetic and molecular clock estimates. *J. Mol. Evol.* 45:392–396.

Purugganan, M. D., S. D. Rounsley, R. J. Schmidt, and M. F. Yanofsky. 1995. Molecular evolution of flower development: Diversification of the plant MADS-box regulatory gene family. *Genetics* 140:345–356.

Rachootin, S. P., and K. S. Thomson. 1981. Epigenetics, paleontology, and evolution. In G. G. E. Scudder and J. L. Reveal (eds.), *Evolution Today*, pp. 181–193. Hunt Institute for Botanical Documentation, Pittsburgh.

Raff, R. A. 1996. *The Shape of Life.* Chicago University Press, Chicago.

Raff, R. A., and G. A. Wray. 1989. Heterochrony: Developmental mechanisms and evolutionary results. *J. Evol. Biol.* 2:409–434.

Raff, R. A., G. A. Wray, and J. J. Henry. 1991. Implications of radical evolutionary changes in early development for concepts of developmental constraint. In L. Warren and H. Koprowski (eds.), *New Perspectives on Evolution*, pp. 189–207. Wiley-Liss, New York.

Ramachandran, V. S., C. W. Tyler, R. L. Gregory, D. Rogers-Ramachandran, S. Duensing, C. Pillsbury, and C. Ramachandran. 1996. Rapid adaptive camouflage in tropical flounders. *Nature* 379:815–818.

Rand, D. A., H. B. Wilson, and J. M. McGlade. 1994. Dynamics and evolution: Evolutionary stable attractors, invasion exponents and phenotype dynamics. *Phil. Trans. R. Soc. Lond. B.* 343:261–283.

Ranganayakulu, G., and A. R. Reddy. 1994. Regulatory differences in developmental expression of alcohol dehydrogenase are related to interspecies differences in ethanol tolerance of *Drosophila. Heredity* 72:374–383.

Rast, J. P., N. A. Hawke, M. K. Anderson, and G. W. Litman. 1996. T-cell receptors and related variable region containing genes from non-tetrapod jawed vertebrates. In M. Nei and N. Takahata (eds.), *Current Topics on Molecular Evolution*, pp. 139–150. Institute of Molecular Evolutionary Genetics, Pennsylvania State University, University Park, PA.

Raup, D. M. 1966. Geometric analysis of shell coiling: General problems. *J. Paleontol.* 40:1178–1190.

Raup, D. M. 1967. Geometric analysis of shell coiling: Coiling in ammonoids. *J. Paleontol.* 41:43–65.

Rausher, M. D. 1992. The measurement of selection on quantitative traits: Biases due to environmental covariances between traits and fitness. *Evolution* 46:616–626.

Rawson, P. D., and T. J. Hilbish. 1991. Genotype-environment interaction for juvenile growth in the hard clam *Mercenaria mercenaria* (L.). *Evolution* 45:1924–1935.

Ray, T. S. 1994. An evolutionary approach to synthetic biology: Zen and the art of creating life. In C. G. Langton (ed.), *Artificial Life: An Overview*, pp. 179–209. MIT Press, Cambridge, MA.

Raybould, A. F., and A. J. Gray. 1994. Will hybrids of genetically modified crops invade natural communities? *Trends Ecol. Evol.* 9:85–88.

Rehfeldt, G. E. 1992. Early selection in *Pinus ponderosa*: Compromises between growth potential and growth rhythm in developing breeding strategies. *For. Sci.* 38:661–677.

Reilly, S. M., E. O. Wiley, and D. J. Meinhardt. 1997. An integrative approach to heterochrony: The distinction between interspecific and intraspecific phenomena. *Biol. J. Linn. Soc.* 60:119–143.

Reiter, R. S., S. A. Coomber, T. M. Bourett, G. E. Bartley, and P. A. Scolnik. 1994. Control of leaf and chloroplast development by the *Arabidopsis* gene *pale cress. Plant Cell* 6:1253–1264.

Ren, P., C.-S. Lim, R. Johnsen, P. S. Albert, D. Pilgrim, and D. L. Riddle. 1996. Control of *C. elegans* larval development by neuronal expression of a TGF-β homolog. *Science* 274:1389–1391.

Rendel, J. M. 1967. *Canalisation and Gene Control.* Logos Press, Academic Press, London.

Rensch, B. 1959. *Evolution above the Species Level.* Columbia University Press, New York.

Resnik, D. 1994. The rebirth of rational morphology: Process structuralism's philosophy of biology. *Acta Biotheor.* 42:1–14.

Reyment, R. A., and W. J. Kennedy. 1991. Phenotypic plasticity in a Cretaceous ammonite analyzed by multivariate statistical methods. *Evol. Biol.* 25:411–426.

Reznick, D. 1985. Costs of reproduction: An evaluation of the empirical evidence. *Oikos* 44:257–267.

Reznick, D. 1992. Measuring the costs of reproduction. *Trends Ecol. Evol.* 7:42–45.

Reznick, D. N., and H. Bryga. 1996. Life-history evolution in guppies (*Poecilia reticulata*: Poeciliidae). V. Genetic basis of parallelism in life histories. *Am. Nat.* 147:339–359.

Reznick, D., and J. Travis. 1996. The empirical study of adaptation in natural populations. In M. A. Rose and G. V. Lauder (eds.), *Adaptation*, pp. 243–289. Academic Press, San Diego.

Reznick, D., and A. P. Yang. 1993. The influence of fluctuating resources on life history: Patterns of allocation and plasticity in female guppies. *Ecology* 74:2011–2019.

Reznick, D. N., H. Bryga, and J. A. Endler. 1990. Experimentally-induced life-history evolution in a natural population. *Nature* 346:357–359.

Reznick, D. N., F. H. Shaw, F. H. Rodd, and R. G. Shaw. 1997. Evaluation of the rate of evolution in natural populations of guppies (*Poecilia reticulata*). *Science* 275:1934–1937.

Rice, S. H. 1997. The analysis of ontogenetic trajectories: When a change in size or shape is not heterochrony. *Proc. Natl. Acad. Sci. USA* 94:907–912.

Richardson, M. K., S. P. Allen, G. M. Wright, A. Raynaud, and J. Hanken. 1998. Somite number and vertebrate evolution. *Development* 125:151–160.

Rickey, T. M., and W. R. Belknap. 1991. Comparison of the expression of several stress-responsive genes in potato tubers. *Plant Mol. Biol.* 16:1009–1018.

Ricklefs, R. E. 1996. Phylogeny and ecology. *Trends Ecol. Evol.* 11:229–230.

Riddle, D., T. Blumenthal, B. Meyer, and J. Priess. 1997. *Caenorhabditis elegans* II. Cold Spring Harbor Laboratory Press, Cold Spring Harbor, NY.

Riechert, S. E., and J. Maynard Smith. 1989. Genetic analyses of two behavioural traits linked to individual fitness in the desert spider *Agelonopsis aperta*. *Anim. Behav.* 37:524–637.

Riedl, R. 1978. *Order in Living Organisms*. John Wiley & Sons, New York.

Rieppel, O., and L. Grande. 1994. Summary and comments on systematic pattern and evolutionary process. In L. Grande and O. Rieppel (eds.), *Interpreting the Hierarchy of Nature: From Systematic patterns to evolutionary process theories*, pp. 227–255. Academic Press, San Diego.

Rieseberg, L. H., C. van Fossen, and A. M. Desrochers. 1995. Hybrid speciation accompanied by genomic reorganization in wild sunflowers. *Nature* 375:313–316.

Riska, B. 1985. Group size factors and geographic variation of morphometric correlation. *Evolution* 39:792–803.

Riska, B. 1986. Some models for development, growth, and morphometric correlation. *Evolution* 40:1303–1311.

Riska, B. 1989. Composite traits, selection response, and evolution. *Evolution* 43:1172–1191.

Riska, B., T. Prout, and M. Turelli. 1989. Laboratory estimates of heritabilities and genetic correlations in nature. *Genetics* 123:865–871.

Rivera-Pomar, R., and H. Jackle. 1996. From gradients to stripes in *Drosophila* embryogenesis: Filling in the gaps. *Trends Genet.* 12:478–483.

Rivera-Pomar, R., X. Lu, N. Perrimon, H. Taubert, and H. Jackle. 1995. Activation of posterior gap gene expression in the *Drosophila* blastoderm. *Nature* 376:253–256.

Roach, D. A. 1986. Life history variation in *Geranium carolinianum*. 1. Covariation between characters at different stages of the life cycle. *Am. Nat.* 128:47–57.

Robertson, A. 1956. The effect of selection against extreme deviants based on deviation or on homozygosis. *J. Genet.* 54:236–248.

Robertson, F. W. 1964. The ecological genetics of growth in *Drosophila*. 7. The role of canalization in the stability of growth relations. *Gen. Res.* 5:107–126.

Roff, D. A. 1992. *The Evolution of Life Histories*. Chapman and Hall, New York.

Roff, D. A. 1994. Optimality modeling and quantitative genetics: A comparison of the two approaches. In C. R. B. Boake (ed.), *Quantitative Genetic Studies of Behavioral Evolution*, pp. 49–66. The University of Chicago Press, Chicago.

Roff, D. A. 1995. Antagonistic and reinforcing pleiotropy: A study of differences in development time in wing dimorphic insects. *J. Evol. Biol.* 8:405–420.

Roff, D. A. 1996. The evolution of threshold traits in animals. *Q. Rev. Biol.* 71:3–35.

Roff, D. A. 1997. *Evolutionary Quantitative Genetics*. Chapman and Hall, New York.

Rogers, B. T., M. D. Peterson, and T. C. Kaufman. 1997. Evolution of the insect body plan as revealed by the *Sex combs reduced* expression pattern. *Development* 124:149–157.

Rogers, C. M., V. J. Nolan, and E. D. Ketterson. 1994. Winter fattening in the dark-eyed junco: Plasticity and possible interaction with migration trade-offs. *Oecologia* 97:526–532.

Rohlf, F. J., and F. D. Bookstein (eds.). 1990. *Proceedings of the Michigan Morphometrics Workshop*. University of Michigan Press, Ann Arbor.

Rohlf, F. J., and L. F. Marcus. 1993. A revolution in morphometrics. *Trends Ecol. Evol.* 8:129–132.

Rollo, C. D. 1994. *Phenotypes: Their Epigenetics, Ecology and Evolution*. Chapman and Hall, London.

Rollo, C. D., and M. D. Hawryluk. 1988. Compensatory scope and resource allocation in two species of aquatic snails. *Ecology* 69:146–156.

Rollo, C. D., and D. M. Shibata. 1991. Resilience, robustness, and plasticity in a terrestrial slug with particular reference to food quality. *Can. J. Zool.* 69:978–987.

Rood, S. B., P. H. Williams, D. Pearce, N. Murofushi, L. N. Mander, and R. P. Pharis. 1990. A mutant gene that increases gibberellin production in *Brassica*. *Plant Physiol.* 93:1168–1174.

Rose, M. R., and G. V. Lauder. 1996. *Adaptation*. Academic Press, San Diego.

Rose, M. R., T. J. Nusbaum, and A. K. Chippindale. 1996. Laboratory evolution: The experimental wonderland and the Cheshire Cat syndrome. In M. R. Rose and G. V. Lauder (eds.), *Adaptation*, pp. 221–241. Academic Press, San Diego.

Roskam, J. C., and P. M. Brakefield. 1996. A comparison of temperature-induced polyphenism in African *Bicyclus* butterflies from a seasonal savannah-rainforest ecotone. *Evolution* 50:2360–2372.

Ross, J. L., P. P. Fong, and D. R. Cavener. 1994. Correlated evolution of the cis-acting regulatory elements and developmental expression of the *Drosophila Gld* gene in seven species from the subgroup *melanogaster*. *Dev. Genet.* 15:38–50.

Rossiter, M. C. 1996. Incidence and consequences of inherited environmental effects. *Annu. Rev. Ecol. Syst.* 27:451–476.

Roth, V. L. 1984. On homology. *Biol. J. Linn. Soc.* 22: 13–29.

Roth, V. L. 1988. The biological basis of homology. In C. J. Humphries (ed.), *Ontogeny and Systematics*, pp. 1–26. Columbia University Press, New York.

Roth, V. L. 1992. Inferences from allometry and fossils: Dwarfing of elephants on islands. *Oxford Surv. Evol. Biol.* 8:261–288.

Roush, W. 1997. A "master control" gene for fly eyes shares its power. *Science* 275:618–619.

Ruddle, F. H., K. L. Bentley, M. T. Murtha, and N. Risch. 1994. Gene loss and gain in the evolution of the vertebrates. *Development Suppl.*:155–161.

Ruiz-Gomez, M., S. Romani, C. Hartmann, H. Jackle, and M. Bate. 1997. Specific muscle identities are regulated by *Krüppel* during *Drosophila* embryogenesis. *Development* 124:3407–3414.

Rusch, J., and M. Levine. 1996. Threshold responses to the dorsal regulatory gradient and the subdivision of primary tissue territories in the *Drosophila* embryo. *Curr. Opin. Genet. Dev.* 6:416–423.

Ruvkun, G., and J. Giusto. 1989. The *Caenorhabditis elegans* heterochronic gene lin-14 encodes a nuclear protein that forms a temporal developmental switch.

Sachs, T. 1988. Ontogeny and phylogeny: Phytohormones as indicators of labile changes. In L. D. Gottlieb and S. K. Jain (eds.), *Plant Evolutionary Biology*, pp. 157–176. Chapman and Hall, London.

Sakai, H., L. J. Medrano, and E. M. Meyerowitz. 1995. Role of SUPERMAN in maintaining *Arabidopsis* floral whorl boundaries. *Nature* 378:199–203.

Sakai, K., and Y. Shimamoto. 1965. Developmental instability in leaves and flowers of *Nicotiana tabacum*. *Genetics* 51:801–813.

Salser, S. J., and C. Kenyon. 1996. A *C. elegans Hox* gene switches on, off, on and off again to regulate proliferation, differentiation and morphogenesis. *Development* 122:1651–1661.

Sanderson, M. J., and L. Hufford. 1996. *Homoplasy: the Recurrence of Similarity in Evolution*. Academic Press, San Diego.

Sandoval, C. P. 1994. Plasticity in web design in the spider *Parawixia bistriata*: A response to variable prey type. *Func. Ecol.* 8:701–707.

Sano, H., I. Kamada, S. Youssefian, M. Katsumi, and H. Wabiko. 1990. A single treatment of rice seedlings with 5-azacytidine induces heritable dwarfism and undermethylation of genomic DNA. *Mol. Gen. Genet.* 220:441–447.

Schaal, B. A., and W. J. Leverich. 1987. Genetic constraints on plant adaptive evolution. In V. Loeschcke (ed.), *Genetic Constraints on Adaptive Evolution*, pp. 173–184. Springer-Verlag, Berlin.

Schaefer, S. A., and G. V. Lauder. 1996. Testing historical hypotheses of morphological change: Biomechanical decoupling in loricarioid catfishes. *Evolution* 50:1661–1675.

Scharloo, W. 1964. Mutant expression and canalization. *Nature* 203:1095–1096.

Scharloo, W. 1987. Constraints in selection response. In V. Loeschcke (ed.), *Genetic Constraints on Adaptive Evolution*, pp. 125–149. Springer-Verlag, Berlin.

Scharloo, W. 1988. Selection on morphological patterns. In G. de Jong (ed.), *Population Genetics and Evolution*, pp. 230–250. Springer-Verlag, Berlin.

Scharloo, W. 1991. Canalization: Genetic and developmental aspects. *Annu. Rev. Ecol. Syst.* 22:65–93.

Scheiner, S. M. 1993a. Genetics and evolution of phenotypic plasticity. *Annu. Rev. Ecol. Syst.* 24:35–68.

Scheiner, S. M. 1993b. Plasticity as a selectable trait: Reply to Via. *Am. Nat.* 142:371–373.

Scheiner, S. M., and C. J. Goodnight. 1984. The comparison of phenotypic plasticity and genetic variation in populations of the grass *Danthonia spicata*. *Evolution* 38:845–855.

Scheiner, S. M., and C. A. Istock. 1991. Correlational selection on life history traits in the pitcher-plant mosquito. *Genetica* 84:123–128.

Scheiner, S. M., and R. F. Lyman. 1991. The genetics of phenotypic plasticity. II. Response to selection. *J. Evol. Biol.* 4:23–50.

Scheiner, S. M., R. L. Caplan, and R. F. Lyman. 1991. The genetics of phenotypic plasticity. III. Genetic correlations and fluctuating asymmetries. *J. Evol. Biol.* 4:51–68.

Schindel, D. E. 1990. Unoccupied morphospace and the coiled geometry of gastropods: Architectural constraint or geometric covariation? In R. M. Ross and W. D. Allmon (eds.), *Causes of Evolution: A Paleontological Perspective*, pp. 270–304. The University of Chicago Press, Chicago.

Schlichting, C. D. 1986. The evolution of phenotypic plasticity in plants. *Annu. Rev. Ecol. Syst.* 17:667–693.

Schlichting, C. D. 1989a. Phenotypic plasticity in *Phlox*. II. Plasticity of character correlations. *Oecologia* 78:496–501.

Schlichting, C. D. 1989b. Phenotypic integration and environmental change. *BioScience* 39:460–464.

Schlichting, C. D., and D. A. Levin. 1984. Phenotypic plasticity of annual *Phlox*: Tests of some hypotheses. *Am. J. Bot.* 71:252–260.

Schlichting, C. D., and D. A. Levin. 1986. Phenotypic plasticity: An evolving plant character. *Biol. J. Linn. Soc.* 29:37–47.

Schlichting, C. D., and D. A. Levin. 1990. Phenotypic plasticity in *Phlox*. III. Variation among natural populations of *P. drummondii*. *J. Evol. Biol.* 3:411–428.

Schlichting, C. D., and M. Pigliucci. 1993. Control of phenotypic plasticity via regulatory genes. *Am. Nat.* 142:366–370.

Schlichting, C. D., and M. Pigliucci. 1995a. Gene regulation, quantitative genetics and the evolution of reaction norms. *Evol. Ecol.* 9:154–168.

Schlichting, C. D., and M. Pigliucci. 1995b. Lost in phenotypic space: Environment dependent morphology in *Phlox drummondii* (Polemoniaceae). *Int. J. Plant. Sci.* 156:542–546.

Schlichting, C. D., B. Devlin, and A. G. Stephenson. Unpublished. The effects of prior fruit dominance and seed number per fruit on male and female flower production in *Cucurbita pepo*.

Schluter, D. 1995. Adaptive radiation in sticklebacks: Trade-offs in feeding performance and growth. *Ecology* 76:82–90.

Schluter, D. 1996. Adaptive radiation along genetic lines of least resistance. *Evolution* 50:1766–1774.

Schluter, D., and J. D. McPhail. 1992. Ecological character displacement and speciation in sticklebacks. *Am. Nat.* 140:85–108.

Schluter, D., and J. D. McPhail. 1993. Character displacement and replicate adaptive radiation. *Trends Ecol. Evol.* 8:197–200.

Schluter, D., and D. Nychka. 1994. Exploring fitness surfaces. *Am. Nat.* 143:597–616.

Schmalhausen, I. I. 1949. *Factors of Evolution*. Blakiston, Philadelphia, PA.

Schmid-Hempel, P. 1990. In search of optima: Equilibrium models of phenotypic evolution. In K. Wohrmann and S. K. Jain (eds.), *Population Biology: Ecological and Evolutionary Viewpoints*, pp. 321–347. Springer-Verlag, Berlin.

Schmidt, K. P., and D. A. Levin. 1985. The comparative demography of reciprocally sown populations of *Phlox drummondii* Hook. I. Survivorships, fecundities, and finite rates of increase. *Am. Nat.* 39: 396–404.

Schmitt, J., and R. D. Wulff. 1993. Light spectral quality, phytochrome and plant competition. *Trends Ecol. Evol.* 8:47–50.

Schmitt, J., J. Niles, and R. D. Wulff. 1992. Norms of reaction of seed traits to maternal environments in *Plantago lanceolata*. *Am. Nat.* 139:451–466.

Schmitt, J., A. C. McCormac, and H. Smith. 1995. A test of the adaptive plasticity hypothesis using transgenic and mutant plants disabled in phytochrome-mediated elongation responses to neighbors. *Am. Nat.* 146:937–953.

Scholl, D. A. 1954. Regularities in growth curves, including rhythms and allometry. In E. J. Boell (ed.), *Dynamics of Growth Processes*, pp. 224–241. Princeton University Press, Princeton, NJ.

Schrag, S. J., and V. Perrot. 1996. Reducing antibiotic resistance. *Nature* 381:120–121.

Schultz, E. A., and G. W. Haughn. 1991. LEAFY, a homeotic gene that regulates inflorescence development in *Arabidopsis*. *Plant Cell* 3:771–781.

Schultz, E. A., and G. W. Haughn. 1993. Genetic analysis of the floral initiation process (FLIP) in *Arabidopsis*. *Development* 119:745–765.

Schwenk, K. 1995. A utilitarian approach to evolutionary constraint. *Zoology* 98:251–262.

Scott, G. E. 1967. Selecting for stability of yield in maize. *Crop Sci.* 7:549–551.

Scott, M. 1966. "Ye canna change the laws of physics, Captain." *Star Trek*.

Scriber, J. M., R. H. Hagen, and R. C. Lederhouse. 1996. Genetics of mimicry in the tiger swallowtail butterflies, *Papilio glaucus* and *P. canadensis* (Lepidoptera: Papilionidae). *Evolution* 50:222–236.

Searcy, K. B. 1992. Developmental selection in response to environmental stress. *Evol. Trends Plants* 6:21–24.

Seger, J., and H. J. Brockmann. 1987. What is bet-hedging? *Oxford Surv. Evol. Biol.* 4:182–211.

Seger, J., and J. W. Stubblefield. 1996. Optimization and adaptation. In M. R. Rose and G. V. Lauder (eds.), *Adaptation*, pp. 93–123. Academic Press, San Diego.

Selzer, J. 1993. (ed.) *Understanding Scientific Prose*, pp. 388. University of Wisconsin Press, Madison.

Semlitsch, R. D. 1987. Paedomorphosis in *Ambystoma talpoideum*: Effects of density, food, and pond drying. *Ecology* 68:994–1102.

Semlitsch, R. D., and H. M. Wilbur. 1989. Artificial selection for paedomorphosis in the salamander *Ambystoma talpoideum*. *Evolution* 43:105–112.

Service, P. M., and M. R. Rose. 1985. Genetic covariation among life-history components: The effect of novel environments. *Evolution* 39:943–945.

Shaffer, H. B., and S. R. Voss. 1996. Phylogenetic and mechanistic analysis of a developmentally integrated character complex: Alternate life history modes in ambystomatid salamanders. *Am. Zool.* 36:24–35.

Shannon, S., and D. R. Meeks-Wagner. 1991. A mutation in the *Arabidopsis TFL1* gene affects inflorescence meristem development. *Plant Cell* 3:877–892.

Shaw, F. H., R. G. Shaw, G. S. Wilkinson, and M. Turelli. 1995a. Changes in genetic variances and covariances: G whiz! *Evolution* 49:1260–1267.

Shaw, R. G., G. A. J. Platenkamp, F. H. Shaw, and R. H. Podolsky. 1995b. Quantitative genetics of response to competitors in *Nemophila menziesii*: A field experiment. *Genetics* 139:397–406.

Sherry, R. A., and E. M. Lord. 1996a. Developmental stability in flowers of *Clarkia tembloriensis* (Onagraceae). *J. Evol. Biol.* 9:911–930.

Sherry, R. A., and E. M. Lord. 1996b. Developmental stability in leaves of *Clarkia tembloriensis* (Onagraceae) as related to population outcrossing rates and heterozygosity. *Evolution* 50:80–91.

Shubin, N., C. Tabin, and S. Carroll. 1997. Fossils, genes and the evolution of animal limbs. *Nature* 388:639–648.

Silander, J. A. 1985. The genetic basis of the ecological amplitude of *Spartina patens*. II. Variance and correlation analysis. *Evolution* 39:1034–1052.

Silvertown, J., and M. Dodd. 1996. Comparing plants and connecting traits. *Phil. Trans. R. Soc. Lond. B* 351:1233–1239.

Simms, E. L., and J. Triplett. 1994. Costs and benefits of plant responses to disease: Resistance and tolerance. *Evolution* 48:1973–1985.

Simon, J., A. Chiang, and W. Bender. 1992. Ten different *Polycomb* group genes are required for spatial control of the *abdA* and *abdB* homeotic products. *Development* 114:493–505.

Simon, R., M. I. Igeno, and G. Coupland. 1996. Activation of floral meristem identity genes in *Arabidopsis*. *Nature* 384:59–62.

Simons, A. M., and D. A. Roff. 1994. The effect of environmental variability on the heritabilities of traits of a field cricket. *Evolution* 48:1637–1649.

Simpson, G. G. 1944. *Tempo and Mode in Evolution*. Columbia University Press, New York.

Simpson, G. G. 1953. The Baldwin effect. *Evolution* 7:110–117.

Simpson, J., M. V. Montagu, and L. Herrera-Estrella. 1986. Photosynthesis-associated gene families: Differences in response to tissue-specific and environmental factors. *Science* 233:34–38.

Sinervo, B., and A. L. Basolo. 1996. Testing adaptation using phenotypic manipulations. In M. R. Rose and G. V. Lauder (eds.), *Adaptation*, pp. 149–185. Academic Press, San Diego.

Sinervo, B., P. Doughty, R. B. Huey, and K. Zamudio. 1992. Allometric engineering: A causal analysis of natural selection on offspring size. *Science* 258:1927–1930.

Sinha, N. R., and E. A. Kellogg. 1996. Parallelism and diversity in multiple origins of C^4 photosynthesis in the grass family. *Am. J. Bot.* 83:1458–1470.

Slatkin, M. 1987. Quantitative genetics of heterochrony. *Evolution* 41:799–811.

Slobodkin, L. B. 1968. Toward a predictive theory of evolution. In R. C. Lewontin (ed.), *Population Biology and Evolution*, Chapter 13. Syracuse University Press, Syracuse, NY.

Smith, A. B. 1997. Echinoderm larvae and phylogeny. *Annu. Rev. Ecol. Syst.* 28:219–242.

Smith, H. 1990. Signal perception, differential expression within multigene families and the molecular basis of phenotypic plasticity. *Plant Cell Envir.* 13:585–594.

Smith, H. 1995. Physiological and ecological function within the phytochrome family. *Annual Review of Plant Physiol. and Plant Mol. Biol.* 46:289–315.

Smith, K. C. 1992. Neo-rationalism versus neo-darwinism: Integrating development and evolution. *Biol. Phil.* 7:431–451.

Smith, L. G., and S. Hake. 1994. Molecular genetic approaches to leaf development: *Knotted* and beyond. *Can. J. Bot.* 72:617–625.

Smith, V., K. N. Chou, D. Lashkari, D. Botstein, and P. O. Brown. 1996. Functional analysis of the genes of yeast chromosome V by genetic footprinting. *Science* 274:2069–2074.

Smith-Gill, S. J. 1983. Developmental plasticity: Developmental conversion versus phenotypic modulation. *Am. Zool.* 23:47–55.

Smits, J. D., F. Witte, and G. D. E. Povel. 1996a. Differences between inter- and intraspecific architectonic adaptations to pharyngeal mollusc crushing in cichlid fishes. *Biol. J. Linn. Soc.* 59:367–387.

Smits, J. D., F. Witte, and F. G. Van Veen. 1996b. Functional changes in the anatomy of the pharyngeal jaw apparatus of *Astatoreochromis alluaudi* (Pisces, Cichlidae), and their effects on adjacent structures. *Biol. J. Linn. Soc.* 59:389–409.

Smouse, P. E., J. C. Long, and R. R. Sokal. 1986. Multiple regression and correlation extensions of the Mantel test of matrix correspondence. *Syst. Zool.* 35:627–632.

Sneath, P. H. A., and R. R. Sokal. 1973. *Numerical Taxonomy*. W.H. Freeman & Co., San Francisco.

Sniegowski, P. D., P. J. Gerrish, and R. E. Lenski. 1997. Evolution of high mutation rates in experimental populations of *E. coli*. *Nature* 387:703–705.

Solomon, B. P. 1985. Environmentally influenced changes in sex expression in an andromonoecious plant. *Ecology* 66:1321–1332.

Somero, G. N. 1978. Temperature adaptation of enzymes: Biological optimization through structure-function comparisons. *Annu. Rev. Ecol. Syst.* 9:1–29.

Somers, K. M. 1986. Multivariate allometry and removal of size with principal components analysis. *Syst. Zool.* 35:359–368.

Somers, K. M. 1989. Allometry, isometry and shape in principal component analysis. *Syst. Zool.* 38:169–173.

Sommer, R. J. 1997a. Evolution and development: The nematode vulva as a case study. *BioEssays* 19:225–231.

Sommer, R. J. 1997b. Evolutionary changes of developmental mechanisms in the absence of cell lineage alterations during vulva formation in the Diplogastridae (Nematoda). *Development* 124:243–251.

Sommer, R. J., and P. W. Sternberg. 1994. Changes of induction and competence during the evolution of vulva development in nematodes. *Science* 265:114–118.

Sommer, R. J., L. K. Carta, and P. W. Sternberg. 1994. The evolution of cell lineage in nematodes. *Development Suppl.*:85–95.

Sondhi, K. C. 1961. Selection for a character with a bounded distribution of phenotypes in *Drosophila subobscura*. *J. Genet.* 57:193–221.

Sondhi, K. C. 1962. The evolution of a pattern. *Evolution* 16:186–191.

Soule, M. E. 1979. Heterozygosity and developmental stability: Another look. *Evolution* 33:396–401.

Spitze, K. 1992. Predator-mediated plasticity of prey life history and morphology: *Chaoborus americanus* predation on *Daphnia pulex*. *Am. Nat.* 139:229–247.

Spofford, J. B. 1956. The relation between expressivity and selection against eyeless in *Drosophila melanogaster*. *Genetics* 41:938–959.

Stace, C. A. 1987. Hybridization and the plant species. In K. M. Urbanska (ed.), *Differentiation Patterns in Higher Plants*, pp. 115–127. Academic Press, London.

Stanton, M. L., C. Galen, and J. Shore. 1997. Population structure along a steep environmental gradient: Consequences of flowering time and habitat variation in the snow buttercup, *Ranunculus adoneus*. *Evolution* 51:79–94.

Stearns, S. C. 1982. The role of development in the evolution of life histories. In J. T. Bonner (ed.), *Evolution and Development*, pp. 237–258. Springer-Verlag, New York.

Stearns, S. C. 1983. The evolution of life-history traits in mosquitofish since their introduction to Hawaii in 1905: Rates of evolution, heritabilities, and developmental plasticity. *Am. Zool.* 23:65–75.

Stearns, S. C. 1984. How much of the phenotype is necessary to understand evolution at the level of the gene? In K. Wohrmann and V. Loeschcke (eds.), *Population Biology and Evolution*, pp. 31–45. Springer-Verlag, Berlin.

Stearns, S. C. 1989a. The evolutionary significance of phenotypic plasticity. *BioScience* 39:436–445.

Stearns, S. C. 1989b. Tradeoffs in life-history evolution. *Func. Ecol.* 3:259–268.

Stearns, S. C. 1992. *The Evolution of Life Histories.* Oxford University Press, New York.

Stearns, S. C. 1994. The evolutionary links between fixed and variable traits. *Acta Palaeon. Polonica* 38:215–232.

Stearns, S. C., and M. Kaiser. 1996. Effects on fitness components of P-element inserts in *Drosophila melanogaster.* Analysis of trade-offs. *Evolution* 50:795–806.

Stearns, S. C., and J. C. Koella. 1986. The evolution of phenotypic plasticity in life-history traits: Predictions of reaction norms for age and size at maturity. *Evolution* 40:893–914.

Stearns, S. C., and T. J. Kawecki. 1994. Fitness sensitivity and the canalization of life history traits. *Evolution* 48:1438–1450.

Stearns, S., G. de Jong, and B. Newman. 1991. The effects of phenotypic plasticity on genetic correlations. *Trends Ecol. Evol.* 6:122–126.

Stearns, S. C., M. Kaiser, and T. J. Kawecki. 1995. The differential genetic and environmental canalization of fitness components in *Drosophila melanogaster. J. Evol. Biol.* 8:539–557.

Stebbins, G. L. 1950. *Variation and Evolution in Flowering Plants.* Columbia University Press, New York.

Stebbins, G. L. 1980. Botany and the synthetic theory of evolution. In E. Mayr and W. B. Provine (eds.), *The Evolutionary Synthesis,* pp. 139–152. Harvard University Press, Cambridge, MA.

Stebbins, G. L., and D. L. Hartl. 1988. Comparative evolution: Latent potentials for anagenetic advance. *Proc. Natl. Acad. Sci. USA* 85:5141–5145.

Stephenson, A. G., B. Devlin, and J. B. Horton. 1988. The effects of seed number and prior fruit dominance on the pattern of fruit production in *Cucurbita pepo* (zucchini squash). *Ann. Bot.* 62:653–661.

Steppan, S. J. 1997a. Phylogenetic analysis of phenotypic covariance structure. I. Contrasting results from matrix correlation and common principal component analyses. *Evolution* 51:571–586.

Steppan, S. J. 1997b. Phylogenetic analysis of phenotypic covariance structure. II. Reconstructing matrix evolution. *Evolution* 51:587–594.

Sternberg, P. W., and M.-A. Felix. 1997. Evolution of cell lineage. *Curr. Opin. Genet. Dev.* 7:543–550.

Stibor, H., and J. Luning. 1994. Predator-induced phenotypic variation in the pattern of growth and reproduction in *Daphnia hyalina* (Crustacea: Cladocera). *Func. Ecol.* 8:97–101.

Stills, S. 1967. For what it's worth. Ten East Springalo Cotillion (BMI).

Stone, J. R. 1996. Computer-simulated shell size and shape variation in the Caribbean land snail genus *Cerion*: A test of geometrical constraints. *Evolution* 50:341–347.

Stone, L. 1993. Period-doubling reversals and chaos in simple ecological models. *Nature* 365:617–620.

Strathmann, R. R., L. Fenaux, and M. F. Strathmann. 1992. Heterochronic developmental plasticity in larval sea urchins and its implications for evolution of nonfeeding larvae. *Evolution* 46:972–986.

Stratton, D. A. 1992. Life-cycle components of selection in *Erigeron annuus*: I. Phenotypic selection. *Evolution* 46:92–106.

Stratton, D. A. 1994. Genotype-by-environment interactions for fitness of *Erigeron annuus* show fine-scale selective heterogeneity. *Evolution* 48: 1607–1618.

Strauss, R. E. 1990. Heterochronic variation in the developmental timing of cranial ossifications in Poeciliid fishes (Cyprinodontiformes). *Evolution* 44:1558–1567.

Strome, S. 1995. (ed.). *C. elegans* lineage map. At http://sunflower.bio.indiana.edu/~sstrome/_photos/photos.html.

Sultan, S. E. 1987. Evolutionary implications of phenotypic plasticity in plants. *Evolutionary Biology* 21: 127–178.

Sultan, S. E. 1992. Phenotypic plasticity and the Neo-Darwinian legacy. *Evolutionary Trends in Plants* 6:61–71.

Sultan, S. E. 1995. Phenotypic plasticity and plant adaptation. *Acta Bot. Neer.* 44:363–383.

Sultan, S. E. 1996. Phenotypic plasticity for offspring traits in *Polygonum persicaria. Ecology* 77:1791–1807.

Sultan, S. E., and F. A. Bazzaz. 1993a. Phenotypic plasticity in *Polygonum persicaria.* I. Diversity and uniformity in genotypic norms of reaction to light. *Evolution* 47:1009–1031.

Sultan, S. E., and F. A. Bazzaz. 1993b. Phenotypic plasticity in *Polygonum persicaria.* II. Norms of reaction to soil moisture and the maintenance of genetic diversity. *Evolution* 47:1032–1049.

Sultan, S. E., and F. A. Bazzaz. 1993c. Phenotypic plasticity in *Polygonum persicaria.* III. The evolution of ecological breadth for nutrient environment. *Evolution* 47:1050–1071.

Sun, H., A. Rodin, Y. Zhou, D. P. Dickinson, D. E. Harper, D. Hewett-Emmett, and W.-H. Li. 1997. Evolution of paired domains: Isolation and sequencing of jellyfish and hydra *Pax* genes related to *Pax-5* and *Pax-6. Proc. Natl. Acad. Sci. USA* 94:5156–5161.

Sundberg, P. 1989. Shape and size-constrained principal components analysis. *Syst. Zool.* 38:166–168.

Sung, Z. R., A. Belachew, B. Shunong, and R. Bertrand-Garcia. 1992. EMF, an *Arabidopsis* gene required for vegetative shoot development. *Science* 258: 1645–1647.

Tabin, C., and E. Laufer. 1993. *Hox* genes and serial homology. *Nature* 361:692–693.

Takhtajan, A. 1943. Correlations of ontogenesis and phylogenesis in higher plants (in Russian with an English summary). *Tr. Erevansk. Gos. Univ.* 22:71–176 (cited in Guerrant, 1988).

Takhtajan, A. 1972. Patterns of ontogenetic alterations in the evolution of higher plants. *Phytomorphology* 22:164–171.

Takhtajan, A. 1976. Patterns of ontogenetic alterations in the evolution of higher plants. In C. B. Beck (ed.),

Origin and Early Evolution of Angiosperms, pp. 207–219. Columbia University Press, New York.

Takimoto, A., S. Kaihara, and M. Yokoyama. 1994. Stress-induced factors involved in flower formation in *Lemna*. *Physiol. Plant.* 92:624–628.

Tamarina, N. A., M. Z. Ludwig, and R. C. Richmond. 1997. Divergent and conserved features in the spatial expression of the *Drosophila pseudoobscura* esterase-5B gene and the esterase-6 gene of *Drosophila melanogaster*. *Proc. Natl. Acad. Sci. USA* 94:7735–41.

Tarasjev, A. 1995. Relationship between phenotypic plasticity and developmental instability in *Iris pumila* l. *Genetika* 31:1655–1663.

Tatusov, R. L., E. V. Koonin, and D. J. Lipman. 1997. A genomic perspective on protein families. *Science* 278:631–637.

Taylor, C., and D. Jefferson. 1994. Artificial life as a tool for biological inquiry. In C. G. Langton (ed.), *Artificial Life: An Overview*, pp. 1–13. MIT Press, Cambridge, MA.

Taylor, D. R., and L. W. Aarssen. 1988. An interpretation of phenotypic plasticity in *Agropyron repens* (Gramineae). *Am. J. Bot.* 75:401–413.

Taylor, E. B., C. J. Foote, and C. C. Wood. 1996. Molecular genetic evidence for parallel life-history evolution within a Pacific salmon (sockeye salmon and kokanee, *Oncorhyncus nerka*). *Evolution* 50:401–416.

te Velde, J. H., and W. Scharloo. 1988. Natural and artificial selection on a deviant character of the anal papillae in *Drosophila melanogaster* and their significance for salt adaptation. *J. Evol. Biol.* 1:155–164.

te Velde, J. H., J. H. Gordens, and W. Scharloo. 1987. The genetic fixation of phenotypic response of an ultrastructural character in the anal papillae of *Drosophila melanogaster*. *Heredity* 61:47–53.

Templeton, A. R. 1982. Why read Goldschmidt? *Paleobiology* 8:474–481.

Templeton, A. R., and J. S. Johnston. 1988. The measured genotype approach to ecological genetics. In G. D. Jong (ed.), *Population genetics and evolution*, pp. 138–146. Springer-Verlag, Berlin.

Thoday, J. M. 1953. Components of fitness. *Symp. Soc. Exp. Biol.* 7:96–113.

Thoday, J. M. 1955. Balance, heterozygosity and developmental stability. *Cold Spring Harb. Symp. Quant. Biol.* 20:318–326.

Thomas, J. H. 1993. Thinking about genetic redundancy. *Trends Genet.* 9:395–399.

Thomas, R. D. K., and W. E. Reif. 1993. The skeleton space: A finite set of organic designs. *Evolution* 47:341–360.

Thomas, R. H., and J. S. F. Barker. 1993. Quantitative genetic analysis of the body size and shape of *Drosophila buzzatii*. *Theor. Appl. Genet.* 85:598–608.

Thompson, D. B. 1992. Consumption rates and the evolution of diet-induced plasticity in the head morphology of *Melanoplus femurrubrum* (Orthoptera, Acrididae). *Oecologia* 89:204–213.

Thompson, D. B. 1998. Different spatial scales of natural selection and gene flow: The evolution of behavioral geographic variation and phenotypic plasticity. In S. Foster and J. Endler (eds.), *Geographic Variation in Behavior: Perspectives in Evolutionary Mechanisms*. Oxford University Press, Oxford.

Thompson, D. W. 1917. *On Growth and Form*. Cambridge University Press, Cambridge.

Thompson, D. W. 1942. *On Growth and Form*. 2nd ed. Cambridge University Press, Cambridge.

Thompson, D. W. 1987. *On Growth and Form*. 2nd ed., abridged. Cambridge University Press, Cambridge.

Thompson, J. N., and J. J. Burdon. 1992. Gene-for-gene coevolution between plants and parasites. *Nature* 360:121–125.

Titus, T. A., and A. Larson. 1996. Molecular phylogenetics of desmognathine salamanders (Caudata: Plethodontidae): A reevaluation of evolution in ecology, life history, and morphology. *Syst. Biol.* 45:449–470.

Tollrian, R. 1995. Predator-induced morphological defenses: Costs, life history shifts, and maternal effects in *Daphnia pulex*. *Ecology* 76:1691–1705.

Tomarev, S. I., P. Callaerts, and J. Piatigorsky. 1997. Squid *Pax-6* and eye development. *Proc. Natl. Acad. Sci. USA* 94:2421–2426.

Tomkins, J. L., and L. W. Simmons. 1996. Dimorphisms and fluctuating asymmetry in the forceps of male earwigs. *J. Evol. Biol.* 9:753–770.

Toquenaga, Y., and M. J. Wade. 1996. Sewall Wright meets Artificial Life: The origin and maintenance of evolutionary novelty. *Trends Ecol. Evol.* 11:478–482.

Trainor, F. R. 1995. The sequence of ecomorph formation in a phenotypicly plastic, multispined *Scenedesmus* species. *Arch. Hydrobiol.* 133:161–171.

Trainor, F. R. 1996. Reproduction in *Scenedesmus*. *Algae (The Korean Journal of Phycology)* 11:183–201.

Travis, J. 1994. Ecological genetics of life-history traits: Variation and its evolutionary significance. In L. Real (ed.), *Ecological Genetics*, pp. 171–204. Princeton University Press, Princeton, NJ.

Travisano, M. 1997. Long-term experimental evolution in *Escherichia coli*. VI. Environmental constraints on adaptation and divergence. *Genetics* 146:471–479.

Travisano, M., F. Vasi, and R. E. Lenski. 1995. Long-term experimental evolution in *Escherichia coli*. III. Variation among replicate populations in correlated responses to novel environments. *Evolution* 49:189–200.

Tsukaya, H. 1995. *Dev. Genet.* of leaf morphogenesis in dicotyledonous plants. *J. Plant Res.* 108:407–416.

Tsuneizumi, K., T. Nakayama, Y. Kamoshida, T. B. Kornberg, J. L. Christian, and T. Tabata. 1997. *Daughters against dpp* modulates *dpp* organizing activity in *Drosophila* wing development. *Nature* 389:627–631.

Tucic, N., M. Milosevic, I. Gliksman, D. Milanovic, and I. Aleksic. 1991. The effects of larval density on genetic variation and covariation among life-history traits in the bean weevil (*Acanthoscelides obtectus* Say). *Func. Ecol.* 5:525–534.

Tucic, B., A. Tarasjev, S. Vujci, S. Milojkovic, and N. Tucic. 1990. Phenotypic plasticity and character differentiation in a subdivided population of *Iris pumila* (Iridaceae). *Plant Syst. Evol.* 170:1–9.

Turelli, M. 1988. Phenotypic evolution, constant covariances, and the maintenance of additive variance. *Evolution* 42:1342–1347.

Turelli, M., and N. H. Barton. 1994. Genetic and statistical analyses of strong selection on polygenic traits: What, me normal? *Genetics* 138:913–941.

Turesson, G. 1922. The genotypical response of the plant species to the habitat. *Hereditas* 3:211–350.

Valentine, J. W. 1997. Cleavage patterns and the topology of the metazoan tree of life. *Proc. Natl. Acad. Sci. USA* 94:8001–8005.

Valentine, J. W., D. H. Erwin, and D. Jablonski. 1996. Developmental evolution of metazoan body plans: The fossil evidence. *Dev. Biol.* 173:373–381.

Van Buskirk, J., and S. A. McCollum. 1997. Natural selection for environmentally induced phenotypes in tadpoles. *Evolution* 51:1983–1192.

Van Delden, W., A. C. Boerema, and A. Kamping. 1978. The alcohol dehydrogenase polymorphism in populations of *Drosophila melanogaster* I. Selection in different environments. *Genetics* 90:161–191.

van Noordwijk, A. J., and G. de Jong. 1986. Acquisition and allocation of resources: Their influence on variation in life history tactics. *Am. Nat.* 128:137–142.

van Tienderen, P. H. 1990. Morphological variation in *Plantago lanceolata*: Limits of plasticity. *Evol. Trends Plants* 4:35–43.

van Tienderen, P. H. 1991. Evolution of generalists and specialists in spatially heterogeneous environments. *Evolution* 45:1317–1331.

van Tienderen, P. H. 1997. Generalists, specialists, and the evolution of phenotypic plasticity in sympatric populations of distinct species. *Evolution* 51:1372–1380.

van Tienderen, P. H., and J. Antonovics. 1994. Constraints in evolution: On the baby and the bath water. *Func. Ecol.* 8:139–140.

van Tienderen, P. H., and G. de Jong. 1994. A general model of the relation between phenotypic selection and genetic response. *J. Evol. Biol.* 7:1–12.

van Tienderen, P. H., and H. P. Koelewijn. 1994. Selection on reaction norms, genetic correlations and constraints. *Genet. Res.* 64:115–125.

van Tienderen, P. H., and J. Vandertoorn. 1991a. Genetic differentiation between populations of *Plantago lanceolata*.I. Local adaptation in three contrasting habitats. *J. Ecol.* 79:43–59.

van Tienderen, P. H., and J. Vandertoorn. 1991b. Genetic differentiation between populations of *Plantago lanceolata*. II. Phenotypic selection in a transplant experiment in three contrasting habitats. *J. Ecol.* 79:43–59.

van Tienderen, P. H., and A. van Hinsberg. 1996. Phenotypic plasticity in growth habit in *Plantago lanceolata*: How tight is a suite of correlated characters? *Plant Species Biol.* 11.

van Valen, L. M. 1982. Homology and causes. *J. Morphol.* 173:305–312.

Vargas-Hernandez, J., and W. T. Adams. 1992. Age-age correlations and early selection for wood density in young coastal Douglas-Fir. *For. Sci.* 38:467–478.

Via, S. 1984a. The quantitative genetics of polyphagy in an insect herbivore. I. Genotype-environment interaction in larval performance on different host plant species. *Evolution* 38:881–895.

Via, S. 1984b. The quantitative genetics of polyphagy in an insect herbivore. II. Genetic correlations in larval performance within and among host plants. *Evolution* 38:896–905.

Via, S. 1987. Genetic constraints on the evolution of phenotypic plasticity. In V. Loeschcke (ed.), *Genetic Constraints on Adaptive Evolution*, pp. 47–71. Springer-Verlag, Berlin.

Via, S. 1991. The genetic structure of host plant adaptation in a spatial patchwork: Demographic variability among reciprocally transplanted pea aphid clones. *Evolution* 45:827–852.

Via, S. 1993a. Adaptive phenotypic plasticity: Target or by-product of selection in a variable environment? *Am. Nat.* 142:352–365.

Via, S. 1993b. Regulatory genes and reaction norms. *Am. Nat.* 142:374–378.

Via, S. 1994a. The evolution of phenotypic plasticity: What do we really know? In L. Real (ed.), *Ecological Genetics*, pp. 35–57. Princeton University Press, Princeton, NJ.

Via, S. 1994b. Population structure and local adaptation in a clonal herbivore. In L. Real (ed.), *Ecological Genetics*, pp. 58–85. Princeton University Press, Princeton, NJ.

Via, S., and J. Conner. 1995. Evolution in heterogeneous environments: Genetic variability within and across different grains in *Tribolium castaneum*. *Heredity* 74:80–90.

Via, S., and R. Lande. 1985. Genotype-environment interaction and the evolution of phenotypic plasticity. *Evolution* 39:505–522.

Via, S., and R. Lande. 1987. Evolution of genetic variability in a spatially heterogeneous environment: Effects of genotype-environment interaction. *Genet. Res.* 49:147–156.

Via, S., R. Gomulkiewicz, G. de Jong, S. M. Scheiner, C. D. Schlichting, and P. H. van Tienderen. 1995. Adaptive phenotypic plasticity: Consensus and controversy. *Trends Ecol. Evol.* 10:212–216.

Vidybida, A. K. 1991. Selectivity and sensitivity improvement in co-operative systems with a threshold in the presence of noise. *J. Theor. Biol.* 152:159–164.

Vlot, E. C., W. H. J. V. Houten, S. Mauthe, and K. Bachmann. 1992. Genetic and nongenetic factors influencing deviations from five pappus parts in a hybrid between *Microseris douglasii* and *M. bigelovii* (Asteraceae, Lactuceae). *Int. J. Plant. Sci.* 153:89–97.

Vogl, C. 1996. Developmental buffering and selection. *Evolution* 50:1343–1346.

von Baer, K. E. 1828. *Entwicklungsgeschichte der Thiere: Beobachtung und Reflexion*. Borntrager, Konegsberg, Germany.

von Bertalanffy, L. 1964. Basic concepts in quantitative biology of metabolism. *Helgolander Wiss. Meeresunters* 9:5–37.

Voss, R. S., and H. B. Shaffer. 1996. What insights into the developmental traits of urodeles does the study of interspecific hybrids provide? *Int. J. Dev. Biol.* 40:885–893.

Waddington, C. H. 1942. Canalization of development and the inheritance of acquired characters. *Nature* 150:563–565.

Waddington, C. H. 1952. Selection of the genetic basis for an acquired character. *Nature* 169:278.

Waddington, C. H. 1953a. Genetic assimilation of an acquired character. *Evolution* 7:118–126.

Waddington, C. H. 1953b. Epigenetics and evolution. *Symp. Soc. Exp. Biol.* 7:186–199.

Waddington, C. H. 1956. Genetic assimilation of the *bithorax* phenotype. *Evolution* 10:1–13.

Waddington, C. H. 1957. *The Strategy of the Genes*. Allen & Unwin, London.

Waddington, C. H. 1959a. Canalization of development and genetic assimilation of acquired characters. *Nature* 183:1654–1655.

Waddington, C. H. 1959b. Evolutionary adaptation. In S. Tax (ed.), *Evolution after Darwin*, pp. 381–402. University of Chicago Press, Chicago.

Waddington, C. H. 1960. Experiments on canalizing selection. *Genet. Res.* 1:140–150.

Waddington, C. H. 1961. Genetic assimilation. *Adv. Genet.* 10:257–290.

Waddington, C. H. 1966. *Principles of Development and Differentiation*. Macmillan, New York.

Waddington, C. H. 1975. *The Evolution of an Evolutionist*. Cornell University Press, Ithaca. NY.

Waddington, C. H. 1979. A catastrophe theory of evolution. *Ann. New York Acad. Sci.* 231:32–42.

Waddington, C. H., and E. Robertson. 1966. Selection for developmental canalisation. *Genet. Res.* 7:303–312.

Wade, M. J. 1992. Sewall Wright: Gene interaction and the shifting balance theory. *Oxford Surv. Evol. Biol.* 8:35–62.

Wade, M. J., and C. J. Goodnight. 1991. Wright's shifting balance theory: An experimental study. *Science* 253:1015–1018.

Wade, M. J., and S. Kalisz. 1990. The causes of natural selection. *Evolution* 44:1947–1955.

Wagner, A. 1994. Evolution of gene networks by gene duplication: A mathematical model and its implications on genome organization. *Proc. Natl. Acad. Sci. USA* 91:4387–4391.

Wagner, A. 1996a. Does evolutionary plasticity evolve? *Evolution* 50:1008–1023.

Wagner, A. 1996b. Genetic redundancy caused by gene duplications and its evolution in networks of transcriptional regulators. *Biol. Cyber.* 74:557–567.

Wagner, A., G. P. Wagner, and P. Similion. 1994. Epistasis can facilitate the evolution of reproductive isolation by peak shifts: A two-locus two-allele model. *Genetics* 138:533–545.

Wagner, G. P. 1986. The systems approach: An interface between development and population genetic aspects of evolution. In D. M. Raup and D. Jablonski (eds.), *Patterns and Processes in the History of Life*, pp. 149–165. Springer-Verlag, Berlin.

Wagner, G. P. 1988a. The influence of variation and of developmental constraints on the rate of multivariate phenotypic evolution. *J. Evol. Biol.* 1:45–66.

Wagner, G. P. 1988b. The significance of developmental constraints for phenotypic evolution by natural selection. In G. de Jong (ed.), *Population Genetics and Evolution*, pp. 222–229. Springer-Verlag, Berlin.

Wagner, G. P. 1989a. The biological homology concept. *Annu. Rev. Ecol. Syst.* 20:51–69.

Wagner, G. P. 1989b. The origin of morphological characters and the biological basis of homology. *Evolution* 43:1157–1171.

Wagner, G. P. 1989c. Multivariate mutation-selection balance with constrained pleiotropic effects. *Genetics* 122:223–234.

Wagner, G. P. 1990. A comparative study of morphological integration in *Apis mellifera* (Insecta, Hymenoptera). *Z. Zool. Syst. Evol.* 28:48–61.

Wagner, G. P. 1995. Adaptation and the modular design of organisms. In F. Moran, A. Moreno, J. J. Merelo, and P. Chacon (eds.), *Advances in Artificial Life*, pp. 317–328. Springer, Berlin.

Wagner, G. P. 1996. Homologues, natural kinds and the evolution of modularity. *Am. Zool.* 36:36–43.

Wagner, G. P. 1998. Complexity matters. *Science* 279:1158–1159.

Wagner, G. P., and L. Altenberg. 1996. Perspective: Complex adaptations and the evolution of evolvability. *Evolution* 50:967–976.

Wagner, G. P., and B. Y. Misof. 1993. How can a character be developmentally constrained despite variation in developmental pathways? *J. Evol. Biol.* 6:449–455.

Wagner, G. P., G. Booth, and H. Bagheri-Chaichian. 1997. A population genetic theory of canalization. *Evolution* 51:329–347.

Waitt, D. E., and D. A. Levin. 1993. Phenotypic integration and plastic correlations in *Phlox drummondii* (Polemoniaceae). *Am. J. Bot.* 80:1224–1233.

Wake, D. B. 1991. Homoplasy: The result of natural selection, or evidence of design limitations? *Am. Nat.* 138:543–567.

Wake, D. B., and J. Hanken. 1996. Direct development in the lungless salamanders: what are the consequences for developmental biology, evolution and phylogenesis? *Int. J. Dev. Biol.* 40:859–869.

Wake, D. B., and G. Roth. 1989. *Complex Organismal Functions: Integration and Evolution in Vertebrates*. John Wiley & Sons, Chichester, UK.

Walker, J. A. 1997. Ecological morphology of lacustrine threespine stickleback *Gasterosteus aculeatus* L. (Gasterosteidae) body shape. *Biol. J. Linn. Soc.* 61:3–50.

Wallace, B. 1982. Phenotypic variation with respect to fitness: The basis for rank-order selection. *Biol. J. Linn. Soc.* 17:269–274.

Wallace, B. 1990. Norms of reaction: Do they include molecular events ? *Perspect. Biol. Med.* 33:323–334.

Walsh, J. B. 1995. How often do duplicated genes evolve new functions? *Genetics* 139:421–428.

Ward, D., and M. K. Seely. 1996. Adaptation and constraint in the evolution of the physiology and behavior of the Namib desert tenebrionid beetle genus *Onymacris*. *Evolution* 50:1231–1240.

Ward, P. J. 1994. Parent-offspring regression and extreme environments. *Heredity* 72:574–581.

Ward, P. S. 1997. Ant soldiers are not modified queens. *Nature* 385:494–495.

Warren, R. W., L. Nagy, J. Selegue, J. Gates, and S. Carroll. 1994. Evolution of homeotic gene regulation and function in flies and butterflies. *Nature* 372:458–461.

Watson, M. A., Geber, M. A., and C. S. Jones. 1995. Ontogenetic contingency and the expression of plant plasticity. *Trends Ecol. Evol.* 10:474–475.

Watt, W. B. 1983. Adaptation at specific loci. II. Demographic and biochemical elements in the maintenance of *Colias* PGI polymorphism. *Genetics* 103:691–724.

Waxman, D., and J. Peck. 1998. Pleiotropy and the preservation of perfection. *Science* 279:1210–1213.

Weber, K. E. 1990. Selection on wing allometry in *Drosophila melanogaster*. *Genetics* 126:975–989.

Weber, K. E. 1992. How small are the smallest selectable domains of form? *Genetics* 130:345–353.

Weber, K. E. 1996. Large genetic change at small fitness cost in large populations of *Drosophila melanogaster* selected for wind tunnel flight: Rethinking fitness surfaces. *Genetics* 144:205–213.

Weeks, S. C., and G. K. Meffe. 1996. Quantitative genetic and optimality analyses of life-history plasticity in the Eastern mosquitofish, *Gambusia holbrooki*. *Evolution* 50:1358–1365.

Weigel, D., and O. Nilsson. 1995. A developmental switch sufficient for flower initiation in diverse plants. *Nature* 377:495–500.

Weigensberg, I., and D. A. Roff. 1996. Natural heritabilities: Can they be reliably estimated in the laboratory? *Evolution* 50:2149–2157.

Weiner, J., and S. C. Thomas. 1992. Competition and allometry in three species of annual plants. *Ecology* 73:648–656.

Weir, B. S., E. J. Eisen, M. M. Goodman, and G. Namkoong. 1988. *Proceedings of the Second International Conference on Quantitative Genetics.* Sinauer Associates, Sunderland, MA.

Weis, A. E., and W. L. Gorman. 1990. Measuring selection on reaction norms: An exploration of the *Eurosta-Solidago* system. *Evolution* 44:820–831.

Weissing, F. J. 1996. Genetic versus phenotypic models of selection: Can genetics be neglected in a long-term perspective? *J. Math. Biol.* 34:533–555.

West-Eberhard, M. J. 1989. Phenotypic plasticity and the origins of diversity. *Annu. Rev. Ecol. Syst.* 20:249–278.

West-Eberhard, M. J. 1992. Behavior and evolution. In P. R. Grant and H. S. Horn (eds.), *Molds, Molecules and Metazoa*. Princeton University Press, Princeton, NJ.

Westoby, M., M. R. Leishman, and J. M. Lord. 1995. On misinterpreting the 'phylogenetic correction.' *J. Ecol.* 83:531–534.

Whiteman, H. H. 1994. Evolution of facultative paedomorphosis in salamanders. *Q. Rev. Biol.* 69:205–221.

Whiteman, H. H. 1997. Maintenance of polymorphism promoted by sex-specific fitness payoffs. *Evolution* 51:2039–2044.

Whitlock, M. C. 1995. Variance-induced peak shifts. *Evolution* 49:252–259.

Whitlock, M. C. 1996. The Red Queen beats the jack-of-all-trades: The limitations on the evolution of phenotypic plasticity and niche breadth. *Am. Nat.* 148, Suppl.:S65-S77.

Whitlock, M. C. 1997. Founder effects and peak shifts without genetic drift: Adaptive peak shifts occur easily when environments fluctuate slightly. *Evolution* 51:1044–1048.

Wilken, D. H. 1977. Local differentiation for phenotypic plasticity in the annual *Collomia linearis* (Polemoniaceae). *Syst. Bot.* 2:99–108.

Wilkens, H. 1988. Evolution and genetics of epigean and cave *Astyanax fasciatus* (Characidae, Pisces); support for the neutral mutation theory. *Evol. Biol.* 23:271–367.

Wilkins, A. S. 1997. Canalization: A molecular genetic perspective. *BioEssays* 19:257–262.

Wilkinson, G. S. 1993. Artificial selection alters allometry in the stalk-eyed fly *Cyrtodiopsis dalmanni*. *Genet. Res.* 62:213–222.

Wilkinson, G. S., K. Fowler, and L. Partridge. 1990. Resistance of genetic correlation structure to directional selection in *Drosophila melanogaster*. *Evolution* 44:1990–2003.

Wilson, A. C. 1976. Gene regulation in evolution. In F. J. Ayala (ed.), *Molecular Evolution*, pp. 225–236. Sinauer Associates, Sunderland, MA.

Wiltshire, R. J. E., I. C. Murfet, and J. B. Reid. 1994. The genetic control of heterochrony: Evidence from developmental mutants of *Pisum sativum* L. *J. Evol. Biol.* 7:447–465.

Wimberger, P. H. 1991. Plasticity of jaw and skull morphology in the neotropical cichlids *Geophagus brasiliensis* and *G. steindachneri*. *Evolution* 45:1545–1563.

Wimberger, P. H. 1992. Plasticity of fish body shape. The effects of diet, development, family and age in two species of *Geophagus* (Pisces: Cichlidae). *Biol. J. Linn. Soc.* 45:197–218.

Windig, J. J. 1994a. Genetic correlations and reaction norms in wing pattern of the tropical butterfly *Bicyclus anynana*. *Heredity* 73:459–470.

Windig, J. J. 1994b. Reaction norms and the genetic basis of phenotypic plasticity in the wing pattern of the butterfly *Bicyclus anynana*. *J. Evol. Biol.* 7:665–695.

Windig, J. J. 1997. The calculation and significance testing of genetic correlations across environments. *J. Evol. Biol.* 10:853–874.

Windig, J. J. 1998. Evolutionary genetics of fluctuating asymmetry in the peacock butterfly (*Inachis io*). *Heredity* 80:382–392.

Windig, J. J., P. M. Brakefield, N. Reitsma, and J. G. M. Wilson. 1994. Seasonal polyphenism in the wild: Survey of wing patterns of five species of *Bicyclus* butterflies in Malawi. *Ecological Entomology* 19:285–298.

Winn, A. A. 1996. The contributions of programmed developmental change and phenotypic plasticity to within-individual variation in leaf traits in *Dicerandra linearifolia*. *J. Evol. Biol.* 9:737–752.

Winn, A. A. In press. Is seasonal variation in leaf traits adaptive for the annual plant *Dicerandra linearifolia*? *J. Evol. Biol.*

Winsor, J. A., L. E. Davis, and A. G. Stephenson. 1987. The relationship between pollen load and fruit maturation and the effect of pollen load on offspring vigor in *Cucurbita pepo*. *Am. Nat.* 129:643–656.

Wolfram, S. 1984. Cellular automata as models of complexity. *Nature* 311:419–424.

Woltereck, R. 1909. Weiterer experimentelle Untersuchungen uber Artveranderung, Speziell uber das Wessen Quantitativer Artunterschiede bei Daphniden. *Versuch. Deutsch Zool. Geselleschaft* 19:110–172.

Wray, G. A. 1994. Developmental evolution: New paradigms and paradoxes. *Dev. Genet.* 15:1–6.

Wray, G. A. 1995. Punctuated evolution of embryos. *Science* 267:1115–1116.

Wray, G. A. 1996. Parallel evolution of nonfeeding larvae in echinoids. *Syst. Biol.* 45:308–322.

Wray, G. A., and A. E. Bely. 1994. The evolution of echinoderm development is driven by several distinct factors. *Development Suppl.*:97–106.

Wray, G. A., and D. R. McClay. 1989. Molecular heterochronies and heterotopies in early echinoid development. *Evolution* 43:803–813.

Wright, S. 1921. Systems of mating, I-V. *Genetics* 6:111–178.

Wright, S. 1931a. Evolution in Mendelian populations. *Genetics* 16:97–159.

Wright, S. 1931b. Statistical theory of evolution. *J. Am. Stat. Assoc.* 26, Suppl.:201–208.

Wright, S. 1932. The roles of mutation, inbreeding, crossbreeding and selection in evolution. *Proc. Sixth Int. Cong. Genet.* 1:356–366.

Wright, S. 1935. The analysis of variance and the correlations between relatives with respect to deviations from an optimum. *J. Genet.* 30:257–266.

Wright, S. 1941. The material basis of evolution (A review of *The Material Basis of Evolution*, by Richard B. Goldschmidt). *Sci. Monthly* 53:165–170.

Wright, S. 1969. *Evolution and the Genetics of Populations, Vol. 2: The Theory of Gene Frequencies.* University of Chicago Press, Chicago.

Wright, S. 1977. *Evolution and the Genetics of Populations, Vol. 3: Experimental Results and Evolutionary Deductions.* University of Chicago Press, Chicago.

Yamada, U. 1962. Genotype by environment interaction and genetic correlation of the same trait under different environments. *Jap. J. Genet.* 37:498–509.

Yampolsky, L. Y., and D. Ebert. 1994. Variation and plasticity of biomass allocation in *Daphnia*. *Func. Ecol.* 8:435–440.

Young, K. A., and J. S. Schmitt. 1995. Genetic variation and phenotypic plasticity of pollen release and capture height in *Plantago lanceolata*. *Func. Ecol.* 9:725–733.

Young, S. S. Y., and H. Weiler. 1960. Selection for two correlated traits by independent culling levels. *J. Genet.* 57:329–338.

Yuh, C.-H., and E. H. Davidson. 1996. Modular *cis*-regulatory organization of *Endo16*, a gut-specific gene of the sea urchin embryo. *Development* 122:1069–1082.

Yuh, C.-H., H. Bolouri, and E. H. Davidson. 1998. Genomic *cis*-regulatory logic: Experimental and computational analysis of a sea urchin gene. *Science* 279:1896–1902.

Zákány, J., C. Fromental-Ramain, X. Warot, and D. Duboule. 1998. Regulation of number and size of digits by posterior *Hox* genes: A dose-dependent mechanism with potential evolutionary implications. *Proc. Natl. Acad. Sci. USA* 94:13695–13700.

Zangerl, A. R., A. M. Arntz, and M. R. Berenbaum. 1997. Physiological price of an induced chemical defense: Photosynthesis, respiration, biosynthesis, and growth. *Oecologia* 109:433–441.

Zelditch, M. L. 1988. Ontogenetic variation in patterns of phenotypic integration in the laboratory rat. *Evolution* 42:28–41.

Zelditch, M. L., and A. C. Carmichael. 1989. Ontogenetic variation in patterns of developmental and functional integration in skulls of *Sigmodon fulviventer*. *Evolution* 43:814–824.

Zelditch, M. L., and W. L. Fink. 1996. Heterochrony and heterotopy: Stability and innovation in the evolution of form. *Paleobiology* 22:241–254.

Zelditch, M. L., F. L. Bookstein, and B. L. Lundrigan. 1992. Ontogeny of integrated skull growth in the cotton rat *Sigmodon fulviventer*. *Evolution* 46:1164–1180.

Zelditch, M. L., F. L. Bookstein, and B. L. Lundrigan. 1993. The ontogenetic complexity of developmental constraints. *J. Evol. Biol.* 6:621–641.

Zeng, Z.-B. 1988. Long-term correlated response, interpopulation covariation, and interspecific allometry. *Evolution* 42:363–374.Zera, A. J., and C. Zhang. 1995. Evolutionary endrocrinology of juvenile hormone esterase in *Gryllus assimilis*: Direct and correlated responses to selection. *Genetics* 141:1125–1134.

Zera, A. J., and C. Zhang. 1995. Evolutionary endrocrinology of juvenile hormone esterase in *Gryllus assimilis*: Direct and correlated responses to selection. *Genetics* 141:1125–1134.

Zera, A. J., S. Mole, and K. Rokke. 1994. Lipid, carbohydrate and nitrogen content of long- and short-winged *Gryllus firmus*: Implications for the physiological cost of flight capability. *J. Insect Physiol.* 40:1037–1044.

Zera, A. J., J. Sall, and K. Grudzinski. 1997a. Flight-muscle polymorphism in the cricket *Gryllus firmus*: Muscle characteristics and their influence on the evolution of flightlessness. *Physiol. Zool.* 70:519–529.

Zera, A. J., T. Sanger, and G. L. Cisper. 1997b. Direct and correlated responses to selection on JHE activity in adult and juvenile *Gryllus assimilis*: Implications for stage-specific evolution of insect endocrine traits. *Heredity* 80:300–309.

Zhang, H., and B. G. Forde. 1998. An *Arabidopsis* MADS box gene that controls nutrient-induced changes in root architecture. *Science* 279:407–409.

Zhang, J., and D. D. Boos. 1993. Testing hypotheses about covariance matrices using bootstrap methods. *Comm. Stat.* 22:723–739.

Zhang, J., and M. Nei. 1996. Evolution of antennapedia-class homeobox genes. *Genetics* 142:295–303.

Zhivotovsky, L. A., A. Bergman, and M. W. Feldman. 1996. A model of individual adaptive behavior in a fluctuating environment. In R. K. Belew and M. Mitchell (eds.), *Adaptive Individuals in Evolving Populations*, pp. 131–152. Santa Fe Institute, Santa Fe, NM.

Zhivotovsky, L. A., M. W. Feldman, and A. Bergman. 1996. On the evolution of phenotypic plasticity in a spatially heterogeneous environment. *Evolution* 50:547–558.

Zwaal, R. R., J. E. Mendel, P. W. Sternberg, and R. H. Plasterk. 1997. Two neuronal G proteins are involved in chemosensation of the *Caenorhabditis elegans* Dauer-inducing pheromone. *Genetics* 145: 715–727.

Index

A

Ab allele, 144–145
Abutilon theophrasti, 152, 262–264
acaulis2 mutant, 248
Acceleration, 113–115
Acer rubrum, 100–101
achaete gene, 282–283
Achillea, 44–45, 195
Adaptation
  and evolution in the field, 17
  and phenotypic plasticity, 9–10
  to variable environments,
    269–270
Adaptive plasticity, 83, 154, 305
Additive genetic model, 4
Adh (Alcohol dehydrogenase)
  gene, 138, 236, 292
  diversification, 286
  induction, 303–304
Alberch, P., 115, 117, 128
Allelic sensitivity, 40
  as genetic control, 80
  genetic control of plasticity,
    72–73, 84
Allelic substitution, 310–312
Allometry, 85–109
  definition, 21–22
  mathematical form, 86–87
  Thompson's work, 194
Alternative synthesis, 29–50

Ambystoma
  paedomorphosis in, 149, 298
  heterochrony in, 122–124
Amphibromus scabrivalvis, 151–152
Andersson, S., 71, 175
Andruss, B. F., 236–237
angustifolia2 mutant, 248
Aniridia gene, 227
Anthocyanin gene, 48–49
Anticipatory plastic responses, 74
Antifreeze proteins (AFGP and
  AFP), 82, 286
Apterous gene, 234
Arabidopsis thaliana
  baupläne in, 164
  cold acclimation polypeptide
    in, 3
  DRN and light, 53
  establishing body plan for,
    244–246
  inflorescences and flowers,
    249–252
  mutagenesis and reproduction,
    107–108
  path analysis of, 105–106
  patterns of gene expression in,
    230
  plasticity in, 175
  roots, 246–247
  sensitivity to UV light, 223–224

  strength of constraints in,
    176–178
  temperature and chlorophyll
    production, 176–177
  transition to reproductive
    phase, 138–139
  vegetative and leaf develop-
    ment, 247–249
Arctic cod (*Boreogadus saida*), 82,
  286
Armbruster, W. S., 278–279
Arnold, S. J., 11–12, 166
Artemia, 38, 234
Artificial life
  cellular automata, 134–135, 336,
    338
  "natural" progressions in, 19–20
Astatoreochromis alluaudi, 146
Asymmetry for paired structures,
  64
Atchley, W., 211–213, 292–293
Australian sheep blowfly (*Lucilia
  cuprina*), 301
Averof, M., 287

B

Bacillus subtilis, 289
Baldwin, J. M., 30–31
Baldwin effect, 30–31, 314
Barton, N. H., 80–81

About the Book

Editor: Andrew D. Sinauer
Project Editor: Nan Sinauer
Production Manager: Christopher Small
Electronic Book Production: Michele Ruschhaupt
Illustration Program: Precision Graphics, Inc.
Copy Editor: Grant Hackett
Indexer: Sharon Hughes
Book Design: Michele Ruschhaupt
Cover Design: Janice M. Holabird
Book Manufacturer: Courier Companies, Inc.